BIPED ROBOTS

Edited by **Armando Carlos de Pina Filho**

Biped Robots
Edited by Armando Carlos de Pina Filho

Published by InTech
Janeza Trdine 9, 51000 Rijeka, Croatia

Publishing Process Manager Ana Nikolic

Technical Editor Teodora Smiljanic

Cover Designer InTech Design Team

Additional hard copies can be obtained from orders@intechopen.com

Biped Robots, Edited by Armando Carlos de Pina Filho
p. cm.
ISBN 978-953-307-216-6

For EU product safety concerns:
IN TECH d.o.o., Prolaz Marije Krucifikse Kozulić 3, 51000 Rijeka, Croatia,
info@intechopen.com or visit our website at intechopen.com.

Meet the editor

Dr. Armando Carlos de Pina Filho, D.Sc. in Mechanical Engineering from COPPE/UFRJ, Brazil is an Associate Professor of the Federal University of Rio de Janeiro. He is Head of the research group ARMS (Automation, Robotics and Modeling of Systems) and is a member of the Brazilian Society of Engineering and Mechanical Sciences (ABCM). He is a member of the Editorial Advisory Board of The Open Cybernetics & Systemics Journal and Editor of the book "Humanoid Robots - New Developments", published by InTech in 2007. Apart from contributions to books related to the study of biped robots, Dr. Armando Carlos de Pina Filho actively publishes research works in international scientific journals, and more than 80 of his works have been published in the proceedings of several congresses.

Contents

Preface

Biped robots represent a very interesting research subject, with several particularities and scope topics. Natural biped locomotion involves a very large number of degrees of freedom. There are more than 200 degrees of freedom in the human locomotion. Besides, we have diverse problems related to kinematics, dynamics, equilibrium, stability, etc.

All these aspects make the study of biped robots a very complex subject, and many times the research results are not totally satisfactory. However, with scientific and technological advances, based on theoretical and experimental works, many researchers have collaborated in the evolution of the biped robots design, looking for to develop autonomous systems, as well as to help in rehabilitation technologies of human beings.

Thus, this book intends to present some works related to the study of biped robots, developed by researchers worldwide.

From the great number of interesting information presented here, I believe that this book can offer some help in new researches, as well as to incite the interest of people for this area of study, and its related topics, such as: mechanical design, gait simulation, patterns generation, kinematics, dynamics, equilibrium, stability, kinds of control, adaptability, biomechanics, cybernetics, and rehabilitation technologies.

Armando Carlos de Pina Filho
Federal University of Rio de Janeiro – UFRJ
Brazil

Part 1

Sensors, Control, and Stability

1

Mobile Sensors for Robotics Research

Tao Liu[1], Yoshio Inoue[1],
Kyoko Shibata[1] and Kouzou Shiojima[2]
[1]*Kochi University of Technology*
[2]*TEC GIHAN Co., LTD*
Japan

1. Introduction

Integrating rehabilitation robots with human motion and force sensors for effective training and positive therapeutic effects is attracting more and more attentions in research and clinic fields (Bonato, 2010; Moreno et al., 2009). In order to control robots at the level of human motor control, the muscular activity of the lower limbs which has been estimated from measurements of joint moments and segment orientations may be useful information for biomedical applications (Wu et al., 2009; Lin et al., 2010). Kinematic and kinetic data have been widely collected using a high-speed camera system and force plate for the estimation of lower limb joint loads in laboratory environments (Shakoor et al., 2010; Wannop et al., 2010). However, these commonly used stationary devices for human dynamics analysis require lots of space, special operators, expensive instruments and complex calibration settings; moreover, the range of measurement is limited to capturing a few strides in a gait laboratory. The main shortcomings restrict the application of these stationary devices to experimental research and it is difficult to find applications of gait evaluation in the daily environment or clinic. As an alternative to these conventional techniques, some inexpensive and easy to use wearable measurement systems which can accurately estimate triaxial ground reaction force (GRF) and three-dimensional (3D) body orientations have been developed to implement human dynamics analysis and gait assessments in different environments (Bachlin et al., 2010; Veltink et al., 2005) .
Recently, some inexpensive in-chip inertial sensors including gyroscopes and accelerometers have been gradually coming into practical application in human motion analysis. To expand the scope of application of a mobile force plate system, a small 3D inertial sensor module can be integrated into the force plate. Liedtke et al. proposed a combination sensor system including six degrees of freedom force and moment sensors and miniature inertial sensors (provided by Xsens Motion Technologies) to estimate the joint moments and powers of the ankle (Liedtke et al., 2007). If 3D orientations of the foot are obtained and integrated with measured triaxial GRF during gait, an inverse dynamic method can be used to implement joint dynamics analysis of the lower limb (Schepers et al., 2007).
We are presently concentrating on the development of some wearable sensors to measure human GRF and segment orientations during gait. A multi-axial force sensor has been developed to measure triaxial GRF and the coordinates of the center of pressure, when fixed under a specially designed shoe (Liu et al., 2007). However, its hard interface and the weight

load on the foot affected normal walking according to our experimental tests. A thin and light force plate based on 3D tactile sensors and using lower-cost materials was proposed in our past research (Liu et al., 2009a), and a sensor matrix will be constructed to directly perform triaxial GRF measurements. Moreover, in order to quantify human movements, we have developed some wearable sensor modules using gyroscopes and accelerometers for ambulatory measurements of human segment orientations (). In this chapter, a mobile force plate and 3D motion analysis system (M3D) is introduced, which have been reported in our former publication (Liu et al., 2010). 3D inertial sensor modules which were designed using lower cost inertial sensors including a triaxial accelerometer and gyroscope were integrated into a newly developed force plate. Verification experiments were conducted to compare the estimation results of M3D with measurements performed on a stationary force plate. Finally, an application experiment is introduced to quantify and evaluate human gait. We measured the 3D GRF and orientations of feet using M3D to evaluate paralysis gait

2. Methods and materials

2.1 Mechanical design of mobile force plate

Small triaxial force sensors (USL06-H5-500N) provided by Tec Gihan Co. Japan can only detect the three-directional force induced on a small circular plate (Φ 6 mm), see Table 1, so it is difficult to apply directly to the measurement of the GRF distributed under a foot. As shown in Fig. 1 (a), a mobile force plate (weight: 156g, size: 80×80×15mm3) was constructed using three small triaxial force sensors, in which two aluminum plates were used as top and bottom plates to accurately fix the three sensors. Each small sensor, when calibrated using data provided by the manufacturer, can measure triaxial forces relative to their slave coordinate systems (∑si) defined on the center of each sensor, where subscript i represents the number of the small sensor in every force plate (i = 1, 2, and 3). The GRF and center of pressure (CoP) measured using the force plate so developed could be expressed in the force plate coordinate system (∑f) which is located at the interface between the force plate and the ground, with the origin of the force plate coordinate system taken as the center of the force plate (see Fig. 1 (b)). The y-axis of the force plate coordinate system was chosen to represent the anterior-posterior direction of human movement on the bottom plate, and the z-axis was made vertical, while the x-axis was chosen such that the resulting force plate coordinate system would be right-handed. We aligned the y-axis of each sensor's slave coordinate to the origin of the force plate coordinate system, while the three origins of the slave coordinates were evenly distributed on the same circle (radius: r = 30 mm), and were fixed 120° apart from each other. Fxi, Fyi and Fzi were defined as the triaxial forces measured using the three triaxial sensors. The triaxial GRF and coordinates of the CoP could be calculated from the following equations:

$$F_x = (F_{x_1} + F_{x_3}) \cdot \cos(60^\circ) - F_{x_2} - (F_{y_3} - F_{y_1}) \cdot \cos(30^\circ) \tag{1}$$

$$F_y = (F_{y_1} + F_{y_3}) \cdot \cos(60^\circ) - F_{y_2} - (F_{x_1} - F_{x_3}) \cdot \cos(30^\circ) \tag{2}$$

$$F_z = F_{z_1} + F_{z_2} + F_{z_3} \tag{3}$$

$$M_x = F_{z_2} \cdot r - (F_{z_1} + F_{z_3}) \cdot \sin(30^\circ) \cdot r \tag{4}$$

$$M_y = (F_{z_1} - F_{z_3}) \cdot \cos(30^\circ) \cdot r \tag{5}$$

$$M_z = (F_{x_1} + F_{x_2} + F_{x_3}) \cdot r \tag{6}$$

$$x_{cop} = M_y / F_z \tag{7}$$

$$y_{cop} = M_x / F_z \tag{8}$$

$$z_{cop} = 0 \tag{9}$$

where *Fx*, *Fy* and *Fz* are defined as the triaxial GRF (F_{GRF}) measured using the force plate in the force plate coordinate system, and *Mx*, *My* and *Mz* indicate triaxial moments estimated from measurements of the three sensors, while x_{COP}, y_{COP} and z_{COP} are the coordinates of the CoP,

Small triaxial force sensors

(a)

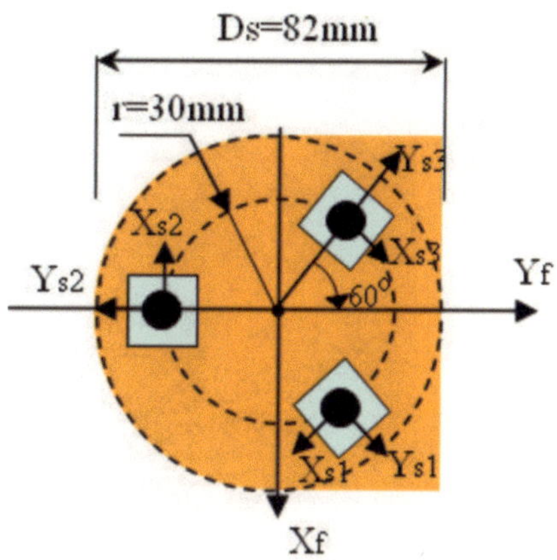

(b)

Fig. 1. (a) Prototype of a mobile force plate, (b) Coordinate systems of the force plate

Type		USL06-H5-500N
Rated Capacity (N)	X- and Y- axis	250
	Z-axis	500
Rated Capacity (με)	X- and Y- axis	900
	Z-axis	1700
Nonlinearity (After calibration of cross effect)		Within 1.0%
Hysteresis (After calibration of cross effect)		Within 1.0%
Size (mm)		20x20x5
Weight (g)		15

Table 1. Main specifications of the small triaxial force sensor used for the mobile force plate

In order to examine the inside force distribution of the mobile force plate, ANSYS FEA software was used to perform a static analysis and to simulate the effects of multi-axial forces which may be distributed over the three contact points of the small sensors on the top aluminum plate.

Fig. 2 shows the finite element mesh and a representative result of the deformation of the top plate which is attached to the small sensors using three M3 screws. When we load the top plate with a z-axial force Fz = 733.57 N (vertical pressure: 0.125 MPa) and y-axial force Fy = 263.5 N (spread over 527 nodes), and x-axial force Fx = 263.5 N (spread over 527 nodes), the induced three-directional forces on the small sensor can be calculated by the finite element method and the results are given in Table 2. The maximum force (274.6 N) on the z-axis of the three sensors, and the maximum x- and y-axis forces of 136.61 N never exceed the measurement capacity of the small sensor.

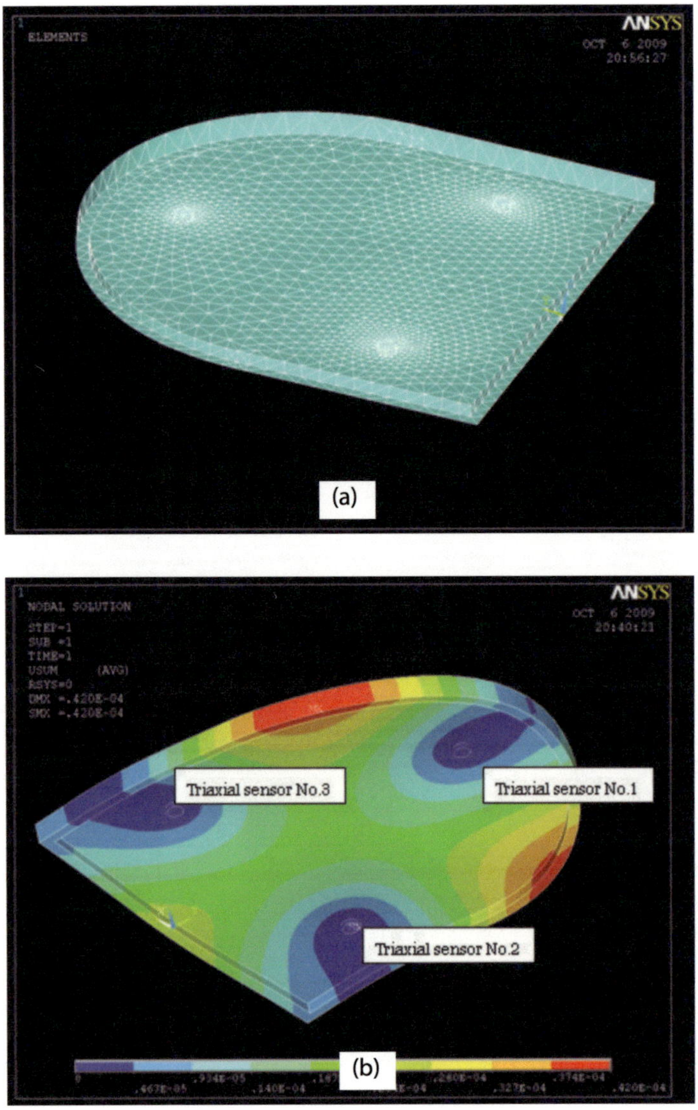

Fig. 2. (a) Finite element mesh, (b) Results of deformation plot

No.	Triaxial Forces (N)		
	X	Y	Z
1	89.48	136.61	274.6
2	59.19	69.03	219.7
3	114.83	58.16	239.27

Table 2. Three-directional forces on the small sensor when we load the force plate by Fz= 733.57N, Fy=263.5N and Fx=263.5N

2.2 Estimation of 3D orientation

As shown in Fig. 3, we constructed a 3D motion sensor module composed of a triaxial accelerometer (MMA7260Q, supplied by Sunhayato Co.) and three uniaxial gyroscopes (ENC-03R, supplied by Murata Co.) on the PCB board inside the mobile force plate. The module can measure triaxial accelerations and angular velocities which can be used to estimate a 3D orientation transformation matrix, so we can implement ambulatory GRF and CoP measurements using the combined system of mobile force plate and 3D motion analysis system (M3D).

Fig. 3. 3D motion sensor module constructed using three uniaxial gyroscopes and a triaxial accelerometer

We defined two local coordinate systems fixed to the two M3Ds under the heel and the forefoot as $\sum_{f\_heel}$ and $\sum_{f\_toe}$ respectively (see Fig. 4). The relative position of the two force plates was aligned using a simple alignment mechanism composed of three linear guides and a ruler to let the origins of $\sum_{f\_toe}$ be on the y-axis of $\sum_{f\_heel}$, and to let the y-axes of the two force plate coordinate systems be collinear, before we mounted them to a shoe. For calculation purposes, such as estimating joint moments and reaction forces of the ankle during loading response and terminal stance phases (Parry, 1992), all vectors including the joint displacement vector, GRF vector and gravity vector have to be expressed in the same coordinate system, that is the global coordinate system ($\sum_g$). Moreover, the origin and orientation of this global coordinate system are renewed for each foot placement to coincide with the heel force plate coordinate system ($\sum_{f\_heel}$), when the heel is flat on the ground.
The integration of the measured angular velocity vector ($\omega = [\omega x, \omega y, \omega z]$) in each force plate coordinate system was defined as $C = [Cx, Cy, Cz]$, which could be used to calculate the 3D

orientation transformation matrix (R) between the global coordinate system and a force plate coordinate system by solving the following equations proposed by Bortz (1970):

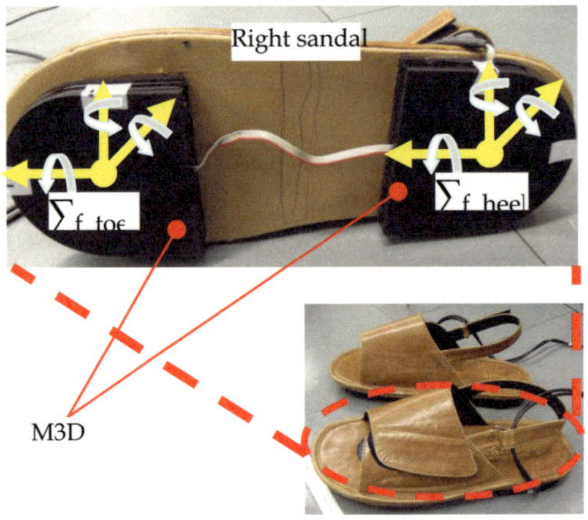

Fig. 4. M3D and the coordinate systems

$$C_i^{i+1} = [\omega x(i) + \omega x(i+1), \omega y(i) + \omega y(i+1), \omega z(i) + \omega z(i+1)] \cdot (\Delta t / 2) \tag{10}$$

$$\left|C_i^{i+1}\right| = \sqrt{(Cx_i^{i+1})^2 + (C_{yi}^{\ i+1})^2 + (Cz_i^{i+1})^2} \tag{11}$$

$$\begin{aligned} R_i^{i+1} = & \frac{C_i^{i+1} \cdot C_i^{i+1^T}}{\left|C_i^{i+1}\right|}(1 - \cos\left|(C_i^{i+1}\right|)) + \begin{bmatrix} \cos\left|(C_i^{i+1}\right|) & 0 & 0 \\ 0 & \cos\left|(C_i^{i+1}\right|) & 0 \\ 0 & 0 & \cos\left|(C_i^{i+1}\right|) \end{bmatrix} \\ & + \frac{\sin(\left|C_i^{i+1}\right|)}{\left|C_i^{i+1}\right|} \begin{bmatrix} 0 & -C_{z_i^{i+1}} & C_{y_i^{i+1}} \\ C_{z_i^{i+1}} & 0 & -C_{x_i^{i+1}} \\ -C_{y_i^{i+1}} & C_{x_i^{i+1}} & 0 \end{bmatrix} \end{aligned} \tag{12}$$

$$R = R_0 \cdot R_0^1 \cdot R_1^2 \cdot\cdot R_i^{i+1} \cdots \tag{13}$$

where [$\omega x(i)$, $\omega y(i)$, $\omega z(i)$] is a sample vector of the triaxial angular velocities of the force plate during a sampling interval Δt, *Cii+1* is an angular displacement vector in the sampling interval, and *R0* is an initial transformation matrix initialized as a unit matrix ($|R_0|=1$). If the force plate is flat on a level ground, we can update R according to $R = R_0$.

2.3 Transformation of triaxial GRF measured by mobile force plates

The triaxial GRF measured by the two M3Ds can be transformed to global coordinates and then combined to calculate the total GRF (F_{FRG}^{g}) and the global coordinate vectors of CoP ($[x,y,z]_{COP}^{g\_heel}$ and $[x,y,z]_{COP}^{g\_toe}$) using the following equations:

$$F_{FRG}^{g} = R_g^{f\_heel} \cdot F_{FRG}^{heel} + R_g^{f\_toe} \cdot F_{FRG}^{toe} \tag{14}$$

$$[x,y,z]_{COP}^{g\_heel} = R_g^{f\_heel} \cdot [x,y,z]_{COP}^{heel} \tag{15}$$

$$[x,y,z]_{COP}^{g\_toe} = R_g^{f\_toe} \cdot [x,y,z]_{COP}^{toe} \tag{16}$$

where F_{FRG}^{heel} and F_{FRG}^{toe} are the triaxial GRF measured by the two M3Ds under the heel and forefoot with their respective coordinate systems; $[x,y,z]_{COP}^{heel}$ and $[x,y,z]_{COP}^{toe}$ are coordinate vectors of CoP measured using the two M3Ds; $R_g^{f\_heel}$ and $R_g^{f\_toe}$ are the orientation transformation matrices of the two M3Ds for transforming the triaxial GRF measured by the two M3Ds in their attached coordinate systems into the measurement results relative to the global coordinate system.

3. Experimental study

3.1 Verification experiment

A stationary TF-4060-A force plate (Tec Gihan Co. Japan) was used as a reference measurement system to verify the measurement results of the M3D system being developed. As shown in Fig. 5, a young volunteer wearing M3D was asked to walk on the stationary force plate and the signals from the two measurement systems were simultaneously sampled at a rate of 100 samples/s, after a trig signal was sent from the data logger of the M3D.

First, a static test experiment was conducted to validate the triaxial force measurement of the M3D without movement. Only one foot wearing the M3D is put on the stationary force plate and the subject arbitrarily moved his center of pressure. As shown in Fig. 6, the triaxial measurement results obtained with the stationary force plate (FP) and M3D almost completely overlap and the maximum errors in the triaxial force measurements were less than 5% of the corresponding maximum forces. Second, in order to verify the M3D ambulatory measurement, a dynamic test was performed on a walking measurement, in which the subject was asked to step on the force plate at a normal speed of about one step/s (see Fig. 7). The verification experiment results indicate that the sensor can measure the triaxial force with high precision (error: less than 6.4% of the maximum measurement force) under static and dynamic working conditions.

3.2 Measurement of paralysis gait using M3D

I As an application of the research, experiments were performed to quantitatively compare and analyze normal walking and paralysis walking using the M3D. The main features of paralysis gait can be summarized as follows: the toe on the paralyzed side rotates to the outside with a larger angle than in normal gait; the knee is stretched to the outside during the swing phase. A healthy subject was trained to imitate the walking feature of paralysis, and we separately measured the imitated paralysis gait of the left leg and right leg. The walking distance of the experimental tests is about 6 m.

Figs. 8 and 9 give the vertical components of GRF on the two feet (Solid blue line: GRF on the left foot; Solid pink line: GRF on the right foot) measured with the M3D system in normal gait and on the right foot imitating paralysis gait, respectively. We note that there are no large differences in the shape of the vertical force (z-axial force) curves induced on

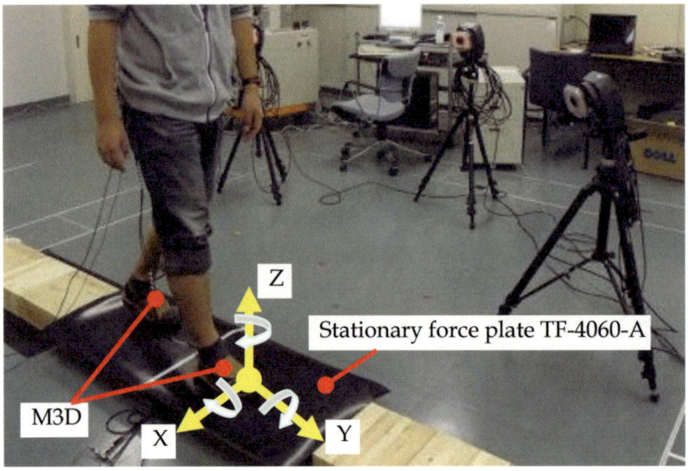

Fig. 5. Verification experiments to validate the measurements of the M3D

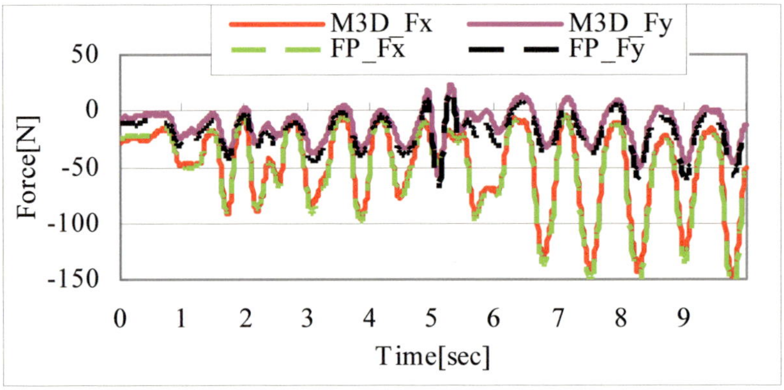

(a) X and Y - axial forces when standing on the stationary force plate

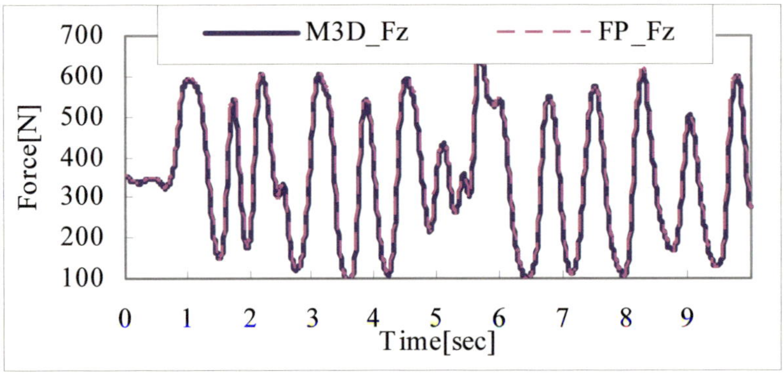

(b) Z-axial force when standing on the stationary force plate

Fig. 6. Experiment results of the static test

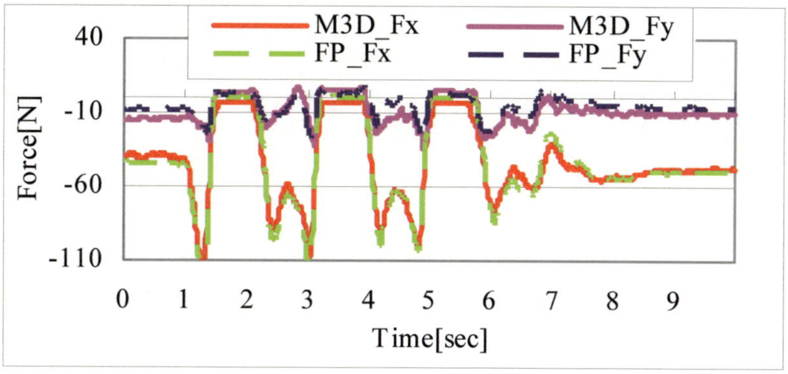

(a) X and Y - axial forces when stepping on the stationary force plate

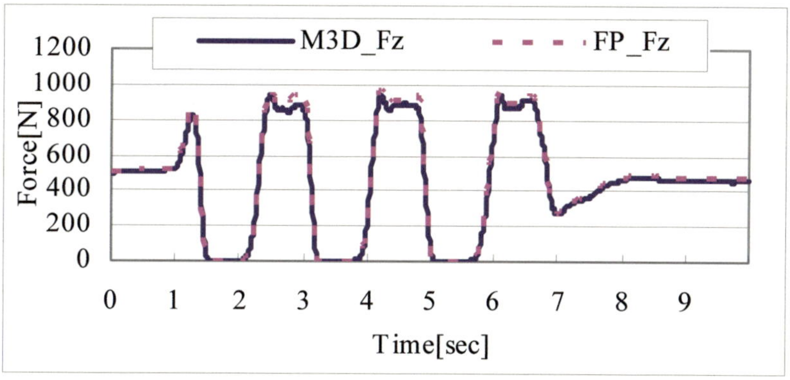

(b) Z-axial force when stepping on the stationary force plate

Fig. 7. Experiment results of the dynamic test

the right and left foot during normal walking, and that the GRF on the two feet have good balance and symmetry during continuous strides. When we compare the curves in Fig. 9 with the normal gait data, we can clearly note that the two peaks of the z-axial force on the left foot which is not on the paralyzed side are depressed. Moreover, it has been understood that the stance phase period of the healthy leg (the left leg) was about 1.6 times longer than the paralyzed side (the right leg) during a stride.

The rotation angles of the toe and heel of the feet around the medial-lateral direction (x-axis, see Fig. 5) are shown in Fig. 10, in which the dotted lines indicate the movements of the right foot and the rotation angles of the left foot are plotted with solid lines. The positive angle values represent the plantar flexion of the foot segments, and the negative values indicate dorsal flexion. The flexion angles of the paralysis foot (the right foot) are reduced significantly and the heel-strike angles and toe-off angles were less than 20 degrees. Moreover, it is noted that the toe joint of the paralysis foot was almost never rotated during the gait, because the heel and toe had the same flexion angles during the entire walking measurements. We also obtained similar results in the measurements of left leg paralysis using the M3D.

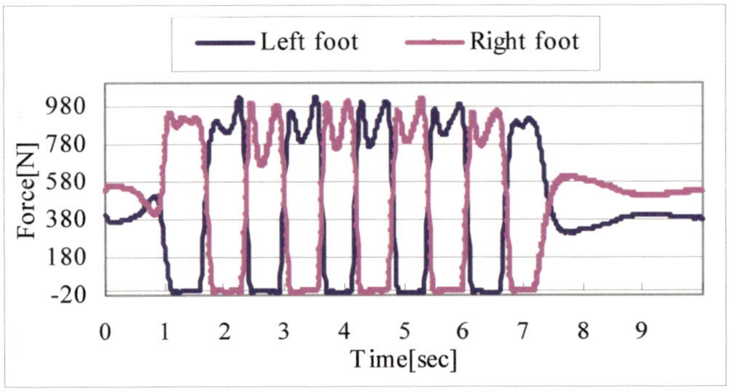

Fig. 8. Z-axis force during normal walking

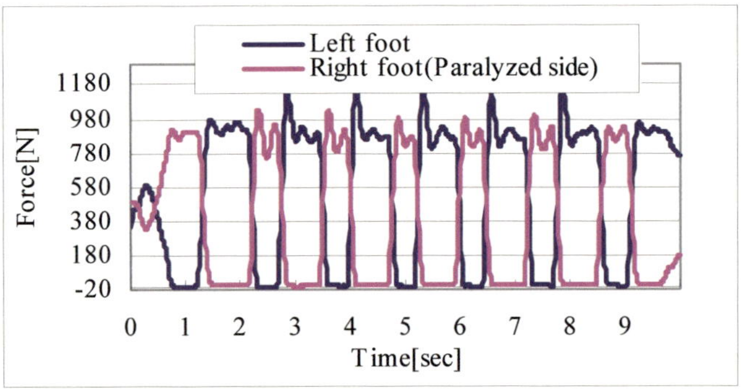

Fig. 9. Z-axis force during paralysis gait

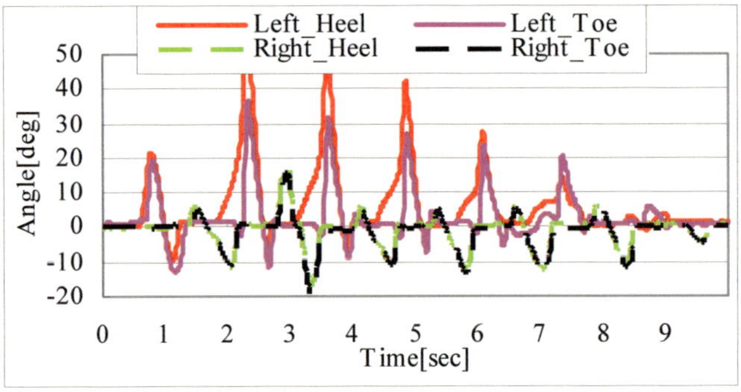

Fig. 10. X-axial angles of the feet during paralysis gait

4. Discussions and conclusions

A mobile force plate and 3D motion analysis system (M3D) was developed using lower cost inertial sensor chips and small triaxial force sensors. In order to apply the system to human gait evaluation, verification experiments were implemented to compare the results estimated by M3D with measurements made with a stationary force plate. In the static tests, the force measurements by the M3D system along three axes were highly correlated in both amplitude and dynamic response to the reference measurements using the stationary force plate (see Fig. 6) and this verifies that the M3D system could measure the triaxial GRF in its fixed local coordinate system with acceptable precision (less than 5% of the corresponding maximum force). However, as shown in Fig. 7, there were larger errors in the triaxial GRF measurements. The most likely source of amplitude error in the triaxial GRF measurement was in the orientation estimate of M3D movements using a triaxial accelerometer which could only implement x- and y-axial angular displacement re-calibration. In the future, we will integrate a triaxial magnetic sensor (Zhu & Zhou, 2009) for estimating the heading angle (z-axial angular displacement) during gait, because the z-axial (vertical) cumulative error induced by the drift effect of the gyroscope sensor could be re-calibrated using measurements from the magnetic sensor. Since only straight level walking was tested with M3D, it is necessary to examine more movements to verify ambulatory measurements obtained with M3D in future gait experiments. Moreover, the sensitivity of results to initial sensor drift and initial orientation fix will be addressed so that the system can be applied to many other applications.

In our research application using the new system, the quantitative differences between paralysis gait and normal gait were analyzed based to the results of z-axial GRF (Fig. 9) and x-axial angular flexions (Fig. 10). In clinical applications, the quantitative analysis of gait variability using kinematic and kinetic characterizations can be helpful to medical doctors in monitoring patient recovery status. Moreover, these quantitative results may help to strengthen their confidence in rehabilitation. Walking speed, stride length, center of mass (CoM) and CoP have been considered as influencing factors in evaluations of human gait (Lee & Chou, 2006). In this paper, only the z-axial GRF (vertical force) and x-axial orientation were analyzed to evaluate different gaits. However, according to one study on slip type falls (Chang et al., 2003), the friction force was used to draw up important safety criteria for detecting safe gait, so the transverse components of GRF may provide important information when quantifying gait variability. The M3D system can be used to obtain multi-dimensional motion and force data on successive gait in non-laboratory environments, so we will develop a new method based on measurements from the mobile system for quantifying gait variability. Moreover, a statistical analysis of the multi-dimensional GRF and orientation data extracted from successive gait measurements will be used to evaluate normal and pathological gait.

5. References

Bonato, P. (2010). Wearable sensors and systems, *IEEE Engineering in Medicine and Biology Magazine*, 29 (3), pp. 25-36.

Bachlin, M.; Plotnik, M.; Roggen, D.; Maidan, I.; Hausdorff, J.M.; Giladi, N. & Troster, G. (2010). Wearable assistant for Parkinsons disease patients with the freezing of gait symptom, *IEEE Transactions on Information Technology in Biomedicine*, 14 (2), pp. 436-446.

Bortz, E. (1970). A new mathematical formulation for strapdown inertial navigation, *IEEE Trans. Aero. Ele.*, 7(1), pp.61-66.

Chang, W.R.; Courtney, T. K. ; Gronqvist, R. & Redfern, M.S. (2003). *Measuring slipperiness: human locomotion and surface factors*, CRC Press, Chapter 1.

Lee, H.J. & Chou, L.S. (2006). Detection of gait instability using the center of mass and center of pressure inclination angles, *Archives of Physical Medicine and Rehabilitation*, 87, pp.569-575.

Liedtke, C.; Fokkenrood, W.; Menger, T.; van der Kooij, H. & Veltink, H. (2007). Evaluation of instrumented shoes for ambulatory assessment of ground reaction forces, *Gait and Posture*, 26(1), pp.39-47.

Lin, Y.C.; Walter, J.P.; Banks, S.A.; Pandy, M.G. & Fregly, B.J. (2010). Simultaneous prediction of muscle and contact forces in the knee during gait, *Journal of Biomechanics*, 43 (5), pp. 945-952.

Liu, T.; Inoue, Y. & Shibata, K. (2007). Wearable Force Sensor with Parallel Structure for Measurement of Ground-reaction Force, *Measurement*, 40(6), pp.644-653.

Liu, T.; Inoue, Y. & Shibata, K. (2009a). A Small and Low-cost 3D Tactile Sensor for a Wearable Force Plate, *IEEE Sensors Journal*, 9(9), pp.1103-1110.

Liu, T.; Inoue, Y. & Shibata, K. (2009b). Development of a Wearable Sensor System for Quantitative Gait Analysis, *Measurement*, 42(7), pp. 978-988.

Liu, T.; Inoue, Y.; Shibata, K.; Hirota, Y. & Shiojima K. (2010). A Mobile Force Plate System and Its Application to Quantitative Evaluation of Normal and Pathological Gait, *Proceedings of IEEE/ASME International Conference on Advanced Intelligent Mechatronics*, Montreal, Canada.

Moreno, J.C.; Brunetti, F.; Navarro, E.; Forner-Cordero, A. & Pons, J.L. (2009). Analysis of the human interaction with a wearable lower-limb exoskeleton, *Applied Bionics and Biomechanics*, 6 (2), pp. 245-256.

Parry, J. (1992). *Gait analysis normal and pathological function, Slack Incorporated*, pp.149-158.

Schepers, M.; Koopman, M. & Veltink, H. (2007). Ambulatory assessment of ankle and foot dynamics, *IEEE Trans. Biomed. Eng.*, 54(5), pp.895-900.

Shakoor, N.; Sengupta, M.; Foucher, K.C.; Wimmer, M.A.; Fogg, L.F. & Block, J.A. (2010). Effects of common footwear on joint loading in osteoarthritis of the knee, *Arthritis Care and Research*, 62 (7), pp. 917-923.

Veltink, H.; Liedtke, C.; Droog, E. & Kooij, H. (2005). Ambulatory measurement of ground reaction forces, *IEEE Trans Neural Syst Rehabil Eng.*, 13(3), pp.423–527.

Wannop, J.W.; Worobets, J.T. & Stefanyshyn, D.J. (2010). Footwear traction and lower extremity joint loading, *American Journal of Sports Medicine*, 38 (6), pp. 1221-1228.

Wu, J.Z.; Chiou, S.S. & Pan, C.S. (2009). Analysis of musculoskeletal loadings in lower limbs during stilts walking in occupational activity, *Annals of Biomedical Engineering*, 37(6), pp.1177-1189.

Zhu, R. & Zhou, Z. (2009). A small low-cost hybrid orientation system and its error analysis, *IEEE Sensors Journal*, 9(3), pp.223-230.

2

Motion Control of Biped Lateral Stepping Based on Zero Moment Point Feedback for Adaptation to Slopes

Satoshi Ito and Minoru Sasaki
Gifu University
Japan

1. Introduction

Biped locomotion consists of both sagittal and lateral (frontal) plane motions. Although the stability of the locomotion must be ensured in both the planes, their natures are different. In the sagittal plane, the main purpose is to move from one place to another; thus, the stability is dynamic – losing static balance is essential in sagittal plane motion; it produces tumble for travel. In the lateral plane, on the other hand, maintaining an upright posture is crucial. Hence, lateral stability is static, and stabilizing a saddle point in the phase plane of the inverted pendulum motion is the main challenge.
In general, the zero moment point (ZMP) criterion is utilized for biped motion control (Kagami et al., 2002; Mitobe et al., 2001; Nagasaka et al., 1999; Suleiman et al., 2009; Yamaguchi & Takanishi, 1997). Although this method is effective and useful, planned motion using this method is not suitable when the environmental conditions change from those considered during motion planning. The literature offers excellent reports on the modification of planned motion (Hirai et al., 1998; Huang et al., 2000; Kulvanit et al., 2005; Lee et al., 2005; Napoleon & Sampei, 2002; Prahlad et al., 2007; Wollherr & Buss, 2004), or online motion generation (Behnke, 2006; Czarnetzki et al., 2009; Héliot & Espiau, 2008; Kajita & Tani, 1996; Nishiwaki et al., 2002; Sugihara et al., 2002) that solve this problem.
Usually, motion planning based on the ZMP criterion is applied to both the sagittal and lateral planes. The concept of this paper is that motion planning in the lateral plane can be skipped because of the difference in the nature of its stability. In the sagittal plane, motion planning is certainly crucial: one cannot proceed without actively generating both leg swing and torso behaviour, followed by the planned motion. The ZMP method was originally proposed to design such co-ordinated motions. However, in the lateral plane, balance is the primary purpose; generating active motion is a secondary problem. Nonetheless, in the ZMP method, the motion is first planned, and balance is maintained as a result of exact tracking of the planned motion. In our opinion, the process should be reversed for motion in the lateral plane, with balance control coming first and motion emerging as a result of balance control. From this viewpoint, trajectory generation for the lateral plane should be eliminated by setting balance as the control object.

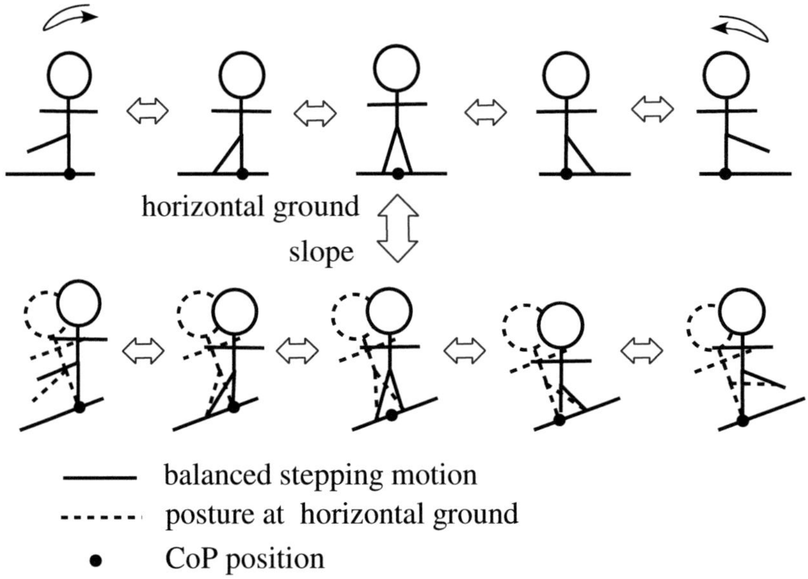

Fig. 1. Lateral stepping motion on different gradients.

To maintain balance without motion planning, we introduce the direct centre of pressure (CoP), i.e. ZMP control (Ito et al., 2003; 2007; 2008) (the CoP is equivalent to the ZMP). Instead of an indirect balancing method, such as tracking the positions of trajectories planned using the ZMP criterion, we select the ZMP directly as the control variable.

Adaptive lateral motion should result without adjusting the controllers or motion pattern generators. This arises from the invariance in the ZMP trajectory in biped lateral motion. Lateral motion on flat and sloped floors is illustrated in Fig. 1. To maintain balance, the motion trajectories of the torso and legs must change adaptively in relation to the angle of the slope. On the other hand, the ZMP trajectory, indicating the time stamp of the load centre, is invariable. Therefore, balance control based on direct ZMP control can naturally produce adaptive motion without re-designing the motion trajectories. In this chapter, we explain a balance control strategy based on direct ZMP feedback and confirm the effectiveness of this method by conducting experiments of improved robot from our previous papers (Ito et al., 2007; 2008).

This chapter is organized as follows: the next section presents the mathematical framework; section 3 describes a control method based on the direct ZMP control; section 4 reports on robot experiments as well as simulation of lateral stepping motion and the section 5 presents our conclusions.

2. Basic Theory of balance control

2.1 Inverted Pendulum model

2.1.1 Assumptions

The CoP is the representative action point of the ground reaction forces and coincides with the ZMP (Goswami, 1999). Because the ZMP contains significant information on balance, the ground reaction forces are also expected to contain the information.

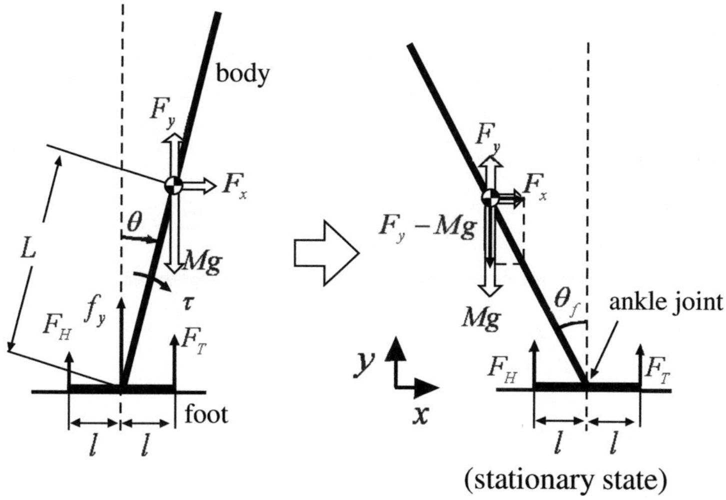

Fig. 2. Inverted pendulum model for biped balance.

From this viewpoint, feedback control of ground reaction forces is introduced for balance control (Ito & Kawasaki, 2005). Because the ankle strategy is dominant for balancing with respect to small disturbances (Horak & Nashner, 1986), the inverted pendulum model illustrated in the left of Fig. 2 is considered with the following assumptions:

- The motion occurs only in the sagittal plane.
- The body (inverted pendulum) and the foot (support) are connected at the ankle joint.
- The foot does not slip on the ground.
- The shape of the foot is symmetrical in the anterior-posterior direction.
- The foot has two ground contact points: the heel and the toe.
- The vertical component of the ground reaction force is measurable.
- The ankle joint is located at the midpoint of the foot with zero height.
- The ankle joint angle and its velocity are detectable.
- An appropriate torque is actively generated at the ankle joint.
- An unknown constant external force is exerted at the centre of gravity (CoG).

The notations are defined as follows: M and m are the mass of the body and foot link, respectively; I is the moment of inertia of the body link around the ankle joint; L is the length between the ankle joint and the CoG of the body link; ℓ is the length from the ankle joint to the toe or the heel; θ is the ankle joint angle; $\dot{\theta}$ is its velocity and τ is the ankle joint torque. F_H and F_T are the vertical components of the ground reaction force at the heel and the toe, respectively. f_y is the vertical component of the internal force between the two links. F_x and F_y are the horizontal and vertical components of constant external force, respectively, and g is the gravitational acceleration.

2.1.2 Control law

The goal of the control is to maintain the postural balance regardless of the constant external forces F_x and F_y. With respect to the stability margin (McGhee & Frank, 1968), F_T and F_H should be kept equal, indicating that the ZMP is held to the centre under the foot. The following control method achieves this.

Theorem 1. *Define an ankle joint torque τ by using adequate feedback gains K_d, K_p, K_f and adequate constant θ_d as*

$$\tau = -K_d\dot{\theta} + K_p(\theta_d - \theta) + K_f \int (F_H - F_T)dt. \tag{1}$$

Then, $\theta = \theta_f$ becomes a locally asymptotically stable posture, and $F_H = F_T$ holds at the stationary state. Here, θ_f is a constant satisfying

$$\sin\theta_f = -\frac{F_x}{A}, \ \cos\theta_f = \frac{Mg - F_y}{A}. \tag{2}$$

where

$$A = \sqrt{(Mg - F_y)^2 + F_x^2} \tag{3}$$

Proof. The motion equation of the body link is described as

$$\begin{aligned} I\ddot{\theta} &= MLg\sin\theta + F_x L\cos\theta - F_y L\sin\theta + \tau. \\ &= AL\sin(\theta - \theta_f) + \tau \end{aligned} \tag{4}$$

On the other hand, the ground reaction forces, with ankle joint torque, are

$$F_T = -\frac{1}{2\ell}\tau + \frac{1}{2}mg + \frac{1}{2}f_y, \tag{5}$$

$$F_H = \frac{1}{2\ell}\tau + \frac{1}{2}mg + \frac{1}{2}f_y, \tag{6}$$

Here, a new state variable τ_f is defined as

$$\tau_f = \int (F_H - F_T)dt. \tag{7}$$

Then, the control law (1) becomes

$$\tau = -K_d\dot{\theta} + K_p(\theta_d - \theta) + K_f\tau_f. \tag{8}$$

which is regarded as a state feedback whose states are θ, $\dot{\theta}$ and τ_f. In addition, differentiating (7) and then substituting (5) and (6) results in

$$\dot{\tau}_f = \frac{1}{\ell}\tau. \tag{9}$$

An equilibrium point $(\bar{\theta}, \bar{\dot{\theta}}, \bar{\tau}_f)$ of the dynamics in (4) and (9) with control law (8) is obtained by setting the time-derivative term as zero. It is given as

$$(\bar{\theta}, \bar{\dot{\theta}}, \bar{\tau}_f) = (\theta_f, 0, \frac{K_p}{K_f}(\theta_f - \theta_d)). \tag{10}$$

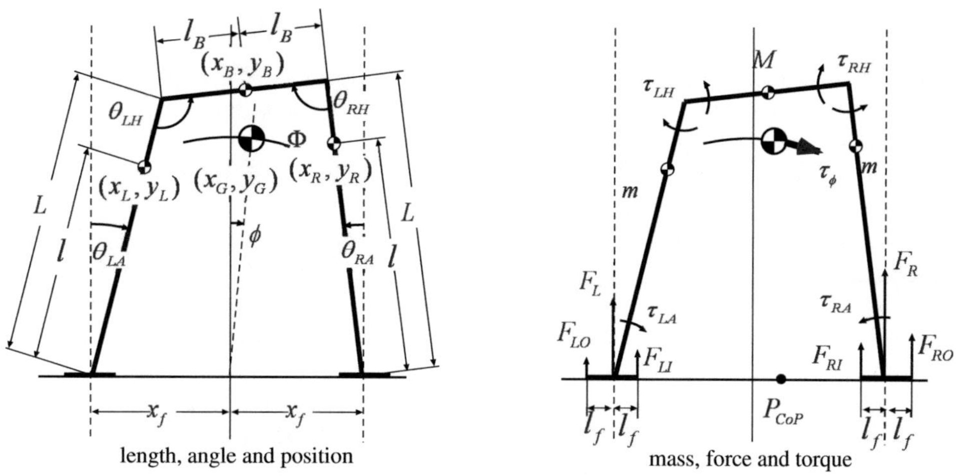

Fig. 3. Lateral sway model for biped balance.

In this state, $\tau = 0$ holds according to (8) and (10), indicating that $F_H = F_T$. The local stability of this equilibrium point is ensured by the controllability of the linearized dynamics around this point. □

2.1.3 Behaviour

The stationary posture from the control law in (1) is illustrated in the right of Fig. 2. This stationary state θ_f depends not on θ_d but the external forces F_x and F_y. It follows that the stationary posture changes adaptively with respect to the environmental conditions expressed as unknown constant external forces. This posture allows the ankle joint torque to be zero in the stationary state, since the moment of the external force is balanced by that of the gravity around the ankle joint. This is an advantage of the control law, in addition to being a model-free property.

2.2 Lateral Sway model

2.2.1 Assumptions

Here, we extend the control law in (1) to active lateral sway with double support. Because the flexion of knee joints in this motion is small, each leg is represented by only one link, without a knee, as shown in Fig. 3. Thus, the following assumptions are introduced:

- The motion is restricted within the lateral plane.
- The lateral motion is approximately represented using a 5-link model consisting of one body, two legs and two feet.
- The foot does not slip on the ground.
- Ankle joints are assumed to be located at the centre of the foot with zero height.
- At the end of both sides, the feet contact the ground.
- The vertical component of the ground reaction forces is measurable.
- The angles and velocities are detectable at the ankle and hip joints.
- Every joint is actively actuated.

- An unknown constant external force is exerted on the CoG of the entire body.

Assume that the feet always maintain contact with the ground. This constraint forces the mechanism to be a closed link constructed by the body and two legs, indicating that the degree of freedom (DoF) of motion is reduced to one.
Here, the following notations are defined: (x_G, y_G) denotes the CoG position in the coordinate frame whose origin is set at the midpoint between two ankle joints, ϕ is a lateral sway angle in this coordinate frame. (x_R, y_R), (x_L, y_L) and (x_B, y_B) are the CoG of the right leg, left leg and body (pelvis), respectively. L is the length of the leg, ℓ is the length from the ground to the CoG of the leg, ℓ_B is the half length of the body, ℓ_f is the length from the ankle joint to the side of the foot and x_f is the distance to the ankle joint from the origin of this coordinate frame. F_{RO}, F_{RI}, F_{LO}, and F_{LI} are the vertical components of ground reaction forces at four contact points, whose subscripts $_{RO}$, $_{RI}$, $_{LI}$ and $_{LO}$ represent the positions of the contact points, indicating the right outside, the right inside, the left inside and the left outside, respectively. $\boldsymbol{F} = [F_x, F_y]^T$ is the external force that is assumed to be constant. $\Theta = [\theta_{RA}, \theta_{RH}, \theta_{LH}, \theta_{LA}]^T$ is a joint angle vector whose elements are the joint angles of the right ankle, right hip, left hip and left ankle, respectively, and $\mathbf{T} = [\tau_{RA}, \tau_{RH}, \tau_{LH}, \tau_{LA}]^T$ is a joint torque vector whose elements are the torque at each joint. τ_ϕ is a generalized force defined in the coordinate frame on the CoG orbit Φ, and P_{ZMP} is the position of the ZMP.

2.2.2 Control law

Under these assumptions, P_{ZMP} is calculated from the magnitude of the ground reaction forces at the four contact points as follows:

$$P_{ZMP} = \frac{F_{RO}}{F_{all}}(x_f + \ell_f) + \frac{F_{RI}}{F_{all}}(x_f - \ell_f) - \frac{F_{LI}}{F_{all}}(x_f - \ell_f) - \frac{F_{LO}}{F_{all}}(x_f + \ell_f), \tag{11}$$

where

$$F_{all} = F_{RO} + F_{RI} + F_{LI} + F_{LO}. \tag{12}$$

The purpose is to control the position of the ZMP at its reference position P_d in the lateral sway model, as shown in Fig. 1. Here, P_d is appropriately planned in advance and may be constant or, alternatively, switched. This is achieved using ZMP feedback obtained by extending theorem 1.

Theorem 2. *Define a generalized force τ_ϕ based on P_{ZMP} as*

$$\tau_\phi = -K_d\dot{\phi} + K_p(\phi_d - \phi) + K_f\int(P_d - P_{ZMP})dt, \tag{13}$$

and assign each joint torque $\mathbf{T}$ so that the following equation holds

$$\tau_\phi = \mathbf{J}^T(\Theta)\mathbf{T}. \tag{14}$$

Here, ϕ_d is a constant, and $\mathbf{J}(\Theta)$ is a Jacobian matrix that relates the deviation of Θ to that of ϕ

$$\Delta\Theta = \mathbf{J}(\Theta)\Delta\phi. \tag{15}$$

Then, P_{ZMP} converges to P_d if it starts in the neighbourhood of P_d.

Proof. Because there is only one DoF of the lateral sway model, the sway angle ϕ uniquely determines each joint angle Θ in the range $0 < \theta_{RH} < \pi, 0 < \theta_{LH} < \pi$. Here, this relationship is described as $\Theta = \Theta(\phi)$. Then, the equation of motion with respect to ϕ is obtained as

$$M(\Theta)\ddot{\phi} + C(\Theta, \dot{\Theta}) + G(\Theta, g, \mathbf{F}) = \tau_\phi. \tag{16}$$

On the other hand, the relationship between P_{ZMP} and τ_ϕ is given as

$$P_{ZMP} = P(\Theta)\tau_\phi + Q(\Theta, \dot{\Theta}) + R(\Theta, g, \mathbf{F}). \tag{17}$$

Here, $M(\Theta) > 0$ is an inertia term, $C(\Theta, \dot{\Theta})$ and $Q(\Theta, \dot{\Theta})$ become the second order terms of the element of $\dot{\Theta}$, G and R contain both the gravity term and external force $\mathbf{F}$. See Appendix 7.3 for the derivation of (16) and (17). Then, a new variable τ_f is introduced:

$$\tau_f = \int (P_{ZMP} - P_d)dt. \tag{18}$$

The differentiation of τ_f provides the relationship

$$\dot{\tau}_f = P_{ZMP} - P_d. \tag{19}$$

And, using (17), it becomes

$$\dot{\tau}_f = P(\Theta)\tau_\phi + Q(\Theta, \dot{\Theta}) + R(\Theta, g, \mathbf{F}) - P_d. \tag{20}$$

In addition, the control law in (13) is described using τ_f

$$\tau_\phi = -K_d\dot{\phi} + K_p(\phi_d - \phi) + K_f\tau_f. \tag{21}$$

Let $[\phi, \dot{\phi}, \tau_f]^T$ be state variables of the dynamics of (16) and (20) with the control law in (21). At the equilibrium point, the derivative terms are forced to zero, indicating that $\dot{\tau}_f = 0$ in (19); thus, $P_{ZMP} = P_d$. To test the stability of the equilibrium point, (16) and (20) are linearized around it.

$$\dot{\xi} = \begin{bmatrix} 0 & 1 & 0 \\ -\bar{G}_\theta \bar{J}/\bar{M} & 0 & 0 \\ (\bar{R}_\theta + \bar{P}_\theta \bar{\tau}_\phi)\bar{J} & 0 & 0 \end{bmatrix} \xi + \begin{bmatrix} 0 \\ 1/\bar{M} \\ \bar{P} \end{bmatrix} \Delta\tau_\phi \tag{22}$$

Here, $\xi = [\Delta\phi, \Delta\dot{\phi}, \Delta\tau_f]^T$ is a deviation from the equilibrium point, $\bar{M} = M(\bar{\Theta})$, $\bar{\Theta} = \Theta(\bar{\phi})$, $\bar{J} = J(\bar{\Theta})$, $\bar{P} = P(\bar{\Theta})$, $\bar{G}_\theta = \dfrac{\partial G(\bar{\Theta})}{\partial \Theta}$, $\bar{R}_\theta = \dfrac{\partial R(\bar{\Theta})}{\partial \Theta}$, $\bar{P}_\theta = \dfrac{\partial P(\bar{\Theta})}{\partial \Theta}$, and $\Delta\tau_\phi$ is a deviation from the input at the equilibrium $\bar{\tau}_\phi = K_p(\phi_d - \bar{\phi}) + K_f\bar{\tau}_f$. The controllability matrix M_c of this linear system becomes

$$M_c = \begin{bmatrix} 0 & 1/\bar{M} & 0 \\ 1/\bar{M} & 0 & -\bar{G}_\theta \bar{J}/\bar{M}^2 \\ \bar{P} & 0 & (\bar{R}_\theta + \bar{P}_\theta \bar{\tau}_\phi)\bar{J}/\bar{M} \end{bmatrix}, \tag{23}$$

whose determinant is calculated as

$$|M_c| = -\frac{1}{\bar{M}^3}(\bar{P}\bar{G}_\theta + \bar{R}_\theta + \bar{P}_\theta \bar{G})\bar{J} = -\frac{1}{\bar{M}^3}\frac{\partial}{\partial \phi}(PG + R)\bigg|_{\phi = \bar{\phi}}. \tag{24}$$

Here, the relationship $\bar{\tau}_\phi = \bar{G} = G(\bar{\Theta})$ from (16) and $\bar{J} = \frac{\partial \Theta}{\partial \phi}$ were applied. To verify $|M_c| \neq 0$, the deviation of the ZMP position is considered. Substituting (16) into τ_ϕ of (17) and linearizing (17) around the equilibrium point results in

$$\Delta P_{ZMP} = PM\Delta\ddot{\phi} + \frac{\partial}{\partial \phi}(PG + R)\bigg|_{\phi=\bar{\phi}} \Delta\phi. \tag{25}$$

This equation implies that the ZMP deviation depends on both the inertial force (the first term) and the gravitational effect (the second term), which varies with the posture, i.e. the CoG position. This is consistent with the definition of the ZMP – it is determined by the inertial and gravitational forces. Now, assume $|M_c| = 0$. Then, $\frac{\partial}{\partial \phi}(PG + R)\big|_{\phi=\bar{\phi}} = 0$ from (24). This produces the conclusion, based on (25), that the ZMP position does not change regardless of the CoG deviation. This contradicts the definition of ZMP; thus, $|M_c| \neq 0$ is ensured. Accordingly, the controllability matrix M_c should be full rank and the linear system is controllable – the equilibrium point can be stabilized by adequate K_d, K_p and K_f in (13). Finally, note that we can find joint torque $\mathbf{T}$ to satisfy the relationship in (14). □

2.2.3 Behaviour

The behaviour of the lateral sway model under control laws (13) - (15) is expected to be similar to that of the inverted pendulum model using control law (1) discussed in the section 2.1.3, i.e., in the stationary state:

- The ZMP is controlled to its reference position P_d.
- The posture changes with the external force.
- The generalized force τ_ϕ becomes zero due to the balance between the gravitational and external forces.

Thus, this control law is a natural extension of control law (1) when there are multiple contact points and active joints.

3. Control of in-place stepping

3.1 Strategy

Here, we focus on in-place stepping motion to achieve it without generating reference trajectories of joint angles, as expected in section 1. The stepping motion is divided into single- and double-support phases. The control law is defined separately in these two phases, and then, two theorems from the previous section are applied, since this task basically involves the stabilization of the inverted pendulum with respect to external forces caused by ground gradients. However, some extensions are needed: definition of the switching conditions between the two control laws and the time-dependent reference for the ZMP position. The local stability of the control laws will ensure tracking of the ZMP position to the time-dependent reference.

3.2 Control

3.2.1 Single-support phase

On a slope, adaptive behaviour is observed – the body tilts around the ankle joint of the supporting leg, as shown in the bottom of Fig. 1. Thus, the ankle joint plays a significant role,

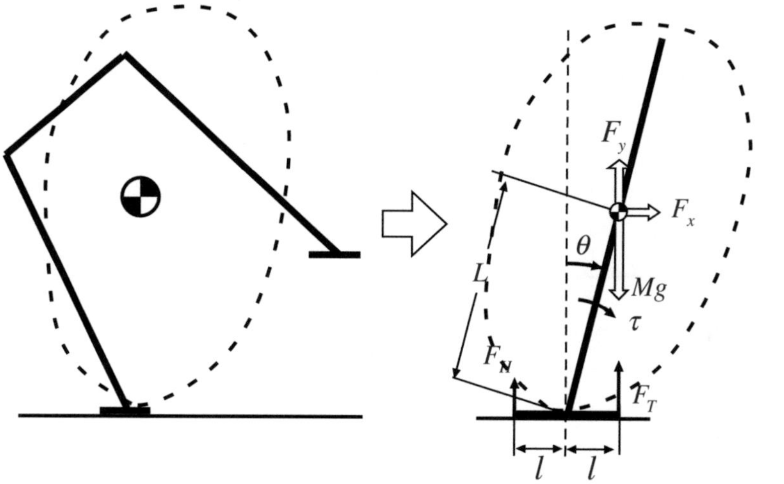

Fig. 4. Single-support phase approximation by using the inverted pendulumn model.

and the dynamics of the single-support phase can be approximated by an inverted pendulum with a foot support, as shown in Fig. 4. Under this approximation, theorem 1 is applicable by regarding the effect of the slope as well as the swing leg dynamics as unknown external forces F_x and F_y. The flow of the balance control is summarized as

1. Detect the angle and its velocity at the ankle joint of the support leg.
2. Detect the ground reaction forces at both ends of the supporting foot.
3. Calculate the ankle joint torque according to (1).
4. Output the ankle joint torque with its actuator.

The trajectory tracking control should be introduced to lift the swing leg.

3.2.2 Switching from single- to double-support phase

Control law (1) is expected to compensate for disturbances caused by the torso and swing leg when stepping. If the torso and swing leg motions are adequately controlled, the posture of the initial state of the single-support phase will be recovered. Thus, the switch condition of the control law is set as the recovery of the initial posture.

3.2.3 Double-support phase

To change to the other support leg, the ZMP position must shift from under the current supporting leg to the other. Control laws (13) - (15) are expected to make the ZMP track such a reference position P_d. Following is the control flow:

1. Detect the angle and its velocity at the ankle and hip joints.
2. Detect the ground reaction forces at both ends of the feet.
3. Calculate the lateral sway angle ϕ by following the next relationship (Appendix 7.1):

$$\phi = \frac{\theta_{LA} - \theta_{RA}}{2}. \tag{26}$$

4. Calculate P_{ZMP} by using (11).

5. Compute the generalized force τ_ϕ according to (13).
6. Distribute the generalized force τ_ϕ to each joint torque $\boldsymbol{\tau}$ so as to satisfy (14). Namely,

$$\mathbf{T} = (\mathbf{J}^T(\Theta))^* \tau_\phi + (I - \mathbf{J}^T(\Theta)(\mathbf{J}^T(\Theta))^*)\mathbf{p}. \quad (27)$$

   Here, $*$ denotes the generalized inverse matrix, and $\mathbf{p}$ is an arbitrary 4-dimensional vector. See Appendix 7.2 for the calculation of $\mathbf{J}(\Theta)$.
7. Output the joint torque by using the actuators.

3.2.4 Switching from double- to single-support phase

According to control law (13), the ZMP position is shifted to the side of the next supporting leg by following P_d. The control law is switched when the ZMP position reaches an area under the next supporting foot.

4. Robot experiment

4.1 Object

In the previous section, we proposed a control method for lateral stepping that does not require motion planning, i.e. the reference trajectory generation of joint angles. This direct ZMP control is expected to allow a robot to naturally change their motion according to the slope. The objective of this experiment is to confirm this effect by using a robot with reduced DoF. The details of the robot are described in section 4.3.

4.2 Simulation

Prior to the experiments, the control method is simulated under the influence of the constant external force, as expressed by

$$F_x = -Mg\sin\alpha \quad (28)$$

$$F_y = -Mg(1 - \cos\alpha). \quad (29)$$

This is equivalent to the gravitational effect on a slope with angle α. The cases where $\alpha = 0$ [rad] (no external force) and $\alpha = 0.2$ [rad] are examined. The parameters are $M = 2.5$ [kg], $m = 1.25$ [kg], $m_f = 0$ [kg], $L = 0.20$ [m], $\ell = 0.1$ [m], $\ell_B = 0.07$ [m], $\ell_f = 0.02$ [m]. The feedback gains of (1) are set to $K_d = 30$, $K_p = 500$ and $K_f = 1$, while those of (13) are $K_d = 5$, $K_p = 10$ and $K_f = 100$. To the hip joint in the single-support phase, the conventional PD control with non-linear compensation is applied with a reference trajectory that lifts up the swing leg – the feedback gains are $K_d = 100$ and $K_p = 500$.

The graphs in Fig. 5(a) and (b) the ZMP position over time. Regardless of the external forces, similar ZMP profiles are obtained, implying that the body weight shifts as expected in both cases. The time-based plot of the horizontal CoG position is depicted in Fig. 5(c): when the external force is exerted, the stepping motion is performed with the posture tilted against it.

4.3 Equipment

Experiments were performed using a biped robot with four DoFs: two in the hip joints, two in the ankle joints and no DoF other than that in the lateral plane. This is an improved version of that in our previous paper (Ito et al., 2007; 2008). The robot is 35 [cm] high and weighs 2.4 [kg]. The sole of the foot is 8.6 [cm] long and the horizontal distance between the right and

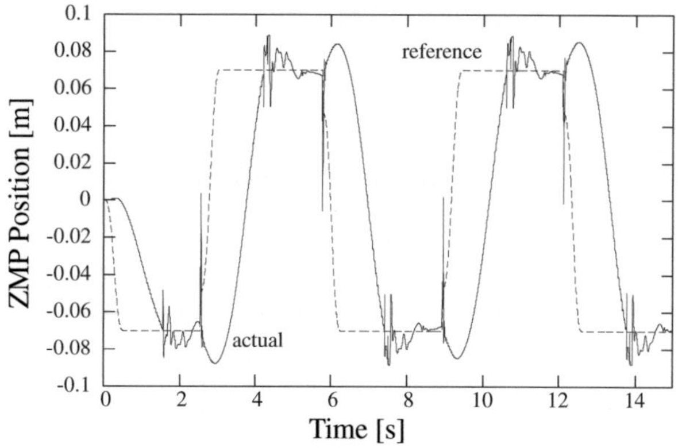

(a) Reference and actual trajectory of ZMP for $\alpha = 0$

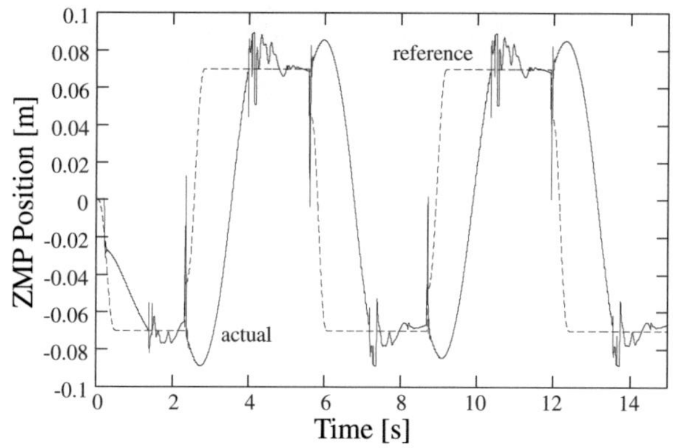

(b) Reference and actual trajectory of ZMP for $\alpha = 0.20$

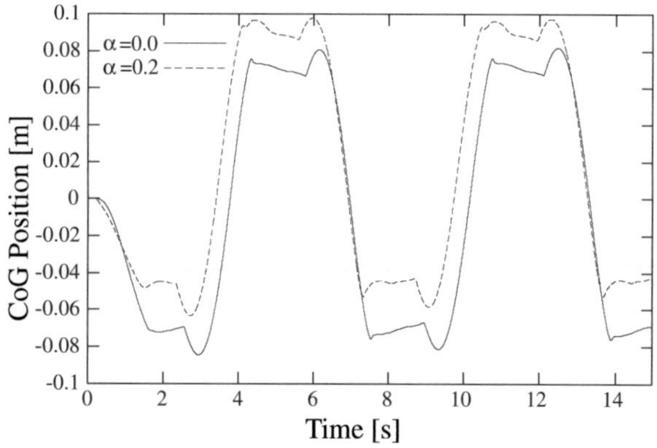

(c) Horizontal position of CoG.

Fig. 5. Simulation results.

left ankles is 13.4 [cm]. Four motors are installed: two drive hip joints, while the others drive ankle joints. A rotary encoder installed in each motor provides information on the joint angles of the robot. Furthermore, three load cells are attached to each sole to provide ZMP detection.

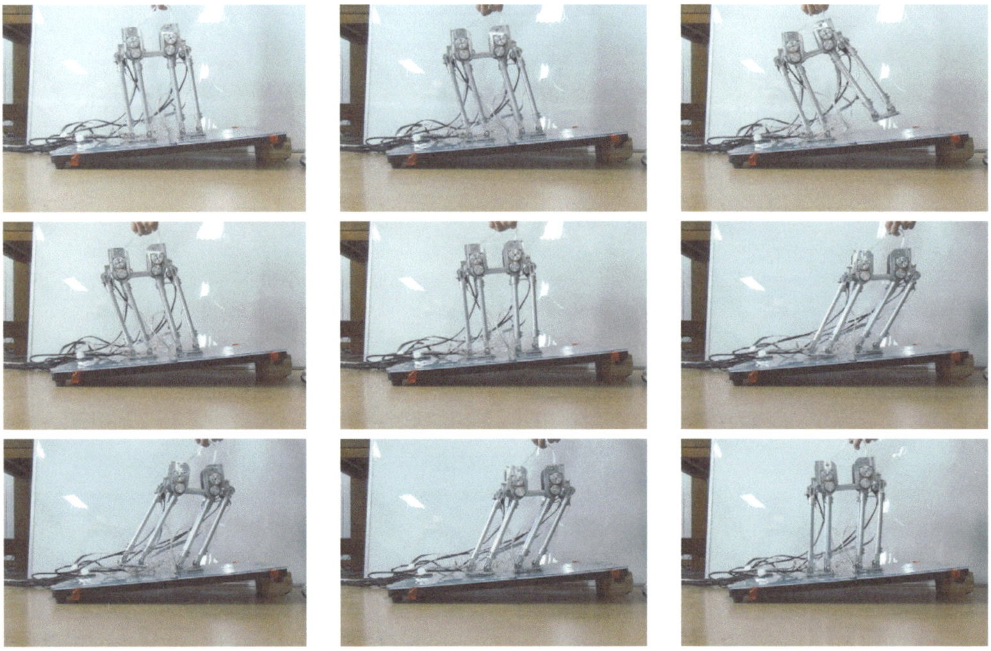

Fig. 6. Snapshots of biped robot experiments on a sloped surface.

The robot controller, operated by ART-LINUX, acquires the sensory information via a pulse counter and A/D converter. It calculates the torque that should be applied at each joint and sends them to the motor driver via a D/A converter. In the experiment, the controller operates at 1 [ms].

4.4 Methods

In the single-support phase, the control law (1) is applied for the ankle joint of the support leg with feedback gains of $K_d = 0.001$, $K_p = 0.005$ and $K_f = 0.0018$. Note that the unit of the angle is set to degrees to allow a simple check of the robot motion in the experiment; thus, the gains are given in the degree unit system. θ in (1) is approximated by the CoG sway angle ϕ, and ϕ at the start of each single-support phase is set to θ_d in (1) so that the ankle joint torque initially becomes zero. The other joint angles are controlled by the PD control. Its reference trajectories are set as follows. The hip joint of the swing leg is held in its neutral position, whereas that of the support leg is extended 30 [deg] from its neutral position in 8 [s], and then, returned to the neutral position again in 8 [s], which is represented by the fifth-order polynomial equation of the time. The ankle joint of the swing leg is controlled so that its sole becomes parallel to the ground at the end of the single-support phase. The control mode is switched when the hip joint angle reaches a neutral position. The feedback gains of the PD control are $K_d = 0.0009$ and $K_p = 0.009$. They are the same for the three joints.

The double-support phase uses control laws (13)-(15). The feedback gains are set to $K_d = 0.001$, $K_p = 0.002$ and $K_f = 0.07$. P_d is set using the fifth-order polynomial equation, to move 18 [cm], i.e. from 8 [cm] (the side of the previous support leg) to 10 [cm] on the reverse side, in 15[s]. To promote ZMP movement, the distance of the ZMP shift is set slightly larger than

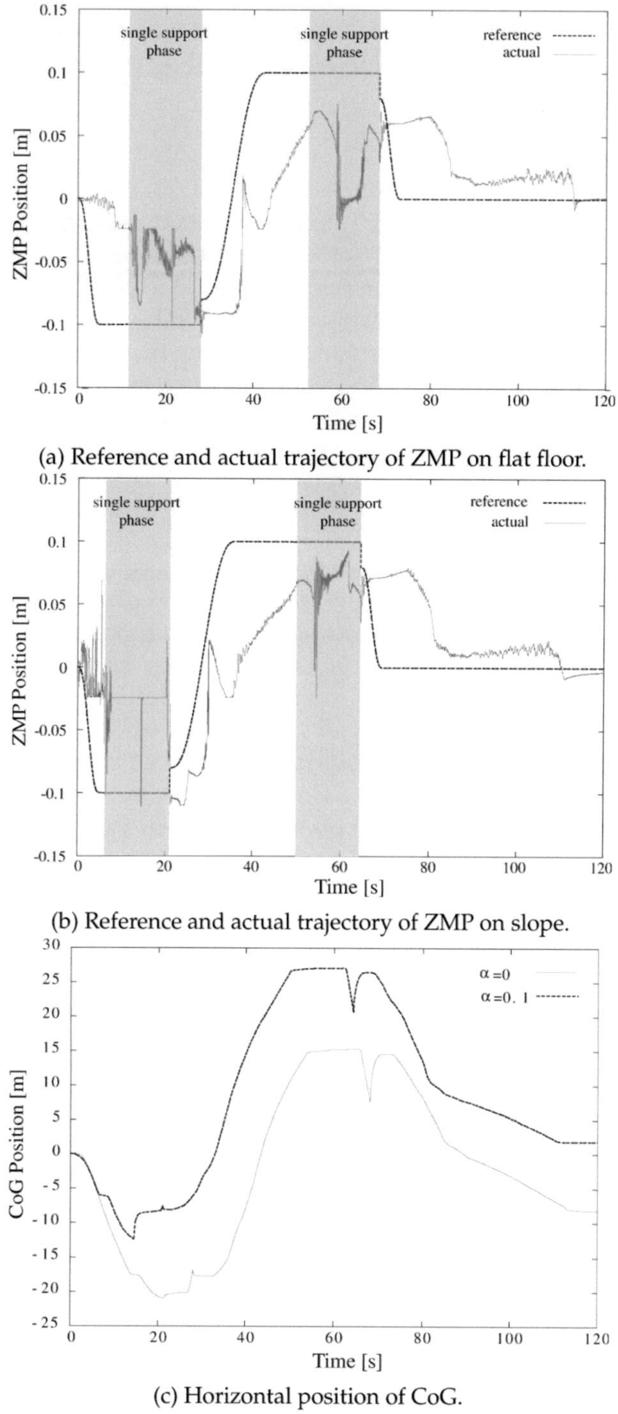

Fig. 7. Experimental results.

the natural width between the two ankle joints (x_f=6.7[cm]). The control mode is switched on the basis of the ZMP position. This threshold is set 7 [cm]. Experiments are executed on both flat ground and an 8 [deg] slope.

4.5 Results and remarks

Snapshots of the robot motion on the slope are shown in Fig. 6. To evaluate the behaviour for both the conditions, time-based plots of the ZMP position in the double-support phase are shown in Fig. 7: (a) is on the flat ground and (b) is on the slope. The ZMP profiles are quite similar, implying that the stepping motion can be achieved regardless of the slope angle. The time based plot of the horizontal CoG position is shown in Fig. 7(c). The profile of the slope condition is shifted up from that on flat ground, indicating that lateral motion is achieved by tilting the entire body adaptively against the slope, as shown in Fig. 1. The slow motion of the robot requires improvement. Correcting mechanical problems, such as backlash at the joints, will improve the motion speed somewhat.

5. Conclusions

The generation of the joint or CoG reference trajectories is a complicated task in biped robot control. By restricting the task to balance control in the lateral plane motion, a control method without the need for generating reference trajectories was proposed. This control method is essentially a feedback control of the ZMP position that makes the most use of the information on the ground reaction forces. Thus, the reference trajectories of both joints and the CoG of the body, which are usually affected by environmental conditions such as the slope, are unnecessary, although those of the ZMP position are required. This approach provides natural adaptive changes in the lateral motion. Applying it to the control of a biped robot, whose DoF of motion were restricted within the lateral plane, experimentally confirmed its effectiveness. Improving the speed of the robot's movements and applying this technique to 3D biped locomotion are considered for future work.

6. Acknowledgement

Authors thank Mr. Shinya Furuta for his co-operation of robot experiments.

7. Appendex

7.1 Definition of ϕ

The lateral sway angle ϕ is calculated as

$$\phi = \arctan \frac{x_G}{y_G}. \tag{30}$$

Here, x_G and y_G are the horizontal and vertical positions of the CoG of the lateral sway model, respectively, and are described as

$$x_G = 2\rho \cos \frac{\theta_{RA} + \theta_{LA}}{2} \sin \frac{\theta_{LA} - \theta_{RA}}{2} \tag{31}$$

$$y_G = 2\rho \cos \frac{\theta_{RA} + \theta_{LA}}{2} \cos \frac{\theta_{LA} - \theta_{RA}}{2}, \tag{32}$$

where

$$\rho = \frac{2m\ell + ML}{2(2m+M)}. \tag{33}$$

Using this relationship, we obtain

$$\frac{x_G}{y_G} = \tan\frac{\theta_{LA} - \theta_{RA}}{2}. \tag{34}$$

According to the definition of the generalized coordinates (30), ϕ is expressed by (26), i.e.,

$$\phi = J_{\phi 1}\mathbf{\Theta} = J_{\phi 2}\mathbf{X}, \tag{35}$$

where

$$J_{\phi 1} = \begin{bmatrix} -\frac{1}{2} & 0 & 0 & \frac{1}{2} \end{bmatrix} \tag{36}$$

$$J_{\phi 2} = \begin{bmatrix} 0 & 0 & 0 & 0 & 0 & \frac{1}{2} & 0 & 0 & -\frac{1}{2} \end{bmatrix}. \tag{37}$$

The definition of $\mathbf{X}$ will be seen later in (47).

7.2 Calculation of the Jacobian matrix

The Jacobian matrix $\mathbf{J}(\Theta)$, which maps $\dot{\phi}$ to $\dot{\Theta}$, is calculated as follows. From (26), we get

$$\dot{\phi} = \frac{\dot{\theta}_{LA} - \dot{\theta}_{RA}}{2}. \tag{38}$$

On the other hand, a kinematic relationship among the joint angles is given as

$$-\theta_{RA} + \theta_{RH} + \theta_{LH} - \theta_{LA} = \pi. \tag{39}$$

Differentiating it, we obtain

$$-\dot{\theta}_{RA} + \dot{\theta}_{RH} + \dot{\theta}_{LH} - \dot{\theta}_{LA} = 0. \tag{40}$$

In addition, the position of the left hip joint (x_{RH}, y_{RH}) can be described in two ways:

$$\begin{bmatrix} x_{RH} \\ y_{RH} \end{bmatrix} = \begin{bmatrix} -x_f + L\sin\theta_{RA} \\ L\cos\theta_{RA} \end{bmatrix} = \begin{bmatrix} x_f - L\sin\theta_{LA} - 2\ell_B\sin(\theta_{LH} - \theta_{LA}) \\ L\cos\theta_{LA} - 2\ell_B\cos(\theta_{LH} - \theta_{LA}) \end{bmatrix}. \tag{41}$$

Differentiating them, the following equations hold.

$$\begin{bmatrix} -L\dot{\theta}_{LA}\cos\theta_{LA} - 2\ell_B(\dot{\theta}_{LH} - \dot{\theta}_{LA})\cos(\theta_{LH} - \theta_{LA}) \\ -L\dot{\theta}_{LA}\sin\theta_{LA} + 2\ell_B(\dot{\theta}_{LH} - \dot{\theta}_{LA})\sin(\theta_{LH} - \theta_{LA}) \end{bmatrix} = \begin{bmatrix} L\dot{\theta}_{RA}\cos\theta_{RA} \\ -L\dot{\theta}_{RA}\sin\theta_{RA} \end{bmatrix}. \tag{42}$$

Solve the three equations (38), (40) and (42) as four variables $\dot{\Theta} = [\dot{\theta}_{RA}, \dot{\theta}_{RH}, \dot{\theta}_{LH}, \dot{\theta}_{LA}]^T$ and the relationship between $\dot{\Theta}$ and $\dot{\phi}$ is represented by

$$\dot{\Theta} = \frac{2}{J_1 + J_3}\begin{bmatrix} -J_1 \\ -J_1 + J_2 \\ J_3 - J_2 \\ J_3 \end{bmatrix}\dot{\phi} = \mathbf{J}(\Theta)\dot{\phi} \tag{43}$$

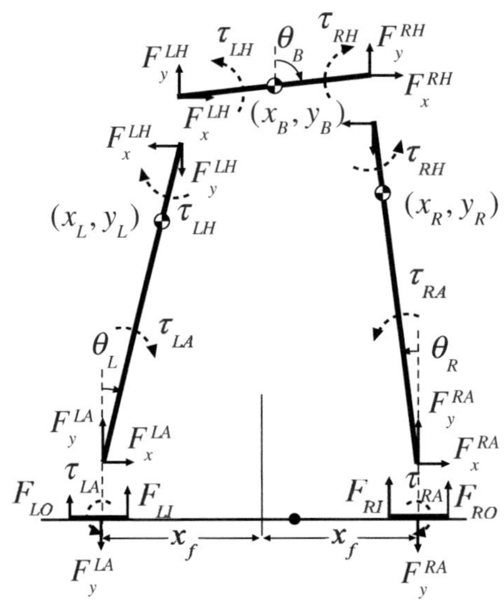

Fig. 8. Notation for the derivation of motion equation.

$$J_1 = 2\ell_B \sin\theta_{LH} \tag{44}$$

$$J_2 = L\sin(\theta_{LH} + \theta_{RH}) \tag{45}$$

$$J_3 = 2\ell_B \sin\theta_{RH}. \tag{46}$$

7.3 Motion equations

We define the vectors $\boldsymbol{X}$ and $\boldsymbol{F}$ as follows:

$$\boldsymbol{X} = [\ x_B \quad y_B \quad \theta_B \quad x_L \quad y_L \quad \theta_L \quad x_R \quad y_R \quad \theta_R\ ]^T \tag{47}$$

$$\boldsymbol{F} = [\ F_x^{LH} \quad F_y^{LH} \quad F_x^{RH} \quad F_y^{RH} \quad F_x^{LA} \quad F_y^{LA} \quad F_x^{RA} \quad F_y^{RA}\ ]^T. \tag{48}$$

The mechanical constraint is described as

$$C_C(\boldsymbol{X}) = 0, \tag{49}$$

where

$$C_C(\boldsymbol{X}) = \begin{bmatrix} x_B - \ell_B\sin\theta_B - x_L - \ell_s\sin\theta_L \\ y_B - \ell_B\cos\theta_B - y_L - \ell_s\cos\theta_L \\ x_B + \ell_B\sin\theta_B - x_R + \ell_s\sin\theta_R \\ y_B + \ell_B\cos\theta_B - y_R - \ell_s\cos\theta_R \\ x_L - \ell\sin\theta_L \\ y_L - \ell\cos\theta_L \\ x_R + \ell\sin\theta_R \\ y_R - \ell\cos\theta_R \end{bmatrix} \tag{50}$$

$$\ell_s = L - \ell. \tag{51}$$

Refer to Fig. 8 for the notations. The motion equation is expressed as

$$M\ddot{\boldsymbol{X}} = J_X^T \boldsymbol{F} + \boldsymbol{G}_0 + J_\theta^T \mathbf{T}. \tag{52}$$

Here,

$$M = diag[M_B, M_B, I_B, M_L, M_L, I_L, M_L, M_L, I_L] \tag{53}$$

$$J_X = \frac{\partial C_C(\boldsymbol{X})}{\partial \boldsymbol{X}} \tag{54}$$

$$\boldsymbol{G}_0 = \boldsymbol{G}_G + J_e^T \boldsymbol{F}_e \tag{55}$$

$$\boldsymbol{G}_G = [\ 0 \quad -M_B g \quad 0 \quad 0 \quad -M_L g \quad 0 \quad 0 \quad -M_L g \quad 0\ ]^T \tag{56}$$

$$J_\theta = \begin{bmatrix} 0 & 0 & 0 & 0 & 0 & 0 & 0 & 0 & 1 \\ 0 & 0 & 1 & 0 & 0 & 0 & 0 & 0 & 1 \\ 0 & 0 & -1 & 0 & 0 & 1 & 0 & 0 & 0 \\ 0 & 0 & 0 & 0 & 0 & 1 & 0 & 0 & 0 \end{bmatrix} \tag{57}$$

$$J_e = \begin{bmatrix} 0 & 0 & 0 & 0 & 0 & \rho\cos\theta_L & 0 & 0 & -\rho\cos\theta_R \\ 0 & 0 & 0 & 0 & 0 & -\rho\sin\theta_L & 0 & 0 & -\rho\sin\theta_R \end{bmatrix} \tag{58}$$

$$\boldsymbol{F}_e = [\ F_x \quad F_y\ ]^T. \tag{59}$$

Differentiating (49) two times, we obtain

$$J_X \ddot{\boldsymbol{X}} + \boldsymbol{C}_0 = 0, \tag{60}$$

where

$$\boldsymbol{C}_0 = \dot{J}_X \cdot \dot{\boldsymbol{X}}. \tag{61}$$

Combining (52) with (60), we can get

$$\begin{bmatrix} M & -J_X^T \\ -J_X & 0 \end{bmatrix} \begin{bmatrix} \ddot{\boldsymbol{X}} \\ \boldsymbol{F} \end{bmatrix} = \begin{bmatrix} \boldsymbol{G}_0 + J_\theta^T \boldsymbol{\tau} \\ \boldsymbol{C}_0 \end{bmatrix}. \tag{62}$$

The matrix of the left hand side has an inverse matrix since M has it. This inverse matrix is put to

$$\begin{bmatrix} M & -J_X^T \\ -J_X & 0 \end{bmatrix}^{-1} = \begin{bmatrix} N_0 & N_1^T \\ N_1 & N_2 \end{bmatrix}. \tag{63}$$

Then, (62) can be solved for $\ddot{\boldsymbol{X}}$ and $\boldsymbol{F}$.

$$\begin{bmatrix} \ddot{\boldsymbol{X}} \\ \boldsymbol{F} \end{bmatrix} = \begin{bmatrix} N_0 & N_1^T \\ N_1 & N_2 \end{bmatrix} \begin{bmatrix} \boldsymbol{G}_0 + J_\theta^T \boldsymbol{\tau} \\ \boldsymbol{C}_0 \end{bmatrix}. \tag{64}$$

From (35),

$$\ddot{\phi} = J_{\phi 2} \ddot{\boldsymbol{X}} = J_{\phi 2}(N_0(\boldsymbol{G}_0 + J_\theta^T J_{\phi 1}^T \tau_\phi) + N_1^T \boldsymbol{C}_0). \tag{65}$$

The dynamics of ϕ is expressed by (16), where

$$M(\Theta) = (J_{\phi 2} N_0 J_{\phi 2}^T)^{-1} \tag{66}$$

$$C(\Theta, \dot{\Theta}) = (J_{\phi 2} N_0 J_{\phi 2}^T)^{-1} J_{\phi 2} N_1^T \boldsymbol{C}_0 \tag{67}$$

$$G(\Theta, g, \mathbf{F}) = (J_{\phi 2} N_0 J_{\phi 2}^T)^{-1} J_{\phi 2} N_0 \mathbf{G}_0. \tag{68}$$

Note that $J_{\phi 1} J_\theta = J_{\phi 2}$ and $\boldsymbol{X}$ is uniquely written by Θ, i.e., $\boldsymbol{X} = \boldsymbol{X}(\Theta)$.
On the other hand, the ground reaction forces are expressed as

$$F_{LO} = \frac{1}{2} F_y^{LA} + \frac{1}{\ell_f} \tau_{LA} \tag{69}$$

$$F_{LI} = \frac{1}{2} F_y^{LA} - \frac{1}{\ell_f} \tau_{LA} \tag{70}$$

$$F_{RO} = \frac{1}{2} F_y^{RA} + \frac{1}{\ell_f} \tau_{RA} \tag{71}$$

$$F_{RI} = \frac{1}{2} F_y^{RA} - \frac{1}{\ell_f} \tau_{RA}. \tag{72}$$

Assume that F_{all} is constant because it corresponds to the total weight. Then, (11) is rewritten as

$$p_{ZMP} = J_{Z1}^T \boldsymbol{F} + J_{Z2}^T \boldsymbol{\tau} \tag{73}$$

$$J_{Z1} = \begin{bmatrix} 0 & 0 & 0 & 0 & 0 & -x_f/F_{all} & 0 & x_f/F_{all} \end{bmatrix}^T \tag{74}$$

$$J_{Z2} = \begin{bmatrix} 2/F_{all} & 0 & 0 & -2/F_{all} \end{bmatrix}^T. \tag{75}$$

From (64), $\boldsymbol{F}$ is expressed as

$$\boldsymbol{F} = N_1(\boldsymbol{G}_0 + J_\theta^T \boldsymbol{\tau}) + N_2 \boldsymbol{C}_0. \tag{76}$$

Substituting this equation to (73), we obtain (17), where

$$P(\Theta) = J_{Z1}^T N_1 J_\theta^T + J_{Z2}^T \tag{77}$$

$$Q(\Theta, \dot{\Theta}) = J_{Z1}^T N_2 \boldsymbol{C}_0 \tag{78}$$

$$R(\Theta, g, \mathbf{F}) = J_{Z1}^T N_1 \boldsymbol{G}_0. \tag{79}$$

8. References

Behnke, S. (2006). Online Trajectory Generation for Omnidirectional Biped Walking, *Proceedings of IEEE International Conference on Robotics and Automation (ICRAAf06), Orlando, Florida* pp. 1597–1603.

Czarnetzki, S., Kerner, S. & Urbann, O. (2009). Observer-based dynamic walking control for biped robots, *Robotics and Autonomous Systems* 57(8): 839–845.

Goswami, A. (1999). Postural Stability of Biped Robots and the Foot-Rotation Indicator (FRI) Point, *The International Journal of Robotics Research* 18(6): 523.

Héliot, R. & Espiau, B. (2008). Online generation of cyclic leg trajectories synchronized with sensor measurement, *Robotics and Autonomous Systems* 56(5): 410–421.

Hirai, K., Hirose, M., Haikawa, Y. & Takenaka, T. (1998). The development of Honda humanoid robot, *Robotics and Automation, 1998. Proceedings. 1998 IEEE International Conference on* 2.

Horak, F. & Nashner, L. (1986). Central programming of postural movements: adaptation to altered support-surface configurations, *Journal of Neurophysiology* 55(6): 1369–1381.

Huang, Q., Kaneko, K., Yokoi, K., Kajita, S., Kotoku, T., Koyachi, N., Arai, H., Imamura, N., Komoriya, K. & Tanie, K. (2000). Balance control of a biped robot combining off-line pattern with real-time modification, *Proc. of the 2000 IEEE International Conference on Robotics and Automation* 4: 3346–3352.

Ito, S., Asano, H. & Kawasaki, H. (2003). A balance control in biped double support phase based on center of pressure of ground reaction forces, *7th IFAC Symposium on Robot Control* pp. 205–210.

Ito, S. & Kawasaki, H. (2005). Regularity in an environment produces an internal torque pattern for biped balance control, *Biological Cybernetics* 92(4): 241–251.

Ito, S., Amano, S., Sasaki, M. & Kulvanit, P. (2007). In-place lateral stepping motion of biped robot adapting to slope change, *Proceedings of the 2007 IEEE International Conference on Systems, Man and Cybernetics* pp. 1274–1279.

Ito, S., Amano, S., Sasaki, M. & Kulvanit, P. (2008). A ZMP Feedback Control for Biped Balance and its Application to In-Place Lateral Stepping Motion, *JOURNAL OF COMPUTERS* 3(8): 23.

Kagami, S., Kitagawa, T., Nishiwaki, K., Sugihara, T., Inaba, M. & Inoue, H. (2002). A Fast Dynamically Equilibrated Walking Trajectory Generation Method of Humanoid Robot, *Autonomous Robots* 12(1): 71–82.

Kajita, S. & Tani, K. (1996). Experimental study of biped dynamic walking, *Control Systems Magazine, IEEE* 16(1): 13–19.

Kulvanit, P., Wongsuwan, H., Srisuwan, B., Siramee, K., Boonprakob, A., Maneewan, T. & Laowattana, D. (2005). Team KMUTT: Team Description Paper, *Robocup 2005: Humanoid League* .

Lee, B., Kim, Y. & Kim, J. (2005). Balance control of humanoid robot for hurosot, *Proc. of IFAC World Congress* .

McGhee, R. & Frank, A. (1968). On the stability properties of quadruped creeping gaits, *Mathematical Biosciences* 3(3-4): 331–351.

Mitobe, K., Capi, G. & Nasu, Y. (2001). Control of walking robots based on manipulation of the zero moment point, *Robotica* 18(06): 651–657.

Nagasaka, K., Inoue, H. & Inaba, M. (1999). Dynamic walking pattern generation for a humanoid robot based on optimal gradient method, *Proc. of 1999 IEEE International Conference on Systems, Man and Cybernetics* 6.

Napoleon, S. & Sampei, M. (2002). Balance control analysis of humanoid robot based on ZMP feedback control, *Proceedings of the 2002 IEEE/RSJ International Conference on Intelligent Robots and Systems* pp. 2437–2442.

Nishiwaki, K., Kagami, S., Kuniyoshi, Y., Inaba, M. & Inoue, H. (2002). Online generation of humanoid walking motion based on a fast generation method of motion pattern that follows desired ZMP, *IEEE/RSJ 2002 International Conference on Intelligent Robots and System* 3.

Prahlad, V., Dip, G. & Meng-Hwee, C. (2007). Disturbance rejection by online ZMP compensation, *Robotica* 26(1): 9–17.

Sugihara, T., Nakamura, Y. & Inoue, H. (2002). Real-time humanoid motion generation through ZMP manipulation based on inverted pendulum control, *Proceedings of 2002 IEEE International Conference on Robotics and Automation* 2.

Suleiman, W., Kanehiro, F., Miura, K. & Yoshida, E. (2009). Improving ZMP-based control model using system identification techniques, *9th IEEE-RAS International Conference on Humanoid Robots*, pp. 74–80.

Wollherr, D. & Buss, M. (2004). Posture modification for biped humanoid robots based on Jacobian method, *Proceedings. 2004 IEEE/RSJ International Conference on Intelligent Robots and Systems(IROS 2004)* 1.

Yamaguchi, J. & Takanishi, A. (1997). Development of a Leg Part of a Humanoid Robot?Development of a Biped Walking Robot Adapting to the Humans' Normal Living Floor, *Autonomous Robots* 4(4): 369–385.

3

Optimal Biped Design Using a Moving Torso: Theory and Experiments

Karthik Kappaganthu[1] and C. Nataraj[2]

[1]*Principal Engineer, R&D Global, Strategic Solution, LLC Virginia, VA 20165*

[2]*Professor & Chair*

Department of Mechanical Engineering, Villanova University, Villanova, PA 19085

USA

1. Introduction

Legged robots have been studied and developed for a long time. The primary advantage of legged motion is that they can traverse terrains inaccessible to wheeled robots. Biped robots are probably the most complicated of legged robots. Despite their complexity there has been a substantial amount of work done in the field including different techniques that have been developed to model, design and control biped robots. There are primarily two kinds of biped robots: active and passive bipeds. Many sophisticated, intelligent bipeds have been built by major companies. Most of the bipeds built are based on active control techniques; these are typically complicated, require high energies and are expensive to build. Also, they are far from mimicking true human motion. Owing to this, reliable biped robots are still elusive even to this day.

McGeer (1990) analyzed the natural dynamics of two-legged systems. He numerically and experimentally analyzed systems with concentrated masses on legs and hip. These bipeds did not have a torso and no external torques were applied. He showed that these systems could walk stably down small slopes and sustain the motion. This class of biped motion is known as Passive Dynamic Walking.

The step length and the velocity of the passive biped depend on parameters such as the masses, lengths and the slope on which the biped is walking. A passive biped is much more efficient than active bipeds and mimics the human motion better. Goswami, Espiau & Keramane (1996); Goswami et al. (1993); Goswami, Thuilot & Espiau (1996); Goswami, Thuilotz & Espiauy (1996) describe the limit cycles and their stability in the passive gait of a biped without a torso. They obtained some numerical solutions of a system with known parameters. In Goswami et al. (1997) and Roussel et al. (98) bifurcation and chaos are studied and the dependence of the gait on the slope is explained.

Asano, Luo & Yamakita (2004) use energy-based control laws to enable the biped to mimic the passive motion; they analyze a biped without torso and knees. Asano, Yamakita, Kamamichi & Luo (2004), Kim & Oh (2004), Paul et al. (2003), Silva & Machado (2001) and Goswami (1999) are some more of the innovative attempts to build a simple yet controllable walking machine. Although passive dynamic walking is efficient, simple to implement and analyze, the step length and the velocity of motion of the passive biped are greatly dependent on the system parameters. The path followed by the biped is uncontrollable and there is no control

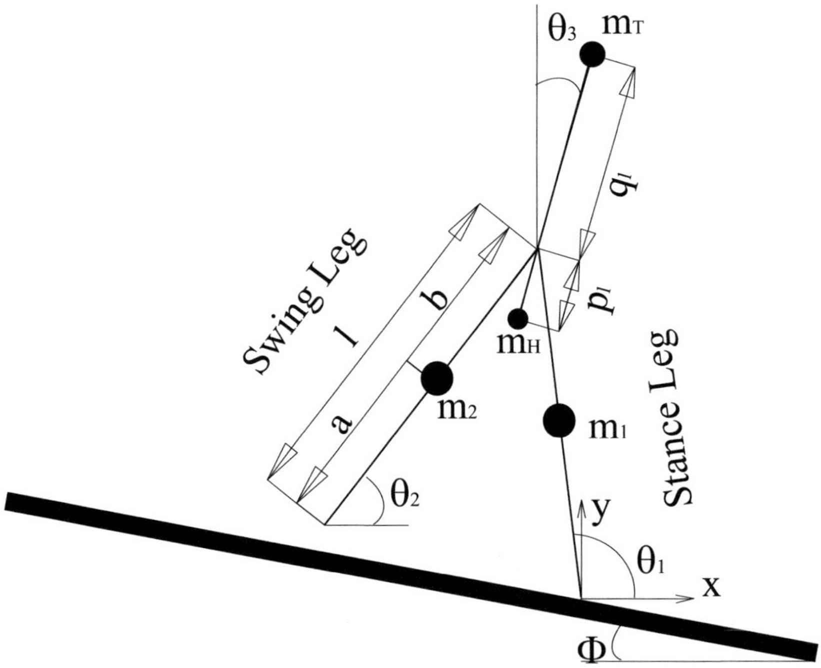

Fig. 1. Two Dimensional Model of the Biped

whatsoever on the system performance. This makes passive bipeds unrealistic for practical applications.

Actively controlled bipeds on the other hand are very adaptable to path planning and stability control at the cost of simplicity. Further they are typically prohibitively costly. Recognizing this, some attempts have been made to incorporate passive dynamics into actively controlled bipeds as it could theoretically result in an efficient method.

The torso plays an important role in the motion of bipeds. Typically most of the mass is concentrated on the torso or on the hip. It can be observed even in human motion that the position of the torso changes with a change in slope. Athletes also seem to use the torso to gain efficiency. Our present work is motivated by this influence the torso has on the biped. Although many bipeds with torsos have been built, little work has been done to utilize the effect of the torso on dynamics. This work tries to incorporate torso dynamics into the control of bipeds.

This paper analyzes the effect of the torso on the step length and velocity of a biped walking down a known slope. This information is used to control the step length and velocity of biped by applying the torque solely at the hip. The legs are passive at all times. The motion of legs is controlled solely by the mathematical coupling between the motion of the torso and the legs. The biped analyzed is shown in Fig. 1.

The goal of this work is to control just the step-length and velocities; clearly, there are infinitely many path profiles that can be executed to achieve this goal. Each path profile has a corresponding torque and external energy associated with it. In this work the path profile that minimizes the external energy input is found and the corresponding torque is applied.

A biped is built to implement and test the effect of torso on the biped motion. The biped walks only down the slopes. The legs are not actuated externally. Different torques are input to the

biped to make it walk with a required step length. The biped weighs 7 kg and is 60 cm tall. A DC motor is used to actuate the torso.
The second section provides the mathematical model for the biped being considered. It explains the phases of biped motion and derives the governing equation of motion. A brief introduction to terminology in bipeds and computational methods in passive bipeds is also provided. The third section presents the simulations of the biped in forward and inverse dynamics. Trajectories for typical parameter values are solved for and the variation of potential energy and kinetic energy is studied. Also the dependence of biped motion on the torso is studied. Section Four defines the optimal problem and its corresponding solution. Optimal trajectories and torques are obtained. Section Five explains the experimental setup, and Section Six, the results of practical application of the principle. Finally, Section Seven summarizes the work and presents a list of future tasks to be accomplished.

2. Modeling

The biped is modeled in two dimensional space. It has two legs without knees, and the mass is assumed to be concentrated at the center of mass of each leg. The lengths of both legs are equal and the two legs are connected by a hip. There is a third mass on the torso which is at the center of the hip. The motor mass is assumed to be concentrated at a known distance from the hip. The degrees of freedom and parameters are as shown in Fig. 1.
The basic motion consists of two phases: the swing phase and the impact phase. The swing phase consists of motion of the swing leg, and the motion is described by a continuous differential equation obtained from Lagrange equations. The impact phase consists of instantaneous impact of the swing leg and the transition to the next step. The following assumptions are made in deriving the equations of motion.

- The biped is modeled only in the lateral plane; the motion in the longitudinal plane is neglected.
- The impact of the swing leg at the end of each step is infinitesimally small.
- The impact of the swing leg is plastic.
- After the impact phase the functionality of each leg is interchanged; i.e., the swing leg becomes the stance leg, and vice versa.
- The angular momentum is conserved during impact.

With these assumptions, the governing equation of the biped can be derived for the swing phase as follows.

$$M(q,\dot{q})\ddot{q} + C(q,\dot{q}) + K(q) = F, \text{ when } \theta_1 + \theta_2 \neq \pi - 2\phi \tag{1}$$

The impact occurs when $\theta_1 + \theta_2 = \pi - 2\phi$; the states after impact are given by

$$q^+ = \begin{pmatrix} 0 & 1 & 0 \\ 1 & 0 & 0 \\ 0 & 0 & 1 \end{pmatrix} q^- \tag{2}$$

$$\dot{q}^+ = \begin{pmatrix} 0 & 1 & 0 \\ 1 & 0 & 0 \\ 0 & 0 & 1 \end{pmatrix} \left\{ q^- + [JM^{-1}J^T]^{-1}\delta\dot{r_e} \right\} \tag{3}$$

r_e is a vector containing the co-ordinates of the end-point of the swing leg. M, C, K, F, and J are system matrices as described by Eqs. 4–8.

$$M(q) = \begin{bmatrix} m_1a^2 + (m_T + m_H + m_2)\ell^2 & -m_2b\ell\cos(\theta_1 - \theta_2) & (m_Hp_l - m_Tq_l)\,\ell\sin(\theta_1 + \theta_C) \\ -m_2b\ell\cos(\theta_1 - \theta_2) & m_2b^2 & 0 \\ (m_Hp_l - m_Tq_l)\,\ell\sin(\theta_1 + \theta_C) & 0 & m_Hp_l^2 + m_Tq_l^2 \end{bmatrix} \quad (4)$$

$$C(\dot{q}, q) = \begin{bmatrix} (m_Hp_l - m_Tq_l)\,\ell\cos(\theta_1 - \theta_2)\,\dot{\theta}_1^2 - m_2b\ell\sin(\theta_1 - \theta_2)\,\dot{\theta}_2^2 \\ m_2b\ell\sin(\theta_1 - \theta_2)\,\dot{\theta}_1^2 \\ (m_Hp_l - m_Tq_l)\,\ell\cos(\theta_1 + \theta_C)\,\dot{\theta}_1^2 \end{bmatrix} \quad (5)$$

$$K(q) = \begin{bmatrix} [m_1a + (m_H + m_T + m_2)\,\ell]\,g\cos\theta_1 \\ -m_2gb\cos\theta_2 \\ (m_Hp_l - m_Tq_l)\,g\sin\theta_C \end{bmatrix} \quad (6)$$

$$J = \frac{\partial r_e}{\partial q} \quad (7)$$

$$F = \begin{bmatrix} 0 \\ 0 \\ \tau \end{bmatrix} \quad (8)$$

The parameters of motion are the step length (λ) and the time taken for each step (t_{end}); t_{end} can be obtained by solving

$$\theta_1(t_{end}) + \theta_2(t_{end}) = \pi - 2\phi \quad (9)$$

The step length λ is then given by

$$\lambda = \sqrt{(x_{ef} - x_{ei})^2 + (y_{ef} - y_{ei})^2} \quad (10)$$

where, x_e, y_e, the co-ordinates of the end point of the swing leg, are given by

$$x_e = \ell\cos\theta_1 - \ell\cos\theta_2 \quad (11)$$

$$y_e = \ell\sin\theta_1 - \ell\sin\theta_2 \quad (12)$$

x_{ei}, y_{ei} are the co-ordinates at $t = 0$, and, x_{ef}, y_{ef} are the co-ordinates at $t = t_{end}$.

3. Simulation

There are two stages of simulation: forward dynamics and inverse dynamics. Forward dynamics refers to the solution of the equations to compute the dynamic response to the imposed excitations, and is typically done using a standard procedure such as the Runge Kutta algorithm (which is the one we used). Inverse dynamics is the computation of the excitation necessary to achieve a certain prescribed motion, and is described in the next subsection.

The initial choice of parameters is such that they are close to the physical system. Later some parameters are varied so that the dependence of the path on the parameters can be verified. The parameter values are chosen considering the final biped to be built and also care is taken that the parameters do not create singularities in the problem. The parameter values chosen are given in Table 1.

Parameters	
Parameter	Value
m_1	0.5 Kg
m_H	2.5 Kg
m_T	3.5 Kg
m_2	0.5 Kg
a	0.14 m
b	0.21 m
p_l	0.1 m
q_l	0.2 m
ℓ	0.35 m
ϕ	3°

Table 1. Parameter Values

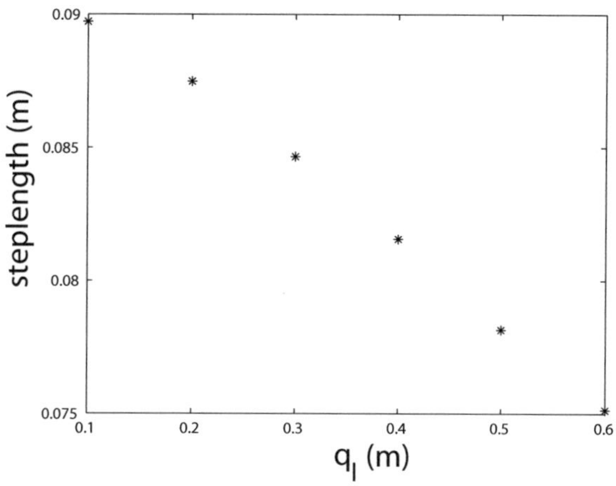

Fig. 2. Dependence of step length on q_l

To check for the dependence of the trajectories on the torso the parameters corresponding to the torso, p_l and q_l, are varied; the torso parameters appearing as moment terms m_T, m_H are held constant.

The results of the forward dynamic simulation are as follows. Mass m_T has a decreasing effect on λ and t_{end}. As shown in Fig. 2, λ decreases from 9 cm to 7.5 cm when q_l changes from 0.1 to 0.6 cm. As expected the decrease in λ causes a decrease in t_{end} from 0.22 sec to 0.16 sec, as shown in Fig. 3. It can be seen from the graphs that the change in the t_{end} is higher near smaller values of q_l.

Mass m_H has an increasing effect on λ and t_{end}. λ increases from 6.6 cm to 9 cm when p_l changes from 0.1 to 0.6 cm. As expected the increase in λ causes an increase in t_{end} from 0.15 sec to 0.23 sec. These are shown in Figs. 4 and 5 respectively.

In order to visualize the change in the shape of the profile, position vector of end point of the swing leg (x_e, y_e) is plotted qualitatively for changes in q_l and p_l. These are shown in Figs. 6 and 7 respectively. In each of the plots the number at the end of the arrow denotes the value of q_l (or p_l) at which the profile was generated. The dotted line represents the slope on which the biped is walking.

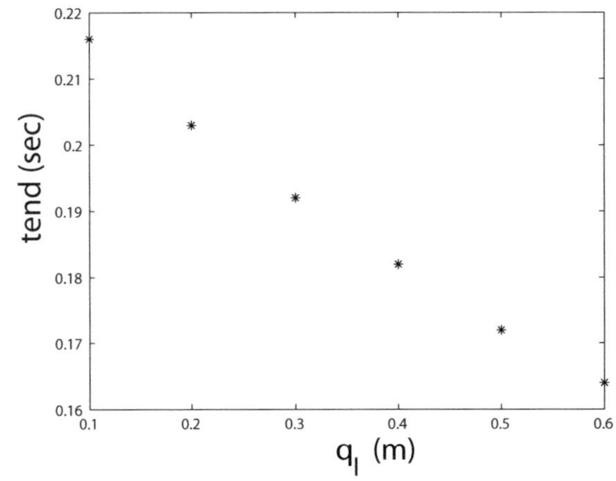

Fig. 3. Dependence of step-end time on q_l

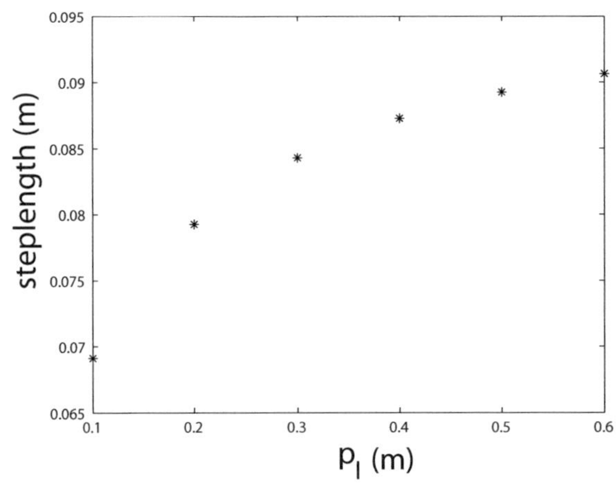

Fig. 4. Dependence of step length on p_l

3.1 Inverse Dynamics

In order to make the biped move with a predetermined step length and velocity, an external torque needs to be applied. The external torque is applied only at the torso. In order to determine the torque we need to solve the inverse dynamics problem. The flowchart to determine the torque by inverse dynamics is shown in Fig. 8.

Figure 9 is the input and Fig. 10 is the output of the algorithm. Note that θ_1 is monotonic in this case. The shape of the feet profile is shown in Fig. 11. The step length and step-end time chosen are 6 cm and 0.43 sec respectively. The maximum torque applied is 4.75 Nm.

4. Optimization

The choice of the profile of θ_1 has been arbitrary in the inverse dynamics problem. Now, we focus on the development of an optimal trajectory for the biped. The objective here is to

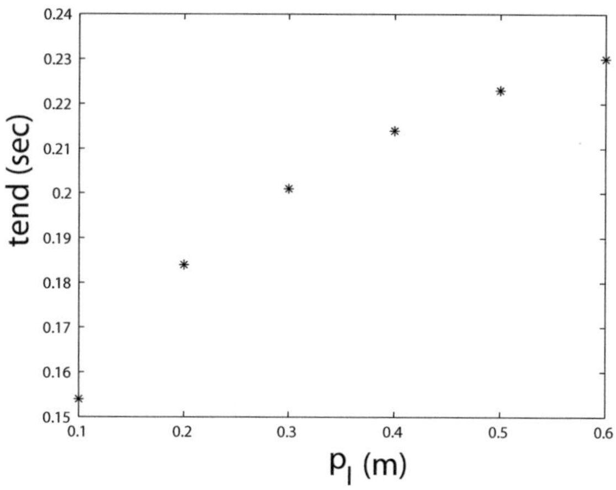

Fig. 5. Dependence of step-end time on q_l

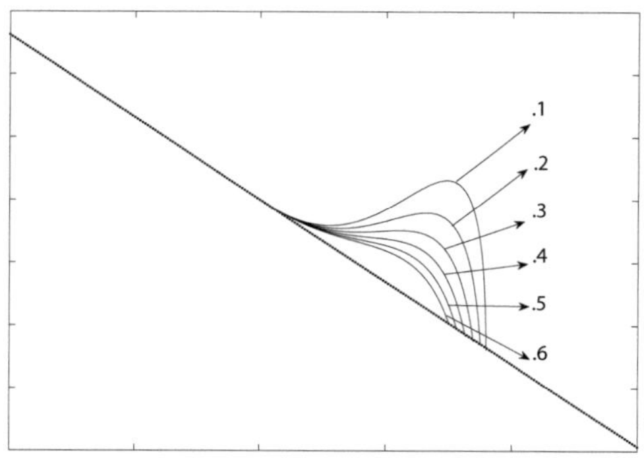

Fig. 6. Qualitative shape of the feet profile for varying q_l as numbered (see text for explanation)

minimize the external energy input to the system given the governing dynamics, step length and the time-period.

The external energy is supplied in the form of a torque which is applied at the torso. Eq. 20 gives the total external energy supplied to the biped. In the optimal problem the profiles of θ_1, θ_2 are reasonably fitted to a cubic curve as shown in Eqs. 13 and 14. There are four unknown coefficients for each variable. Since the initial conditions, λ and t_{end} are known and since the profiles should satisfy these conditions, the number of independent coefficients for each profile is two. These coefficients are the design variables for the optimization problem.

Since the profiles of both θ_1 and θ_2 are chosen they should satisfy the governing equation, hence the second equation in the governing equation that couples θ_1 and θ_2 becomes the nonlinear equality constraint for the optimization shown in Eq. 22. Also, it should be taken care that the feet-profile is consistent with the slope; i.e., during the swing phase it should always be above the ground. This can be modeled using geometry by ensuring that each point of the feet profile when substituted in the equation of the slope should be non-negative.

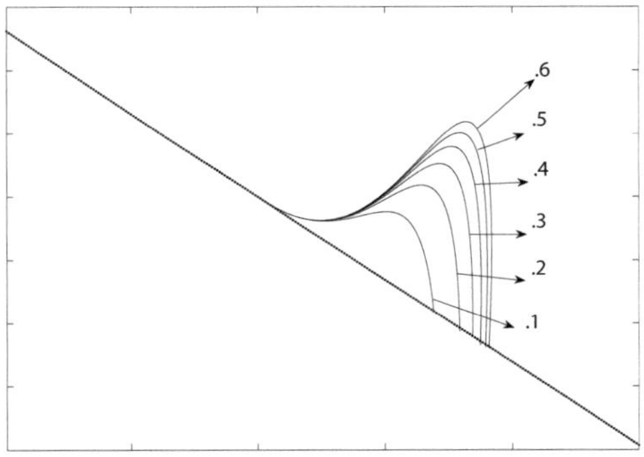

Fig. 7. Qualitative shape of the feet profile for varying p_l as numbered (see text for explanation)

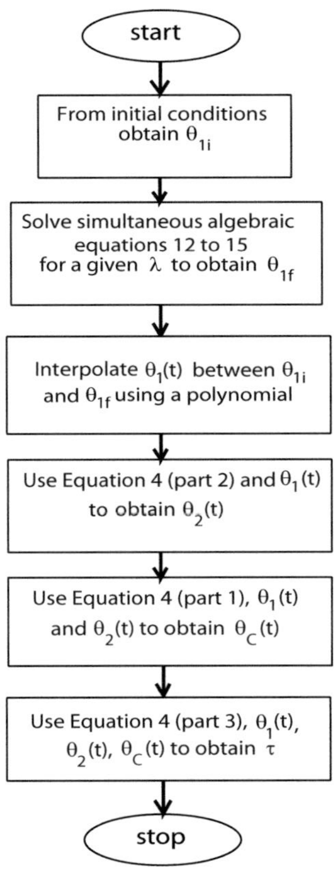

Fig. 8. Flowchart for inverse dynamics (λ)

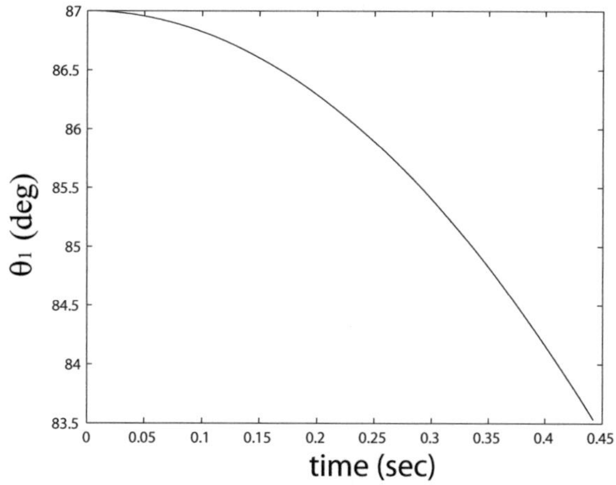

Fig. 9. θ_1^{inv} Obtained for the chosen step length by interpolation

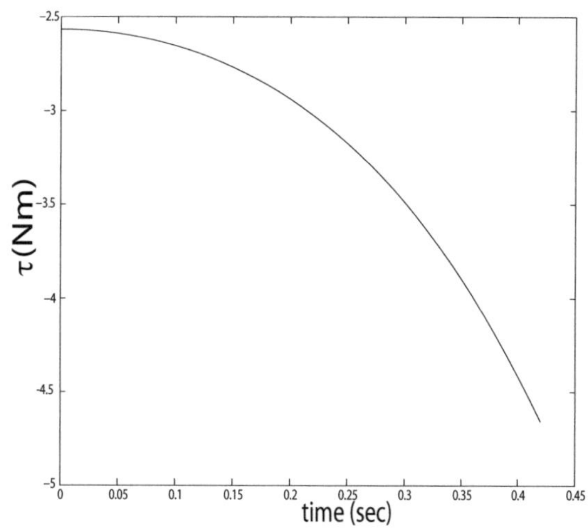

Fig. 10. Torque obtained for the chosen step length from inverse dynamics

This becomes the nonlinear inequality constraint shown in Eq. 23. The optimal problem formulation is shown in the next subsection.

4.0.1 Problem definition

Let

$$\theta_{1_{\text{opt}}}(t) = a_0 + a_1 t + a_2 t^2 + a_3 t^3 \quad (13)$$

$$\theta_{2_{\text{opt}}}(t) = b_0 + b_1 t + b_2 t^2 + b_3 t^3 \quad (14)$$

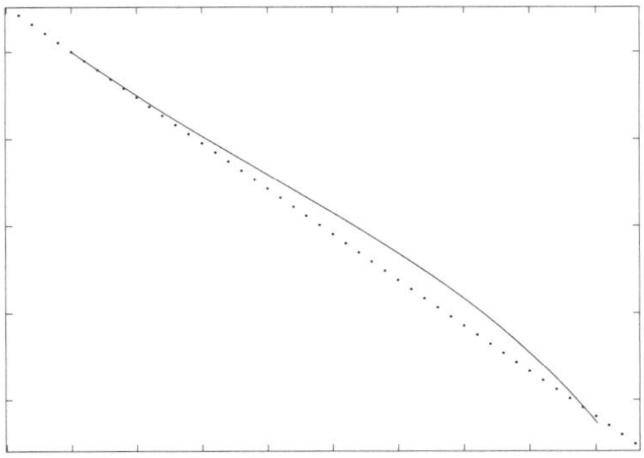

Fig. 11. Qualitative shape of feet profile for the chosen step length from inverse dynamics

Since $\theta_{1_{opt}}$ and $\theta_{2_{opt}}$ satisfy the initial and end condition

$$a_0 = \theta_{1_i} \tag{15}$$

$$b_0 = \theta_{2_i} \tag{16}$$

$$a_1 = \frac{\theta_{1_f} - \theta_{1_i} - a_2 t_{\text{end}}^2 - a_3 t_{\text{end}}^3}{t_{\text{end}}} \tag{17}$$

$$b_1 = \frac{\theta_{2_f} - \theta_{2_i} - b_2 t_{\text{end}}^2 - a_b t_{\text{end}}^3}{t_{\text{end}}} \tag{18}$$

The design variables for the optimization problem are defined as

$$x_{\text{opt}} = \begin{bmatrix} a_3 \\ a_2 \\ b_3 \\ b_2 \end{bmatrix} \tag{19}$$

The total external energy supplied per step is

$$E_{\text{ext}}(x_{\text{opt}}) = \int_{t_0}^{t_{\text{end}}} \tau \dot{\theta_C} dt \tag{20}$$

The optimal problem is hence (minimization of energy):

$$\text{Min } E_{\text{ext}}(x_{\text{opt}}) \tag{21}$$

such that

$$-m_2 \ell b \cos(\theta_{1_{\text{opt}}} - \theta_{2_{\text{opt}}})\ddot{\theta}_{1_{\text{opt}}} + m_2 b^2 \ddot{\theta}_{1_{\text{opt}}} + m_2 \ell b \sin(\theta_{1_{\text{opt}}} - \theta_{2_{\text{opt}}})\dot{\theta}_{1_{\text{opt}}}^2 - m_2 g b \cos\theta_{2_{\text{opt}}} = 0 \tag{22}$$

$$-y_e - \tan\phi x_e \leq 0 \tag{23}$$

where, y_e and x_e are the co-ordinates of the end point of the swing leg. Sequential Quadratic Programming is used to solve the optimal problem. Solving the optimal problem with $\lambda = 6$ cm and $t_{\text{end}} = 0.35$, the maximum torque obtained here is less than half that obtained by

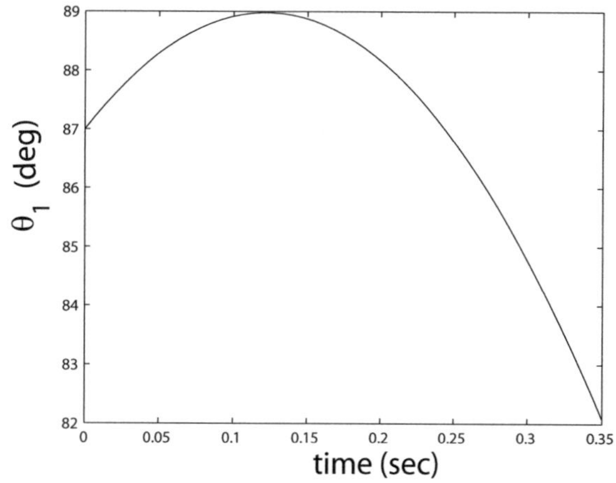

Fig. 12. Optimal trajectory for θ_1 for the chosen step length

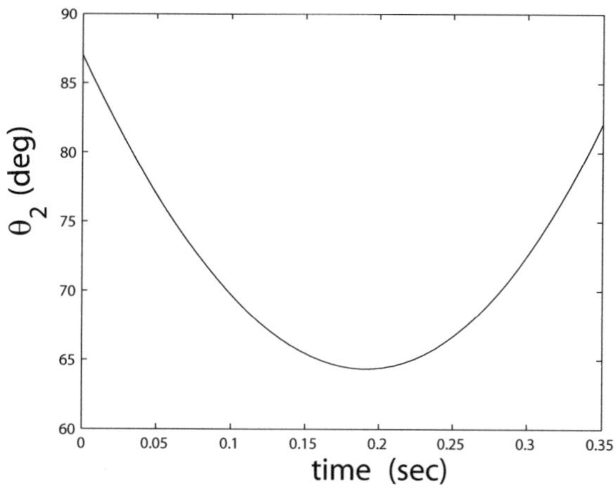

Fig. 13. Optimal trajectory for θ_2 for the chosen step length

choosing a quadratic interpolation of θ_1. The cubic interpolations of θ_1 and θ_2 from the optimal solution are shown in Fig. 12 and 13 respectively. Unlike the inverse dynamics case the angles do not decrease monotonically. θ_1 first increases to 89° and then decreases to 82°. This indicates a substantial difference in the behavior of the system. The optimal torque necessary for this motion is shown in Fig. 14. More extensive simulations are documented in Kappaganthu (2007).

5. Experimental setup

The aim of the experiment is to build a two dimensional biped that can be controlled with the torso. It is used to verify the controllability of the step length and velocity using a torso. The third dimension is neglected, steel guides are used to balance the biped in the 3rd dimension. Prismatic joints actuated by solenoids are used to provide sufficient clearance during the step take off. The solenoids are timed using the 555 timer circuit whose switches are placed at the

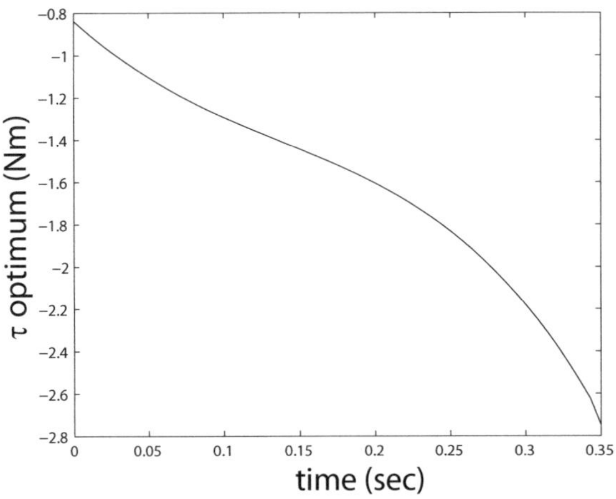

Fig. 14. Optimum torque obtained for the chosen step length

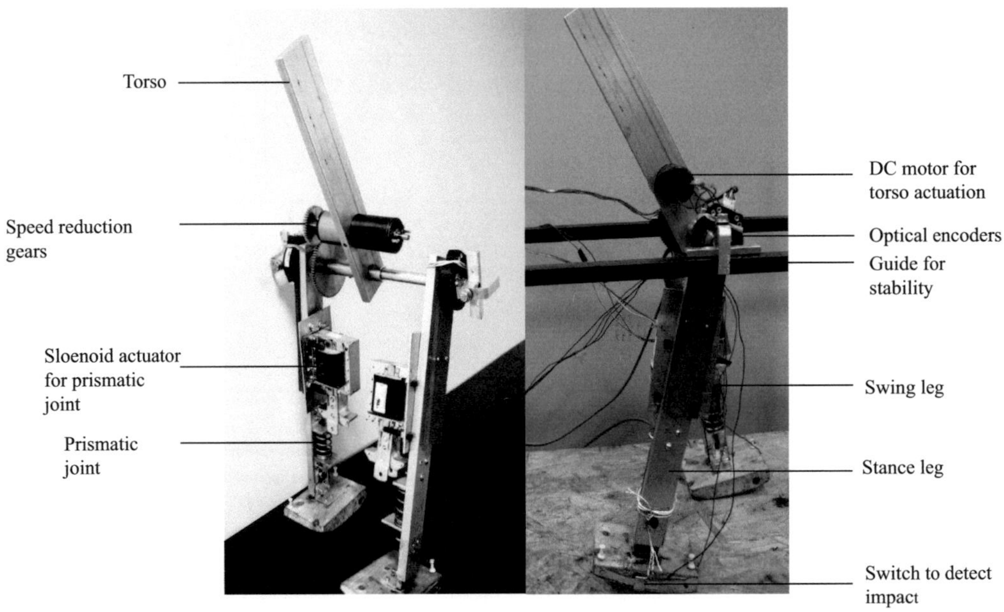

Fig. 15. The Villanova biped robot

feet. A motor is used to control the torso. dSpace is used to interface the hardware with the software. The biped robot built is shown in Fig. 15. The components of the biped and the principle of operation are explained in the following sections.

5.1 Components

5.1.1 Legs

Two aluminium C channels .125 inches thick are used as legs. The length of the aluminum channels is 14 inches. Each leg is 1.5 inches wide, the length of the channelsŠ leg is .75 inches. A hole .25 inches in diameter is drilled at the top of each leg to connect the hip. A small hole

.125 inches in diameter is drilled near the bottom to assemble the prismatic joint. Mounting bases, 5 inches long and 2.75 inches wide are provided at 3 inches from the top to mount the solenoids. An aluminium L angle support is attached at the 10 inches to load the spring of the prismatic joint.

5.1.2 Prismatic joint

A slider with a slot 1 inch long in a rectangular slab of 2.5X1.26 inches is attached to the leg with a screw. L angles of 1 inch are fixed at either end of the slider. The top angle is used to load a spring and the bottom angle is attached to the foot. A spring, 1.5 inch long, is placed between the angles on the leg and slider.

5.1.3 Solenoid

An AC intermittent solenoid is used to actuate the prismatic joint. The solenoid has a stroke length of 1 inch; it runs on 120VAC, 60 Hz current. The solenoid is mounted on the mounting base provided on the leg. The solenoid is 3 inches long and 2.33 inches wide. The pull force at 1 inch stroke length is 6 Lb. The solenoid weighs 2.7 Lb.

5.1.4 Hip

The hip made from a cylindrical aluminium rod of diameter .25 inches. The hip is 10 inches long. Three hex threads are machined at each end of the hip. Bearings 1 inch long made of PVC are attached at each end. The bearings are step turned to prevent play.

5.1.5 Feet

Feet are probably the most complicated parts to design. The feet design obtained after a lot of experimentation is 5 inches long and 3.25 inches wide and are carved from wood. Choosing the feet profile is a difficult task. It should be such that it does not interfere with the motion. As there is no actuation at the feet, it should be sufficiently curved to allow free rotation about the point of contact, however it should not be too steep as this would topple the biped. The profile has been obtained by trial and error. Each foot has a switch which turns on when the feet hit the slope.

5.1.6 Torso, motor and gears

The torso is made up of a 2X14 inches aluminium block. A hole .25 inches in diameter is drilled at the end to connect it to the hip. The torso is fixed in place on the hip using cylindrical restraints. A DC servomotor is used to actuate the torso. The motor has an internal gear train for speed reduction. The motor has a nominal voltage of 24VDC and a stall torque of 1250 mNm. It weighs 400 grams and is 3 inches long. The planetary gear train has a reduction ratio of 531:1. The maximum torque output in intermittent operation is 20Nm. The length of motor gear train combination is 7 inches. The total weight is 1.5 Kg. Additionally a 1.75 inches diameter brass gear is attached to the shaft; this meshes with a gear 3 inches in diameter. The larger gear is fixed to the hip.

5.1.7 555 Time circuit

The 555 timer is used to switch the relay, which actuates the solenoid. The 555 timer is a circuit whose input is a trigger and the output is a step signal of known period. The period can be adjusted by changing the resistance and conductance in the circuit. A description of the circuit can be found in any popular book on circuits. The time period of the output pulse is given by

Eq. 24, where R_v and C_v are the resistance and capacitance respectively.

$$T_{out} = R_v C_v \tag{24}$$

5.1.8 Power sources and relays

A total of four power sources are used. Two power sources are used to drive the motors in forward and backward direction. Agilent 3614A power sources are used. These can be remotely controlled using dSpace and have a gain of 20 and a range of 0-20VDC. One power source is connected to the 555 timer circuit and the fourth powers the 12VDC relay which switches the solenoids. A total of six relays are used. Two 12VDC relays switch the solenoids. Four 7VDC SPDT relays are used to switch the two power sources controlling the forward and backward motion of the torso.

5.1.9 Guideways

When the solenoid is actuated the prismatic joint causes the feet to lift up; however because of the weight of leg the biped tilts sideways. In order to prevent this, guideways are used. These guideways support the biped at the hip. A better solution would be to use lighter actuators for the prismatic joint and provide a reactive force at the stance legs ankle. This has not been implemented as guideways are easier and cheaper to build and at the same time satisfy the requirement of validating the qualitative effect of the torso on walking in 2 dimensions. An 'L' hook at the hip is used for safety to prevent the biped from deviating from the path and falling.

5.2 Working principle

There are three different processes happening at the same time, the up and down motion of the prismatic joints, the swinging of legs causing the biped to walk and the controlled motion of the torso. Data is collected from and transferred only in the latter two processes. The first process is autonomous to most extent. Fig. 17 and 16 explain the processes involved.

When the left leg hits the slope, the switch on it sends a signal to the first 555 timer circuit which sends a timed pulse to the solenoid attached to the right leg, this gives clearance to the leg, due to the dynamics of the system the leg moves forward and takes a step. The timer is set such that it is less than 50% of step period, this ensures that the leg gets back to its original length before the end of the step. When the right leg hits the slope the second 555 timer sends a signal to the left leg and the process repeats itself to create an obstructed motion.

The motor is actuated by an Agilent E3615A power source. The range of this power source is 0-20VDC; however, negative voltage needs to be applied to drive the motor in the reverse direction. This problem has been overcome by using two power sources. The power sources drive the motor in two opposite direction. This too has a problem; the power sources go into overdrive due to a short circuit between the independent power sources. To completely open the circuit when the power source is not in use, two relays are used. These relays are closed when the power source is to be inactive and open when the power source is to be active.

6. Experimental results

Data is collected and analyzed for three sets of data on a slope of 3°. In the first two cases the torso is fixed and no external torque is applied. The torso is fixed at two different positions, once leaning forward and once leaning backward. In the third case optimal torque is applied.

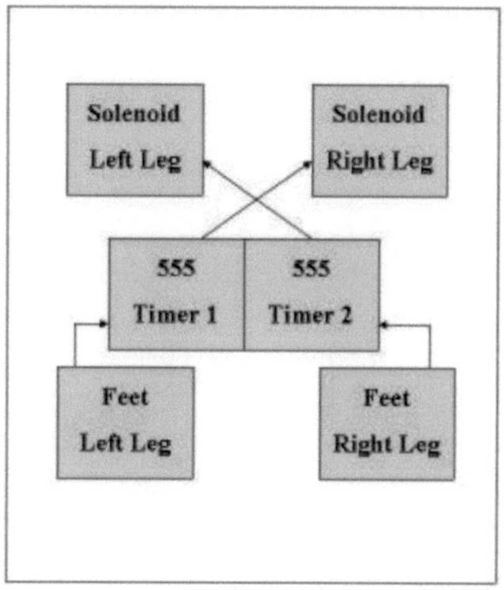

Fig. 16. Solenoid Control Schematic

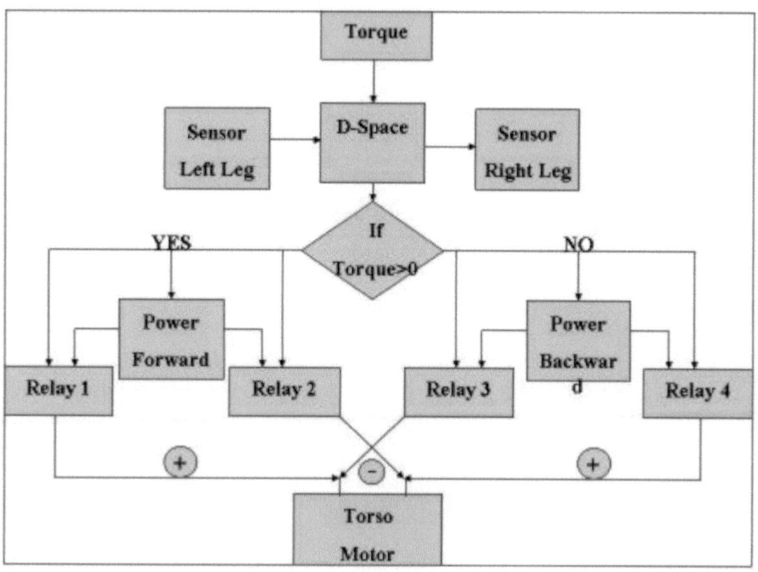

Fig. 17. Torso Control Schematic

6.1 Fixed torso

The profiles of θ_1 and θ_2 are obtained without a torque on the torso. In the first case the torso is held at an angle of 10° and in the second case at an angle of -10°. Fig. 18 show the raw data obtained with $\theta_c = 10^o$ and Fig. 19 show the raw data obtained with $\theta_c = -10^o$. It can be seen that there is no periodicity in the variation. Various initial conditions were tried but no limit cycle was found. Also it is not possible to wait for the biped to converge to a nearest limit cycle if it exists, since only 10 to 15 steps are possible on the ramp. The step profiles for a single step are shown in Figs. 20 and 21 for $\theta_c = 10^o$ and $\theta_c = -10^o$ respectively.

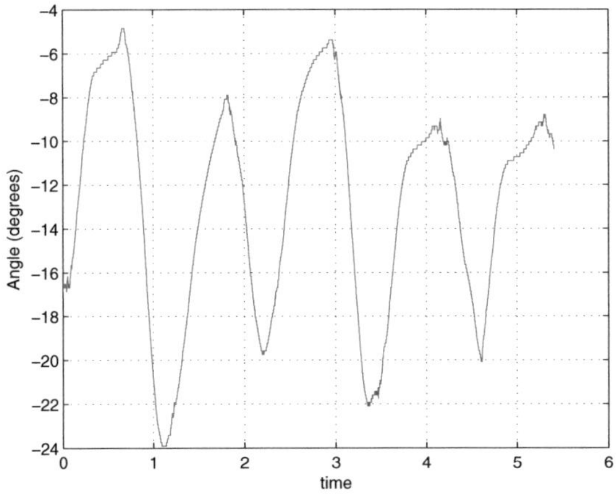

Fig. 18. Sensor output of angle subtended by the leg with vertical when $\theta_C = 10^o$

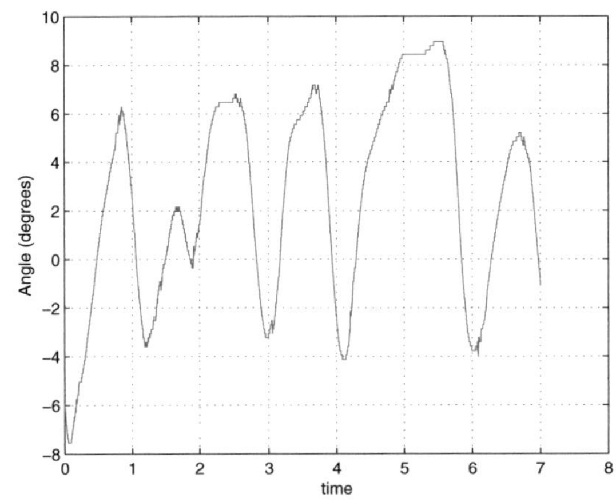

Fig. 19. Sensor output of angle subtended by the leg with vertical when $\theta_C = -10^o$

6.2 Optimal torque

The optimal torque is calculated for a step length of 6 cm and a time period of .35 sec. However, again because of the the inherent differences between the mathematical model and the real system, there is a substantial difference between the time periods. The optimal torque obtained is stretched evenly over the real time period. After many iterations the torque's time period and the real time period were made to match at .595 sec. The torque applied is shown in Fig. 22. Figure 23 show the raw data obtained from the sensor. This is more periodic and indicative of a useful walking robot.

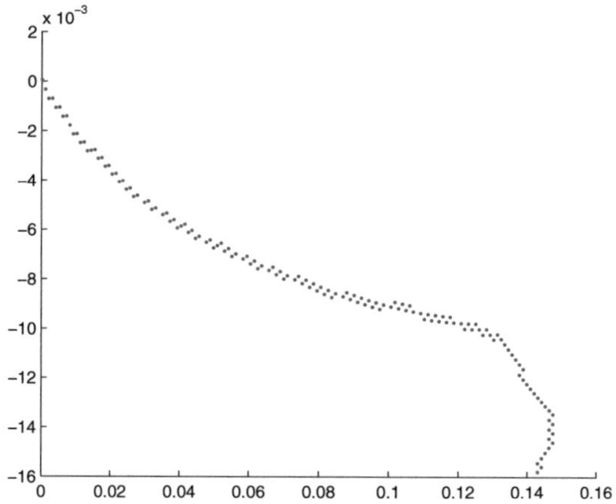

Fig. 20. Step profile when θ_C is held constant at 10°

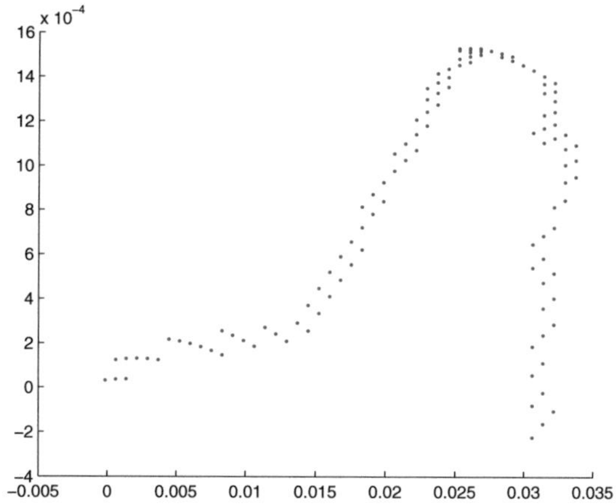

Fig. 21. Step profile when $\theta_C = 10^o$

The comparison of experimental and numerical feet profiles is shown in Fig. 24. It can be seen that there is a marked qualitative similarity between the experimental and numerical results. Also it can be observed that there is more repetitiveness in this case when compared to the case with no torque. The experimental step length was 4.5 cm.

7. Conclusion

This research has explored a novel aspect of biped robots with torsos. It has shown the importance and utility of the torso in the dynamics of the biped. The torso has been used efficiently to make the biped walk with the specified step length and velocity. The use of the torso has reduced the number of actuators required; further, the use of optimal torque has greatly reduced the external energy required for walking. The biped we analyzed, designed and constucted effectively uses the natural dynamics of the system; at the same time, the external excitation at torso has increased the practicability of the biped.

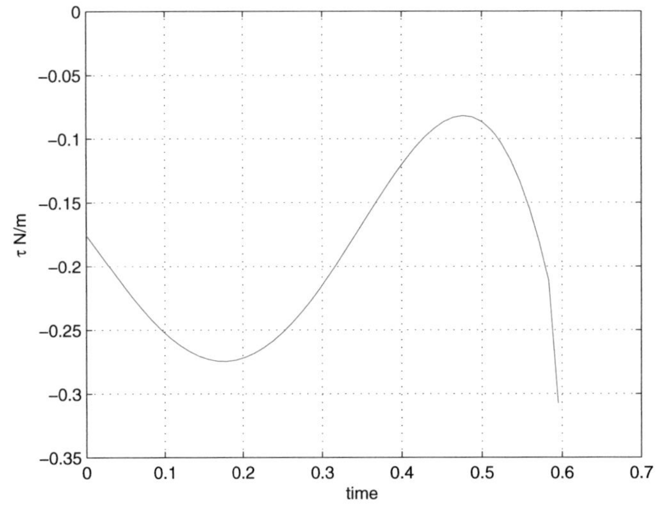

Fig. 22. Applied Optimal Torque

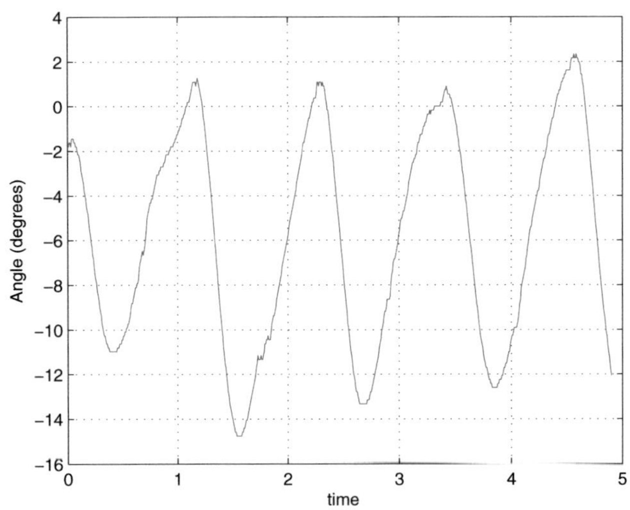

Fig. 23. Sensor output of the angle subtended by the leg with vertical when optimal torque is applied

A working model of the biped has been built. This biped has shown impressive qualitative agreement with numerical results. The Villanova biped walks on slopes, and the step-length and velocity can be controlled. When a torque acts on the biped it shows a higher repeatability of step-length than when no torque is applied. Optimal torque has been computed and applied, and the biped exhibits satisfactory performance.

Currently efforts are underway to add knees and feet to the analysis. Since the method is computationally expensive, work is being done to implement control using more efficient algorithms.

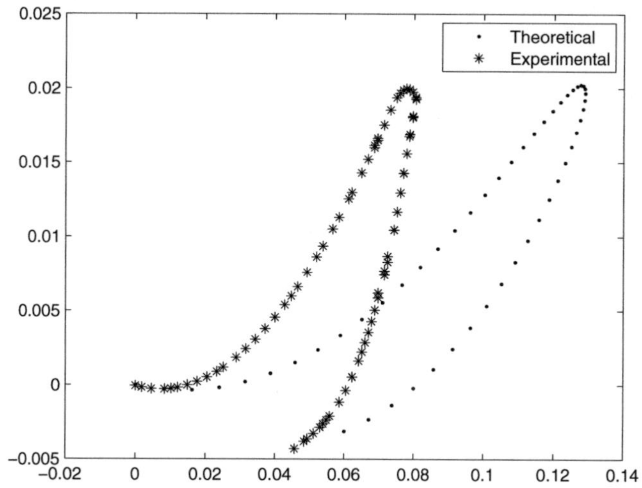

Fig. 24. Comparison of experimental and numerical step profiles when $\tau = \tau_{optimal}$

8. References

Asano, F., Luo, Z.-W. & Yamakita, M. (2004). Some extensions of passive walking formula to active biped robots, *Proceedings If 2004 IEEE, International Conference in Robotics and Automation* pp. 3797–3802.

Asano, F., Yamakita, M., Kamamichi, N. & Luo, Z.-W. (2004). A novel gait generation for biped walking robots based on mechanical energy constraint, *IEEE Transactions on Robotics and Automation* 20(3): 565–573.

Goswami, A. (1999). Postural stability of biped robots and the foot rotation Indicator(FRI) point, *International Journal of Robotics Research* .

Goswami, A., Espiau, B. & Keramane, A. (1996). Limit cycles and their stability in a passive bipedal gait, *IEEE International Conference onRobotics and Automation* .

Goswami, A., Espiau, B. & Keramane, A. (1997). Limit cycles in a passive compass gait biped and passivity mimicking control laws, *International Journal of Robotics* .

Goswami, A., Quaid, A. & Peshkin, M. (1993). Complete parameter identification of a robot from partial pose information, *IEEE International Conference on Robotics and Automation* .

Goswami, A., Thuilot, B. & Espiau, B. (1996). Compass-like biped robot part i: Stability and bifurcation of passive gaits, *Technical report*, Institut National De Recherche En Informatique Et En Autimatique.

Goswami, A., Thuilotz, B. & Espiauy, B. (1996). A study of the passive gait of a compass like biped robot: Symmetry and chaos, *International Journal of Robotics Research* .

Kappaganthu, K. (2007). *Design and development of a biped with a moving torso*, Master's thesis, Villanova University.

Kim, J.-H. & Oh, J.-H. (2004). Walking control of the humanoid platform KHR-1 based on torque feedback control, *Proceedings of the 2004 IEEE, International Conference in Robotics and Automation* pp. 622–625.

McGeer, T. (1990). Passive dynamic walking, *International Journal of Robotics Research* 9(2): 62–82.

Paul, C., Yokoi, H. & Matushita, K. (2003). Design and control of humanoid robot locomotion with passive legs and upper body actuation, *International Symposium on Robotics* .

Roussel, L., de Wit, C. C. & Goswami, A. (98). Generation of energy optimal complete gait cycles for BipedRobots, *IEEE International Conference on Robotics and Automation* pp. 2036–2041.

Silva, F. M. & Machado, J. A. T. (2001). Goal-oriented biped walking based on force interaction control, *Proceedings of the 2001 IEEE, International Conference of Robotics and Automation* pp. 4122–4127.

4

Effect of Circular Arc Feet on a Control Law for a Biped

Tetsuya Kinugasa[1], Christine Chevallereau[2], Yannick Aoustin[3] and Koji Yoshida[4]
[1,4] *Okayama University of Science,*
[2,3]*Institut de Recherche en Communications et Cybernétique de Nantes, Ecole Centrale de Nantes, Centre National de la Recherche Scientifique, Université de Nantes*
[1,4]*JAPAN*
[2,3]*FRANCE*

1. Introduction

Over the past several years a considerable amount of studies have been proposed on biped walking. The choice of type of feet such as a contact points, flat feet and circular arc feet is important, because walking stability is essentially affected by the contact with the ground. Control methods of many traditional humanoids with flat foot are based on zero moment point (ZMP) that remains inside the convex hull of the foot support using the ankle torque. There are lots of successful results, but the gaits seem not to be so natural. On the other hand, for a biped with point contact a geometric tracking method for biped walking using input-output linearization (Aoustin & Formalsky, 1999; Grizzle et al., 2001; Aoustin& Formalsky, 2003; Chevallereau et al., 2003) produces stable gait that seems quite natural. (The idea of the geometric tracking can be seen in the previous studies of Furusho (Furusho et al., 1981) and Kajita (Kajita & Tani, 1991).) Grizzle, et al. (Grizzle et al., 2001) proposed the method for a three-link model, only two outputs are controlled, the reference are expressed as a function of the biped state. Zero dynamics with an impact event of the controlled system were analyzed by Poincaré method. The effectiveness of geometric tracking has been verified on a platform called 'Rabbit' (Chevallereau et al., 2003) (Fig.1 left) with point feet. Westervelt, et al. (Westervelt et al., 2005) gave some additional results to show capability for robustness, changing average walking rate, and rejecting a perturbation by 'one-step transition control' and 'event-based control'.
In the field of passive dynamic walking mechanisms (McGeer, 1990), it is shown that a biped with large radius circular arc feet can take easily a lot of steps. The prototype Emu (Fig.1 right) can be equipped with various arc feet with different radii (Kinugasa et al., 2003; 2007). In previous walking experiments the biped Emu is excited by gravity or forced oscillation of the length of legs. If the feet radius is 10% of leg length, the biped could only take few steps (Kinugasa et al., 2003) excited by the effect of gravity because of the sensitivity to disturbances produced by the cables, the guide to avoid lateral motion and so on. The biped could not walk by the forced oscillation. In the case of a radius which is 97% of leg length, the biped

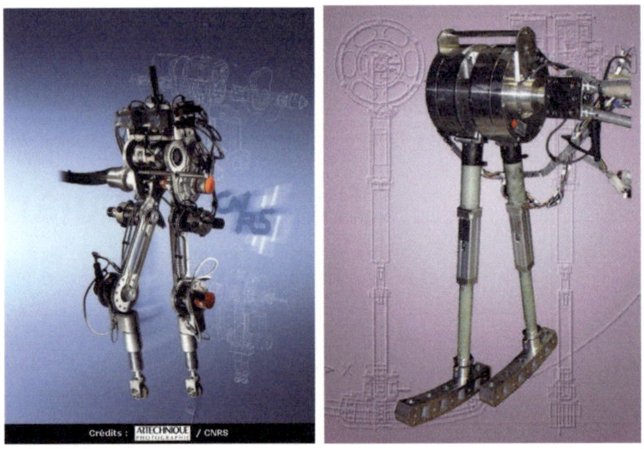

Fig. 1. Biped bipeds, "Rabbit" (left) and "Emu" (right).

Emu (Fig.1 right) can take easily few dozen of steps (Kinugasa et al., 2007) by the gravity and the leg oscillations. The step number is limited only by the space of our laboratory. The effect of the radii of circular feet was significant for our results, but the change of radius is also accompanied by other difference in physical parameters, thus a direct conclusion on the experimental study is not obvious and a more rigorous study must be done. In fact, the same results are well known in the field of passive dynamic walking as it is mentioned in Section 2. The geometric tracking method that was used for the underactuated biped Rabbit can be extended to the case of underactuated biped with circular arc feet. If the biped has the circular arc feet, the analytical stability study given by Chevallereau, et al. (Chevallereau et al., 2003) can not be applied directly. The contact point between the supporting foot and the ground moves forward during the step in this case. The same difficulty appears also in a flat feet model. For this problem, Djoudi and Chevallereau (Chevallereau & Djoudi, 2006) gave a solution to analyze the stability with a chosen evolution of the ZMP.

The purpose of the paper is to show the effects of the circular arc feet for an underactuated planar biped controlled by a geometric tracking method. The effect of the feet shape on the control properties is obviously depending on the walking strategies. Therefore it is significant to clarify the effect of the feet shape on the geometric tracking even if it is well known in the passive dynamic walking field.

A model of our biped is composed of five links. Prismatic knee joints are employed to avoid the foot clearance problem which occurs in association with large foot, not actuated ankle and rotational knee joint. A geometric evolution of the biped configuration is controlled, instead of a temporal evolution. The input-output linearization with a PD control law and a feed forward compensation is used for geometric tracking. The temporal evolution is analyzed using Poincaré map. The map is given by an analytic expression based on the angular momentum about the mobile contact point. The effect of the radius of the circular arc feet on stability and the basin of attraction is revealed by analytic calculation. It is compared to the effect of radius of the circular arc feet on passive dynamic walking. Section 2 presents an overview of previous studies on the circular arc feet. Section 3 gives the biped model. It is composed of a dynamic model and the impact model (instantaneous double support). Section 4 presents the control method. Section 5 gives the stability analysis. Some simulation results are shown and some discussion on the effects of the feet radius is developed in Section 6. Section 7 concludes the paper.

2. Previous studies on biped with circular arc feet

A circular arc feet for the biped are often treated in the field of passive dynamic walking McGeer (1990). It is well known that a passive dynamic walking gives an extremely natural gait. McGeer showed that an eigenvalue of the "speed mode" came to unit when the radius of a circular arc foot approaches the length of legs, and the eigenvalue becomes unit for synthetic wheel which has the foot radius equals to the leg length. The "speed mode" was related to dissipation of energy at the impact.

Wisse, et al. Wisse & van Frankenhuyzen (2003) showed that the larger feet radius, the larger amount of disturbances is accepted in experiments. The robustness against disturbances is connected to the size of a basin of attraction for walking. Wisse explained in the other paper Wisse et al. (2005) that "The walker will fall backward if it has not enough velocity to overcome the vertical position. Circular feet smoothen the hip trajectory and thus relax the initial velocity requirement. As the result, the basin of attraction is enlarged." However a decisive study on the effect of circular arc feet on the basin of attraction has yet to be performed. Recently, Wisse, et al. Wisse et al. (2006) presented a stability analysis of passive dynamic walking with flat feet and passive ankles. The effect of the flat feet was analogous to the effect of the circular arc feet for many properties in the sense that ZMP smoothly and monotonically moves forward from heel to toe. However he pointed out the need of validation for a more accurate model of the heel strike transition. Asano and Luo Asano & Luo (2007) discussed similar effect between the circular arc feet and the flat feet with actuated ankles.

Adamczy, Collins and Kuo Adamczyk et al. (2006) studied the centre of mass (CoM) mechanical work per step with respect to foot radius for various simple models of biped powered by an instantaneous push-off impulse under the stance foot just before contralateral heel strike Kuo (2001). They also showed relationships between foot radius and metabolic costs from measured via respiratory gas exchange. The data are collected through human walking with feet attached to rigid arc, and they conclude that the most effective walking is obtained when the foot radius equals to 30% of leg length. Geometrically speaking, feet length should be at least twice of the product of the coxa angle between two legs and the radius of feet McGeer (1990). Therefore one might choose the radius as 1/3 of a leg length with an angle 0.3 rad between two legs, in order to make an anthropomorphic biped, as McGeer wrote.

Thus for anthropomorphic models, 1/3 of leg length seems to be desirable in the sense of geometry between step length and feet lengths McGeer (1990), "foot clearance problem" Wisse & van Frankenhuyzen (2003) and energy costs Adamczyk et al. (2006).

3. The biped modeling

A biped presented in Fig.2 is composed of a torso and two symmetric legs which consist of the prismatic frictionless knees and the circular arc feet. The hips are rotational frictionless joints. We assume that the contact point does not slip and the biped walks in a vertical sagittal plane. The vector $\theta = [l_1, l_2, \theta_1, \theta_2, \theta_3]'$ (" $'$ " means transpose) of configuration variables (see Fig. 2, left) describes the shape of the biped during single support, l_i is the length of leg i, $\theta_i, i = 1,2$ is the angle between the torso and the leg i, θ_3 is the absolute angle of the supporting leg. The contact point between the biped and the ground is N_1. The lowest point of the swing leg tip is noted N_2. The actuator torques and forces are expressed by a vector $\Gamma = [\Gamma_1, \Gamma_2, \Gamma_3, \Gamma_4]'$. The absolute orientation of the biped θ_3 is not directly actuated. Thus, in a single support (SS), the biped is an under-actuated system. The walking gait consists of single support phases

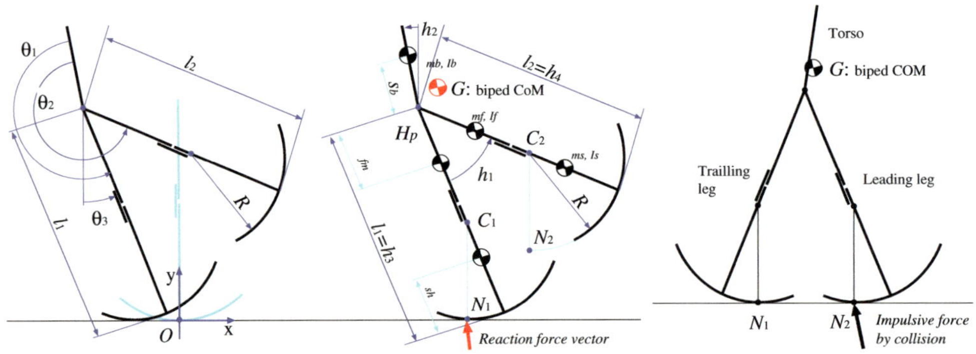

Fig. 2. The biped model: Left: coordinate of the model. Middle: physical parameters. Right: impact model.

separated by impacts, which are instantaneous double supports where a leg exchange takes place.

3.1 Dynamic model for single support phase

The dynamic model can be written as follows:

$$D(\theta)\ddot{\theta} + H(\theta, \dot{\theta}) = B\Gamma, \tag{1}$$

where $D \in \Re^{5\times5}$ is the inertia matrix, the vector $H \in \Re^5$ contains Coriolis, centrifugal and gravity terms. $B \in \Re^{5\times4}$ defines how the inputs Γ enter the model. Due to the choice of joint coordinates, the matrix B is written as: $B = [I_4, O_{4\times1}]'$.

3.2 Impact model

To derive an impact model, an general dynamic model is written:

$$D_e(\theta)\ddot{\theta}_e + H_e(\theta_e, \dot{\theta}) = B_e\Gamma + D_{R_i}(\theta)R_i. \tag{2}$$

where $\theta_e = [\theta', x_H, y_H]'$, and x_H and y_H are the Cartesian coordinates of the hip position H_p shown in Fig.2 (right), $D_e \in \Re^{7\times7}$ is the inertia matrix, the vector $H_e \in \Re^7$ contains Coriolis, centrifugal and gravity terms. $R_i = [R_{x_i}, R_{y_i}]'$ is a ground reaction force vector applied at the contact point. $B_e \in \Re^{7\times4}$ and $D_{R_i} \in \Re^{7\times2}$ defines how the inputs Γ and R_i enter the model, i is the number of the leg in contact with the ground, $i = 1$, $i = 2$, or $i = 1, 2$.

When the leg i rolls on the ground, the contact with the ground occurs in N_i. If leg i touches the ground and since, we assume that no sliding occurs, the position of N_i is $ON_i = [-R\theta_3, 0]'$, where O is defined such that for the current step, the point contact is in 0 when θ_3 is zero. This position can also be calculated by : $ON_i = OH_p + H_pC_i + C_iN_i$ (Fig. 2, middle). Thus, we have :

$$\begin{bmatrix} -R\theta_3 \\ 0 \end{bmatrix} = \begin{bmatrix} x_H + (l_i - R)\sin\theta_3 \\ y_H - (l_i - R)\cos\theta_3 - R \end{bmatrix}. \tag{3}$$

Therefore, the following constraint equation is obtained:

$$\Psi_i := \begin{bmatrix} x_H + R\theta_3 + (l_i - R)\sin\theta_3 \\ y_H - R - (l_i - R)\cos\theta_3 \end{bmatrix} = 0. \tag{4}$$

Equation (4) is differentiated twice with respect to time, to obtain a constraint on the joint acceleration:

$$D'_{R_i}\ddot{\theta}_e + C_{R_i}(\theta_e, \dot{\theta}_e)\dot{\theta}_e = 0. \tag{5}$$

where $D'_{R_i} = \partial\Psi_i/\partial\theta_e$ and C_{R_i} comes from the derivation.
We assume that the impact is inelastic and instantaneous without sliding. Let $\dot{\theta}_e^-$ and $\dot{\theta}_e^+$ be the angular velocities just before and just after the impact, respectively. Let $I_{m_i} = [I_{mx_i}, I_{my_i}]'$, for $i = 1,2$ be the vector of magnitudes of the impulsive reaction at the contact point of the stance and the swing leg. During the impact, the previous supporting leg can stay on the ground or take-off. If the leg takes off, the velocity of N_1 after the impact is positive. The impulsive ground reaction associated to a leg that stays on the ground must be positive and be in the friction cone. If the supporting leg takes off, the associated impulsive ground reaction is zero. The impact occurs when the leg tip of the swing leg contacts to the ground. To take into account the two cases, the following impact equation can be written:

$$\begin{cases} D_e(\theta)(\dot{\theta}_e^+ - \dot{\theta}_e^-) = D_R(\theta)I_m \\ D'_R(\theta)\dot{\theta}_e^+ = 0 \end{cases}, \tag{6}$$

where,

$$D_R(\theta) = \begin{cases} D_{R_2}(\theta), & \dot{y}_{N_1}^+ > 0 \\ D_{R_{12}}(\theta), & I_{my_1} > 0,\ I_{my_2} > 0 \end{cases}, \quad I_m = \begin{cases} I_{m_2}, & \dot{y}_{N_1}^+ > 0 \\ I_{m_{12}}, & I_{my_1} > 0,\ I_{my_2} > 0 \end{cases},$$

$$D_{R_{12}}(\theta) = \begin{bmatrix} D_{R_1}(\theta) & 0 \\ 0 & D_{R_2}(\theta) \end{bmatrix}, \quad I_{m_{12}} = \begin{bmatrix} I_{m_1} \\ I_{m_2} \end{bmatrix}.$$

From Eq. (6), we obtain:

$$\dot{\theta}_e^+ = (I_{7\times 7} - D_e^{-1}D_R(D'_R D_e^{-1} D_R)^{-1} D'_R)\cdot \dot{\theta}_e^-. \tag{7}$$

Before and after the impact, the biped is in contact with the ground on at least one leg, thus x_H, y_H can be calculated as function of θ, and $\dot{x}_H, \dot{y}_H$ can be calculated as function of $\dot{\theta}$. Equation (7) can be transformed into an equation of $\theta, \dot{\theta}$ only.

$$\dot{\theta}^+ = \Delta(\theta)\dot{\theta}^-, \tag{8}$$

where $\Delta(\theta) \in \Re^{5\times 5}$ is the impact matrix. This matrix depends on the foot radius R. In the gait studied, the legs swap their roles from one step to the next, thus since the biped is symmetric, the dynamic model is derived only for the support on leg 1. And the leg exchange is taken into account just after the impact. The state of the biped to begin the next step is :

$$\theta_i = T_{LS}\theta_f, \quad \dot{\theta}_i = T_{LS}\dot{\theta}^+,\ \dot{\theta}^+ = \Delta(\theta_f)\dot{\theta}_f, \tag{9}$$

where $T_{LS} \in \Re^{5\times 5}$ is the permutation matrix describing the leg exchange, the indexes $_{i,\ f}$ denoted the initial and final states of the biped for one step.

4. Control law

Since the studied biped is underactuated, and since some significant results have been obtained for the control of underactuated biped with point contact Chevallereau et al. (2003); Westervelt et al. (2005), our strategy for walking is to control four variables, such that they track the reference defined with respect to the monotonic variable θ_3. The four variables that are controlled are grouped in vector $h = [h_1, h_2, h_3, h_4]' = [\theta_2 - \theta_1, \theta_3 - \theta_1 + \pi, l_1, l_2]'$, composed of the angle between two legs, the absolute angle of the torso, and the leg lengths, (shown in Fig. 2, middle). This vector h, plus θ_3 defines the configuration of the biped. The relation with vector θ is the following:

$$\theta = \begin{bmatrix} h_3 \\ h_4 \\ -h_2 + \theta_3 \\ h_1 - h_2 + \theta_3 \\ \theta_3 \end{bmatrix} = \begin{bmatrix} 0 & 0 & 1 & 0 \\ 0 & 0 & 0 & 1 \\ 0 & -1 & 0 & 0 \\ 1 & -1 & 0 & 0 \\ 0 & 0 & 0 & 0 \end{bmatrix} h + \begin{bmatrix} 0 \\ 0 \\ 1 \\ 1 \\ 1 \end{bmatrix} \theta_3 \tag{10}$$

$$\theta = \frac{\partial \theta}{\partial h} h + \frac{\partial \theta}{\partial \theta_3} \theta_3. \tag{11}$$

where $\frac{\partial \theta}{\partial h}$ and $\frac{\partial \theta}{\partial \theta_3}$ are the constant matrices given in (10). Thus we have also:

$$\ddot{\theta} = \frac{\partial \theta}{\partial h} \ddot{h} + \frac{\partial \theta}{\partial \theta_3} \ddot{\theta}_3. \tag{12}$$

The control law is based on a computed torque control law and is such that the behavior of the controlled variables are:

$$\ddot{h} = \ddot{h}^d - K_p(h - h^d) - K_d(\dot{h} - \dot{h}^d). \tag{13}$$

But the reference to follow is a function of the variable θ_3 thus the reference is:

$$h^d = h^d(\theta_3) \tag{14}$$

$$\dot{h}^d = \frac{dh^d}{d\theta_3}(\theta_3)\dot{\theta}_3 \tag{15}$$

$$\ddot{h}^d = \frac{dh^d}{d\theta_3}(\theta_3)\ddot{\theta}_3 + \frac{d^2h^d}{d\theta_3^2}(\theta_3)\dot{\theta}_3^2, \tag{16}$$

Thus the desired behavior in closed loop is given by:

$$\ddot{h} = \frac{dh^d}{d\theta_3}(\theta_3)\ddot{\theta}_3 + \frac{d^2h^d}{d\theta_3^2}(\theta_3)\dot{\theta}_3^2 - K_p(h - h^d(\theta_3)) - K_d(\dot{h} - \frac{dh^d}{d\theta_3}(\theta_3)\dot{\theta}_3). \tag{17}$$

This expression is denoted:

$$\ddot{h} = \frac{dh^d}{d\theta_3}(\theta_3)\ddot{\theta}_3 + v(\theta, \dot{\theta}). \tag{18}$$

The dynamic model (1) can be expressed as function of $\ddot{h}$ and $\ddot{\theta}_3$ using (12)

$$D(\theta)(\frac{\partial\theta}{\partial h}\ddot{h} + \frac{\partial\theta}{\partial\theta_3}\ddot{\theta}_3) + H(\theta,\dot{\theta}) = B\Gamma, \tag{19}$$

The torques will be calculated in order to have in closed loop the behavior given in (18), thus the torques must satisfy:

$$D(\theta)((\frac{\partial\theta}{\partial h}\frac{dh^d}{d\theta_3}(\theta_3) + \frac{\partial\theta}{\partial\theta_3})\ddot{\theta}_3 + \frac{\partial\theta}{\partial h}v(\theta,\dot{\theta})) + H(\theta,\dot{\theta}) = B\Gamma, \tag{20}$$

Since the biped is underactuated, all the motion are not possible and based on the expression of matrix B, the admissible acceleration $\ddot{\theta}_3$ can be deduced. The dynamic model is decomposed into two sub-models. The first sub-model is composed of the first four lines and allows to calculate the torque. The second sub-model is composed of the fifth line and allows to calculate $\ddot{\theta}_3$. This sub-system gives:

$$\ddot{\theta}_3 = \frac{-D_5(\theta)\frac{\partial\theta}{\partial h}v(\theta,\dot{\theta}) - H_5(\theta,\dot{\theta})}{D_5(\theta)(\frac{\partial\theta}{\partial h}\frac{dh^d}{d\theta_3}(\theta_3) + \frac{\partial\theta}{\partial\theta_3})}, \tag{21}$$

where the index 5 refers to the 5th line of matrix D and vector H.
Finally, the control law is obtained:

$$\Gamma = D_{1,4}(\theta)((\frac{\partial\theta}{\partial h}\frac{dh^d}{d\theta_3}(\theta_3) + \frac{\partial\theta}{\partial\theta_3})\ddot{\theta}_3 + \frac{\partial\theta}{\partial h}v(\theta,\dot{\theta})) + H_{1,4}(\theta,\dot{\theta}), \tag{22}$$

where the indexes $1,4$ refer to the first four lines of matrix D and vector H.

5. Stability analysis

With the control, the output vector h converges to the reference path $h^d(\theta_3)$, and if the reference function is such that the impact condition is satisfied, the output is zero step after step for convenient choice of the control gains K_p, K_d Morris & Grizzle (2005).

5.1 Reference path

Since the initial and final configurations for a single support are double support configurations, when h^d is given, θ_3 can be deduced from geometrical relations. Thus the initial and final values of θ_3 on one step are known and denoted θ_{3i} and θ_{3f}. Since the condition of the impact is a geometrical condition, if the control law has converged and if θ_3 has a monotonic evolution, the configuration at the impact is the desired one. The reference function is designed such that the impact condition is satisfied. According to equations (8), (9), and (11), the reference path must be such that:

$$\theta(\theta_{3i}) = T_{LS}\theta(\theta_{3f}). \tag{23}$$

$$(\frac{\partial\theta}{\partial h}\frac{\partial h^d}{\partial\theta_3}(\theta_{3i}) + \frac{\partial\theta}{\partial\theta_3})\dot{\theta}_{3i} = T_{LS}\Delta(\theta_{3f})(\frac{\partial\theta}{\partial h}\frac{\partial h^d}{\partial\theta_3}(\theta_{3f}) + \frac{\partial\theta}{\partial\theta_3})\dot{\theta}_{3f}, \tag{24}$$

Equality (24) is composed of five scalar equations, thus $\frac{\partial h^d}{\partial\theta_3}(\theta_{3i})$ and $\frac{\dot{\theta}_{3i}}{\dot{\theta}_{3f}}$ can be calculated as function of $\frac{\partial h^d}{\partial\theta_3}(\theta_{3f})$. The ration of velocities is denoted $\delta_{\dot{\theta}_3}$:

$$\delta_{\dot{\theta}_3} = \frac{\dot{\theta}_{3i}}{\dot{\theta}_{3f}}. \tag{25}$$

5.2 Principle of the stability analysis

With the control law, the output vector h converges to the reference path $h^d(\theta_3)$. In the following section we assume that $h = h^d(\theta_3)$, that is, the system tracks the reference path. The five degrees of freedom (DoF) of the biped can be reduced to one DoF of a virtual equivalent pendulum under the condition, and we will hence analyze stability of the pendulum instead of the original biped.

This condition does not mean that the biped motion is cyclic with respect to time since the temporal evolution of θ_3 is the result of integration of Eq. (21), and thus depends on the reference path $h^d(\theta_3)$. For a SS phase θ_3 must evolve monotonically from θ_{3i} to θ_{3f}. The temporal evolution of the biped during a SS phase is completely defined by the velocity $\dot{\theta}_3$ for one particular value θ_3. The stability analysis is based on the Poincaré return map, and this return map will be built just before the impact, when the biped is in the configuration $h^d(\theta_{3f}), \theta_{3f}$. The variable that is effective to study the convergence to a cyclic motion is $\dot{\theta}_{3f}$. Since the angular momentum is proportional to $\dot{\theta}_{3f}$, the angular momentum (or its square value) can also be used in the stability analysis

5.3 SS phase

According the Newton-Euler second law, as the gravity is the only external force that produces a torque around N_1, the equilibrium of the biped in rotation around the mobile contact point N_1 gives:

$$\dot{\sigma}_{N_1} + MV_{N_1} \times V_G = r_{N_1G} \times M\vec{g}, \tag{26}$$

where V_{N_1} and V_G are the velocities at the points $N_1 = [-R\theta_3, 0]'$ and the center of mass, $G = [x_G, y_G]'$, M is the total mass of the biped, the gravity vector is $\vec{g} = [0, -g]'$, and σ_{N_1} is the angular momentum about N_1. The general expression of σ_{N_1} is:

$$\sigma_{N_1} = \sum_i m_i r_{N_1 G_i} \times V_{Gi} + \sum_i I_i w_i \tag{27}$$

where G_i is the center of mass for the link i, m_i and I_i are the mass and the inertia of link i, w_i is the angular velocity of link i, and V_{Gi} is the linear velocity of G_i. This quantity is linear with respect to the joint velocity component and can be written:

$$\sigma_{N_1} = S(\theta)\dot{\theta} \tag{28}$$

We assume that the biped follows reference path thus we have:

$$\theta = \frac{\partial\theta}{\partial h} h^d(\theta_3) + \frac{\partial\theta}{\partial\theta_3}\theta_3. \tag{29}$$

$$\dot{\theta} = \frac{\partial\theta}{\partial h}\frac{\partial h^d}{\partial\theta_3}(\theta_3)\dot{\theta}_3 + \frac{\partial\theta}{\partial\theta_3}\dot{\theta}_3. \tag{30}$$

Thus the angular momentum σ_{N_1} (28) is rewritten:

$$\sigma_{N_1} = S(\theta)\left(\frac{\partial\theta}{\partial h}\frac{\partial h^d}{\partial\theta_3}(\theta_3) + \frac{\partial\theta}{\partial\theta_3}\right)\dot{\theta}_3 = I_{\theta_3}(\theta_3)\dot{\theta}_3. \tag{31}$$

Equation (26) can be developed using the expression of r_{N_1G}, V_G, V_{N_1} as:

$$\dot{\sigma}_{N_1} = -Mg(x_G(\theta_3) + R\theta_3) + MR\frac{dy_G(\theta_3)}{d\theta_3}\dot{\theta}_3^2. \tag{32}$$

Equation (31) is combined to Eq. (32) to express the derivative of σ_{N_1} with respect to θ_3, under the assumption that θ_3 is monotonic:

$$\frac{d\sigma_{N_1}}{d\theta_3} = -Mg(x_G + R\theta_3)\frac{I_{\theta_3}}{\sigma_{N_1}} + MR\frac{dy_G}{d\theta_3}\frac{\sigma_{N_1}}{I_{\theta_3}}. \quad (33)$$

A new variable $\xi = \sigma_{N_1}^2/2$ is introduced, to transform Eq. (33) into an equation that can be integrated analytically:

$$\begin{aligned}\frac{d\xi}{d\theta_3} &= \kappa_1(\theta_3) + 2\kappa_2(\theta_3)\xi, \\ \kappa_1(\theta_3) &= -Mg(x_G + R\theta_3)I_{\theta_3}, \\ \kappa_2(\theta_3) &= \frac{MR}{I_{\theta_3}}\left(\frac{\partial y_G(\theta)}{\partial\theta}\right)'\frac{d\theta^d}{d\theta_3}.\end{aligned} \quad (34)$$

Equation (34) is a first order ordinary differential equation linear in ξ. Therefore, a general solution can be obtained, for a step that begins with θ_{3i} as a initial value:

$$\xi(\theta_3) = \delta_{SS}^2(\theta_3)\xi(\theta_{3i}) + V(\theta_3), \quad (35)$$

$$\delta_{SS}(\theta_3) = \exp\left(\int_{\theta_{3i}}^{\theta_3}\kappa_2(\tau_2)d\tau_2\right), \quad (36)$$

$$V(\theta_3) = \int_{\theta_{3i}}^{\theta_3}\exp\left(\int_{\tau_1}^{\theta_3}2\kappa_2(\tau_2)d\tau_2\right)\kappa_1(\tau_1)d\tau_1. \quad (37)$$

ξ and V are a pseudo-kinetic and a pseudo-potential energies of the virtual equivalent pendulum, respectively.
As a consequence if $\dot\theta_{3i}$ is known $\dot\theta_3$ can be deduced for the current step as a function of V and δ_{SS} without integration of (26). To be able to deduce from this equation the evolution of ξ (and in consequence of σ_{N_1} and $\dot\theta_3$) step after step, the evolution of ξ at the impact must be taken into account. In the following section, the index k will be added to denote the number of the current step

5.4 Impact phase

Let us consider the impact between steps k and $k+1$. Using (31), ξ at the end of step k is:

$$\xi_k(\theta_{3f}) = \frac{1}{2}(I_{\theta_{3f}}(\theta_{3f})\dot\theta_{3f,k})^2 \quad (38)$$

and ξ at the beginning of the step $k+1$ is:

$$\xi_{k+1}(\theta_{3i}) = \frac{1}{2}(I_{\theta_{3i}}(\theta_{3i})\dot\theta_{3i,k+1})^2 \quad (39)$$

Using (25), and defining δ_I by,

$$\delta_I = I_{\theta_3}(\theta_{3i})/I_{\theta_3}(\theta_{3f}), \quad (40)$$

we obtain:

$$\xi_{k+1}(\theta_{3i}) = \delta_I^2\delta_{\dot\theta_3}^2\xi_k(\theta_{3f}). \quad (41)$$

5.5 Poincaré map

Combining (35) and (41), the final value of ξ from the kth step to the $(k+1)^{\text{th}}$ step is as follows:

$$\xi_{k+1}(\theta_{3f}) = \delta^2(\theta_{3f})\xi_k(\theta_{3f}) + V(\theta_{3f}), \quad (42)$$
$$\delta(\theta_{3f}) = \delta_{SS}(\theta_{3f})\delta_I\delta_{\dot{\theta}_3}, \quad (43)$$

where θ_{3f} is the value of θ_3 just before the impact. This equation describes the Poincaré map. If a cyclic motion exists, then $\xi_{k+1}(\theta_{3f})$ corresponds to $\xi_k(\theta_{3f})$. Thus, a fixed point $\xi_c(\theta_{3f})$ is given using (42) as follows:

$$\xi_c(\theta_{3f}) = \frac{V(\theta_{3f})}{1-\delta^2(\theta_{3f})}. \quad (44)$$

Since $\xi_c(\theta_{3f})$ is positive, $V(\theta_{3f})$ and $1-\delta^2(\theta_{3f})$ must have the same sign. The following cases can occur:

Case 1: From (42), the fixed point is stable, if $\delta^2(\theta_{3f}) < 1$. Therefore, if $\delta^2(\theta_{3f}) < 1$ and $V(\theta_{3f}) > 0$, then an asymptotically stable cyclic motion exists.

Case 2: If $\delta^2(\theta_{3f}) = 1$ and $V(\theta_{3f}) = 0$, from (42), $\xi_{k+1}(\theta_{3f}) = \xi_k(\theta_{3f})$, namely, all motions are cyclic.

Case 3: From (42), the fixed point is unstable, if $\delta^2(\theta_{3f}) > 1$. Therefore, if $\delta^2(\theta_{3f}) > 1$ and $V(\theta_{3f}) < 0$, then an unstable cyclic motion exists.

Case 4: $V(\theta_{3f})(1-\delta^2(\theta_{3f})) < 0$, no cyclic motion exists.

Since by definition $\xi \geq 0$, from Eq. (35) for the complete step, ξ_c must satisfy the following inequality:

$$\xi_c(\theta_{3f}) \geq \xi_{\min} = \max_{\theta_3} \frac{-V(\theta_3)}{\delta^2(\theta_3)}. \quad (45)$$

to have $\xi(\theta_3) > 0$ for θ_3 between θ_{3i} and θ_{3f}.
Since a product of the two variables $(\delta_I \cdot \delta_{\dot{\theta}_3})$ is the ratio of momentum σ_{N_1} at the contact point N_1 before and after the impact, the speed of convergence is mainly associated with this ratio (This point will be detailed in the following sections), and connected to the distance between the contact points and velocity of the mass center before the impact Chevallereau et al. (2004). The contact point before the impact, at the end of the single support phase, is denoted N_1, the contact point after the impact, at the beginning of the next single support phase, is denoted N_2. Using equilibrium relation it is possible to compute the change of angular momentum around the contact point at impact as function of the value of the radii.
The distance d between the N_1 and N_2 is (see Fig.2)

$$N_1N_2 = d = 2(l-R)\sin(h_1/2). \quad (46)$$

The angular momentum before the impact denoted $\sigma^-_{N_1}$ is calculated around N_1 and can also be calculated around N_2, it is then denoted $\sigma^-_{N_2}$, the angular momentum transfer gives:

$$\sigma^-_{N_2} = \sigma^-_{N_1} - M \cdot d \cdot \dot{y}^-_G. \quad (47)$$

m_s	1 [kg]	I_s	0.05[kgm^2]	s_h	0.4 [m]	l_1	0.8~0.85 [m]
m_f	1 [kg]	I_f	0.05[kgm^2]	f_m	0.2 [m]	l_2	0.75~0.8 [m]
m_b	15 [kg]	I_b	3[kgm^2]	s_b	0.1 [m]	R	0~1.0 [m]

Table 1. Physical parameters for the dynamic model

At the impact, considering the vertical component I_{my_1} of the impulsive ground reaction I_{m_1} in the point N_1, the equilibrium in rotation around N_2 gives:

$$\sigma_{N_2}^{+} = \sigma_{N_2}^{-} - d \cdot I_{my_1}, \tag{48}$$

where I_{my_1} is the vertical component of the impulsive ground reaction I_{m_1} applied by the ground in N_1. The vertical equilibrium of the biped at the impact is :

$$I_{my_1} + I_{my_2} = M(\dot{y}_G^{+} - \dot{y}_G^{-}), \tag{49}$$

where I_{my_1} and I_{my_2} are the vertical components of the impulsive ground reactions I_{m_1} and I_{m_2} respectively in the points N_1 and N_2. The impact are such that the two legs stay on the ground, thus $I_{my_1} > 0$ and $I_{my_2} > 0$ and we have:

$$0 < I_{my_1} < M(\dot{y}_G^{+} - \dot{y}_G^{-}). \tag{50}$$

As a consequence, combining (47), (48), and (50), we have:

$$\sigma_{N_1}^{-} - M \cdot d \cdot \dot{y}_G^{+} < \sigma_{N_2}^{+} < \sigma_{N_1}^{-} - M \cdot d \cdot \dot{y}_G^{-}, \text{ if } d > 0, \tag{51}$$

$$\sigma_{N_2}^{+} = \sigma_{N_1}^{-}, \quad \text{if } d = 0, \tag{52}$$

$$\sigma_{N_1}^{-} - M \cdot d \cdot \dot{y}_G^{-} < \sigma_{N_2}^{+} < \sigma_{N_1}^{-} - M \cdot d \cdot \dot{y}_G^{+}, \text{ if } d < 0. \tag{53}$$

When $I_{\theta_3} > 0$ (see Fig.7) and $\dot{\theta}_3 < 0$ (see Fig.4), $\sigma_{N_1}^{-} < 0$. Considering (25), (31) and (40), the ratio $\delta_I \delta_{\dot{\theta}_3}$ is bounded:

$$1 - M \cdot d \cdot \frac{\dot{y}_G^{-}}{\sigma_{N_1}^{-}} < \delta_I \delta_{\dot{\theta}_3} < 1 - M \cdot d \cdot \frac{\dot{y}_G^{+}}{\sigma_{N_1}^{-}}, \ (d > 0), \tag{54}$$

$$\delta_I \delta_{\dot{\theta}_3} = 1, \quad (d = 0), \tag{55}$$

$$1 - M \cdot d \cdot \frac{\dot{y}_G^{+}}{\sigma_{N_1}^{-}} < \delta_I \delta_{\dot{\theta}_3} < 1 - M \cdot d \cdot \frac{\dot{y}_G^{-}}{\sigma_{N_1}^{-}}, \ (d < 0) \tag{56}$$

6. Simulation

In simulations, the physical parameters of the biped shown in Fig.2 are used (see Table 1). The gains of the control law are chosen so that tracking errors can be smaller than 10^{-4} for all walking gaits.

$$\begin{cases} K_p = \text{diag}([10^5, 10^4, 10^5, 5 \times 10^4]) \\ K_d = \text{diag}([5 \times 10^2, 5 \times 10^2, 10^3, 5 \times 10^2]) \end{cases} \tag{57}$$

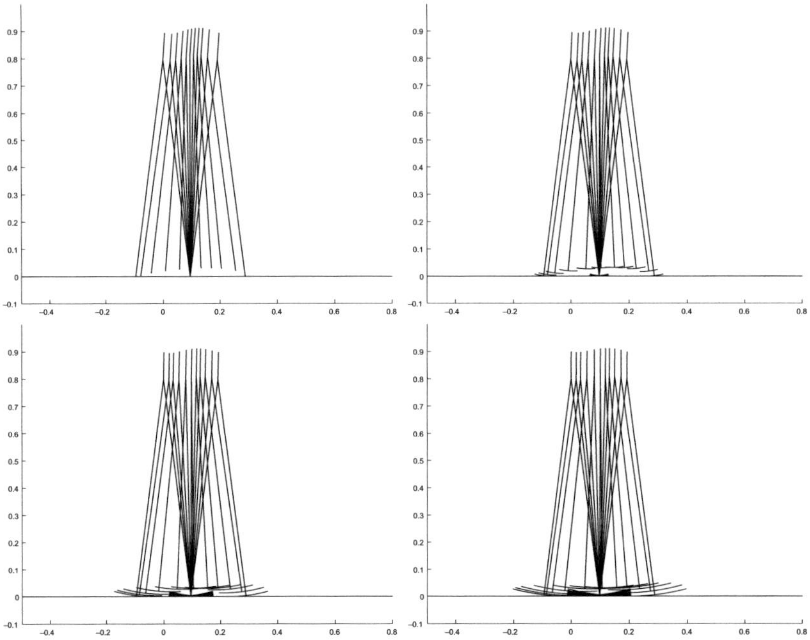

Fig. 3. The stick diagrams of walking. The foot radii $R = 0$ [m] (above left), 0.2 [m] (above right), 0.5 [m] (bellow left) and 0.7 [m] (bellow right).

Foot radius [m]	0	0.1	0.2	0.3	
Angle of torso [rad]	-0.060	-0.051	-0.043	-0.034	
Foot radius [m]	0.4	0.5	0.6	0.7	0.8
Angle of torso [rad]	-0.026	-0.018	-0.011	-0.004	0.002

Table 2. Torso angles. The angles are chosen such that cyclic motions have the same value $\xi_c(\theta_{3f}) = \xi(-0.12) = 16.27$.

6.1 Design of reference path

The reference path h^d is defined by a fourth order polynomial function such that:

$$h_d(\theta_3) = a[1,\ \theta_3^1,\ \theta_3^2,\ \theta_3^3,\ \theta_3^4]', \tag{58}$$

where $a \in \Re^{4\times 5}$ is a coefficient matrix for the reference h_d. An intermediate position of SS phase, positions and derivative with respect to θ just before the impact are given in order to calculate the coefficients of the reference paths (see Fig.3). Position and derivative with respect to θ after the impact are calculated by equations (23) and (24) .

Walking is depending on not only the radii of feet but also of the reference path of the length of the legs. The foot radius reduces the velocity of the CoM before the impact. The reference paths of the legs are chosen to smoothen the vertical variation of the CoM. However the references of the legs are affected by the impact, and the choice of the reference paths is limited accordingly. The radius mainly smoothens the vertical CoM motion.

The initial and the final length for the both legs are chosen as the same value. The final velocity for the biped are arbitrary fixed. The intermediate configuration for the legs is chosen such that the swing leg length decreases 0.02 m and the stance leg length increases 0.01 m during

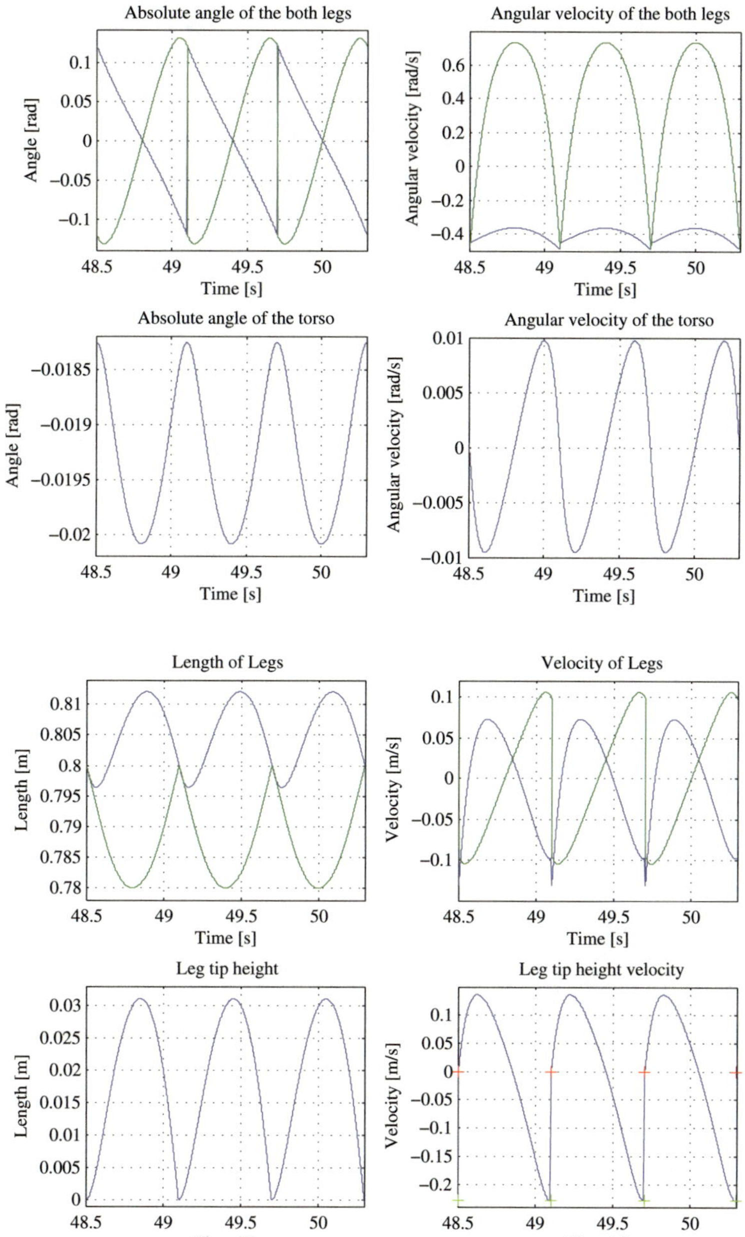

Fig. 4. Time responses at the cyclic motion with $R = 0.5$ [m] of the angle of the both legs, the torso, the length of legs and the leg tip. The reference paths are very well tracked.

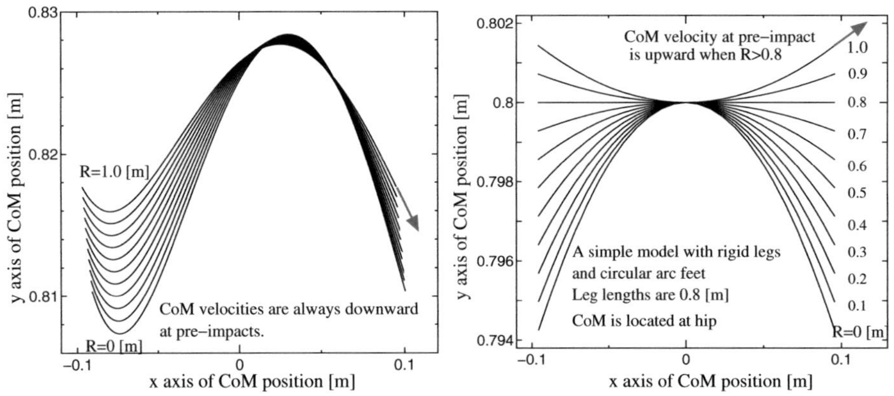

Fig. 5. CoM positions with respect to R. Left: the case of our biped shown in Fig.2. Tangent vectors of right ends of lines are expressing a post-impact velocity of CoM. Right: the case of a simple model with rigid legs and circular arc feet. CoM is located at hip position. When R > 0.8 [m], CoM velocities are upward. It gives a contradiction at the impact or there would be a flight phase.

the step to avoid that the swing leg tip touches the ground and the length of the legs is 0.8 [m] at the impact. Therefore the top position of the CoM is almost the same for each foot radius as shown in Fig.5. For one value R, we choose the angle of the torso at the impact arbitrary. The angle of the torso at the intermediate configuration is equal to 110% of the value of the torso angle at the impact. The corresponding value $\xi_c(\theta_{3f})$ is deduced. For example, the coefficient matrix in Eq.(58) for $R = 0.5$ is obtained as follows:

$$a|_{R=0.5} = \begin{bmatrix} 0 & -3.02 & -0.158 & 70.8 & 10.9 \\ -0.0201 & 0.0002 & 0.255 & -0.0106 & -8.89 \\ 0.810 & -0.122 & -1.58 & 8.50 & 61.2 \\ 0.780 & -0.0037 & 1.91 & 0.254 & -36.5 \end{bmatrix} \tag{59}$$

Then from this reference motion we deduced the reference motion for the other value of the radius R. The angle of the torso at the impact $h_2(\theta_{3f})$ is adjusted such that the cyclic motions for all foot radii R have the same value $\xi_c(\theta_{3f})$ as shown in Table 2.
Fig.3 shows examples of stick diagrams of walking for one step with the foot radii R =0 [m], 0.2 [m], 0.5 [m] and 0.7 [m] and the step angle =0.24 [rad]. A cyclic motion for $R = 0.5$ [m] is given in Fig.4. CoM positions with respect to R are shown in Fig.5. Tangent vectors of right ends of lines are expressing a post-impact velocity of CoM. The variation of CoM velocities at the impact are presented in Fig.6.
Energy excitation for continuous walking with smaller feet radius is mainly done by the asymmetric mass distribution due to the torso forward inclination. Leg swing also provides a way of putting energy. For small feet radii, the energy for walking is produced by the weight of the torso that is inclined forward. For larger feet radii, the energy for walking is produced by the motion of the swing leg.
Since the impact equation changes, the initial configuration and velocity are changed accordingly. During the impact, for the chosen reference path, the two legs stay on the ground.

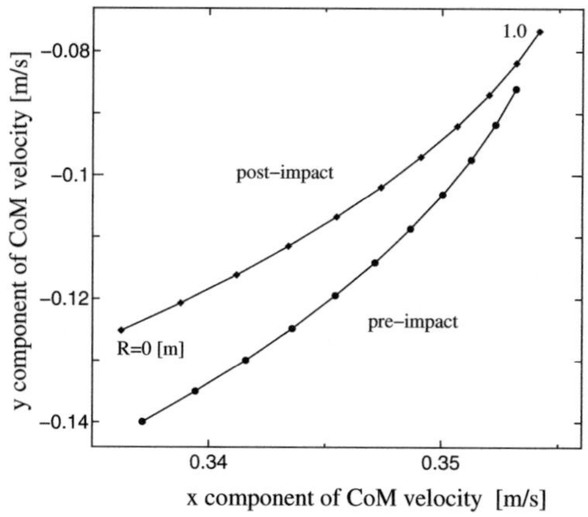

Fig. 6. CoM velocities at the impact with respect to R. The point corresponds to different value of R from 0 to 1, the abscissa of the point gives the horizontal velocity $\dot{x}_G$ respectively before and after the impact, the ordinates gives the vertical velocity $\dot{y}_G$ respectively before and after the impact. The vertical velocities before the impact are always directed downward.

6.2 Stability analysis

The variables in the analytic solution (35) are shown in Fig.7 with respect to the monotonic variable θ_3 for various values of the foot radius R. It should be noted that the monotonic variable is evolving from a positive value to a negative value, $\theta_3 : 0.12$ [rad] $\rightarrow -0.12$ [rad]. In Fig.7, $\xi_c(\theta)$ is given for all the cyclic motions. It can be observed that $\xi_c(\theta_{3f}) = \xi(-0.12) = 16.27$. The figure of $\delta^2_{SS}(\theta_3)$ is given by Eq. (36). The convergence of Poincaré map, as shown in Eq. (43), is function of $\delta^2_{SS}(\theta_{3f}) = \delta^2_{SS}(-0.12)$. However the values of $\delta^2_{SS}(-0.12)$ are very close to unit thus the convergence of Poincaré map is essentially defined by the impact map : $\delta(\theta_{3f}) \approx \delta_I \delta_{\dot{\theta}_3}$. The second figure from the left of Fig.7 represents the evolution of V defined by Eq. (37). These functions are essentially affected by the evolution ξ. The third figure of Fig.7 shows the term I_{θ_3} given by Eq. (31), I_{θ_3} is always positive and has not large variation. This first study concerns reference path with an interlink angle at the impact equals to 0.24 [rad]. For this value, the evolution of $\delta^2_{SS}(\theta_{3f})$,$\delta_I$, $\delta_{\dot{\theta}_3}$ and $\delta(\theta_{3f})$ are given in solid line in Fig.8, as function of the R. The cyclic motion is stable for $R < 0.8$.

In order to determine if the radius $R = 0.8$ is a limit of stability only for one specific reference path or if this limit is more physical, different kinds of reference motion are considered in the following. Only the interlink angle $h_1(\theta_{3f})$ at the impact is changed. For different values of h_1 and radii R, the coefficient involves in the convergence condition are drawn in Fig.8.

$\delta_{\dot{\theta}_3}$ and δ_I increase when R increases and $h_1(\theta_{3f})$ decreases from Fig.8. δ^2 also increases at the same time. The term δ^2 comes to unit when $R = 0.8$ [m] which means that R has the same values as the length of legs at the impact.

Remark: We confirmed in another simulations that variations of the torso angle had small influences on δ_I and $\delta_{\dot{\theta}_3}$ although it essentially affects ξ. The variables V, δ_{SS}, I_{θ_3} and ξ in the analytic solution for SS phase change for the torso angle. However the variation of δ_{SS} is smaller than the variations of δ_I and $\delta_{\dot{\theta}_3}$ with respect to the foot radii. △

Fig.9 presents the stability property with respect to the foot radii. Two black rigid lines show V and $\delta^2 - 1$. V and $\delta^2 - 1$ have opposite sign thus a cyclic motion may exist such that (45) is satisfied for any value of radii R. For $R < 0.8$ [m], the motion is stable. For $R > 0.8$ [m], the

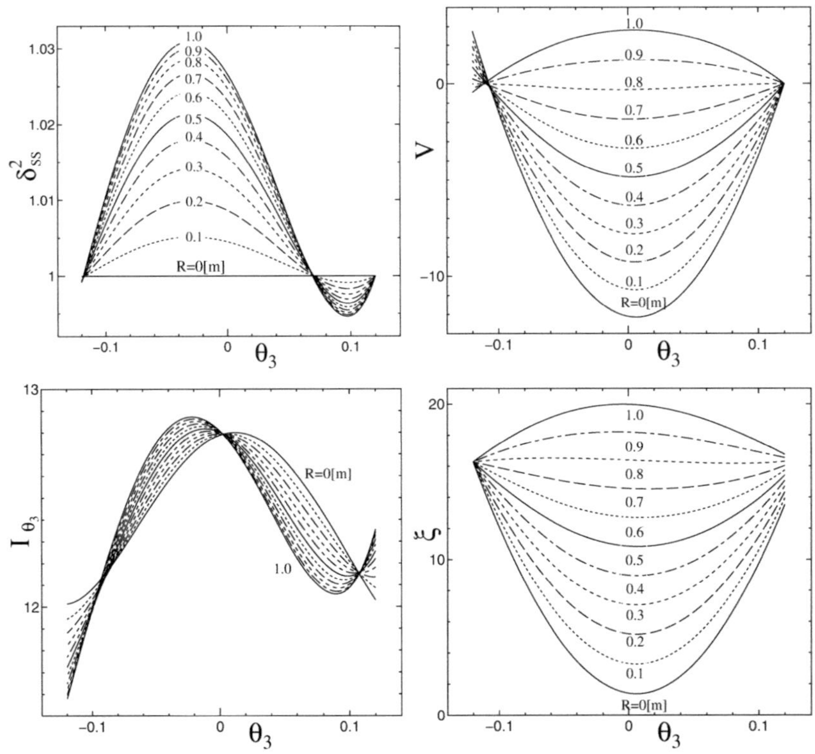

Fig. 7. Analytic solutions for SS phase. The figures are δ^2_{SS} by Eq. (36) (above left), V by Eq. (37) (above right), the function I_{θ_3} by Eq.(31) (bellow left)and ξ by Eq. (35) (bellow right). θ_3 evolves from positive (0.12) to negative (-0.12).

motion is unstable. For $R = 0.8$ [m], the motion is neutral, in this case any value ξ_c produces cyclic motions.

Case corresponding to a radius superior to the length of each leg, ($R > 0.8$ [m]) can be studied if we consider that the motions of feet are not in the same sagittal plane to avoid collisions. In the leg exchange, at the impact, the contact point moves back but the contact point has a large forward progression during the single support phase, the biped goes forward.

The gradient δ^2 (Eq. (43)) of Poincaré map (Eq. (42)) depends on the SS phase (δ_{SS}) and the impact phase ($\delta_I \cdot \delta_{\dot{\theta}_3}$). δ_{SS} was close to unit at the impact. Since $\dot{y}_G^- < \dot{y}_G^+ < 0$ (see Fig.6), we obtain that the foot radius R and the sign of d defined the position of the ratio $\delta_I \delta_{\dot{\theta}_3}$ with respect to 1 from Eq. (54) to Eq. (56).

- if $R < l, d > 0$, and $\delta_I \delta_{\dot{\theta}_3} < 1$
- if $R = l, d = 0$, and $\delta_I \delta_{\dot{\theta}_3} = 1$
- if $R > l, d < 0$, and $\delta_I \delta_{\dot{\theta}_3} > 1$

The property of the gradient δ^2 agrees with "speed mode" of passive dynamic walking obtained by McGeer McGeer (1990). Wisse Wisse et al. (2006) finds results that are different from our results. For passive walking he finds that for stability point of view the best radius is 14% of leg length, this value corresponds to a case where two monotonic lines of eigenvalues are crossing. The increasing one is represented 'Speed mode', and the decreasing one is

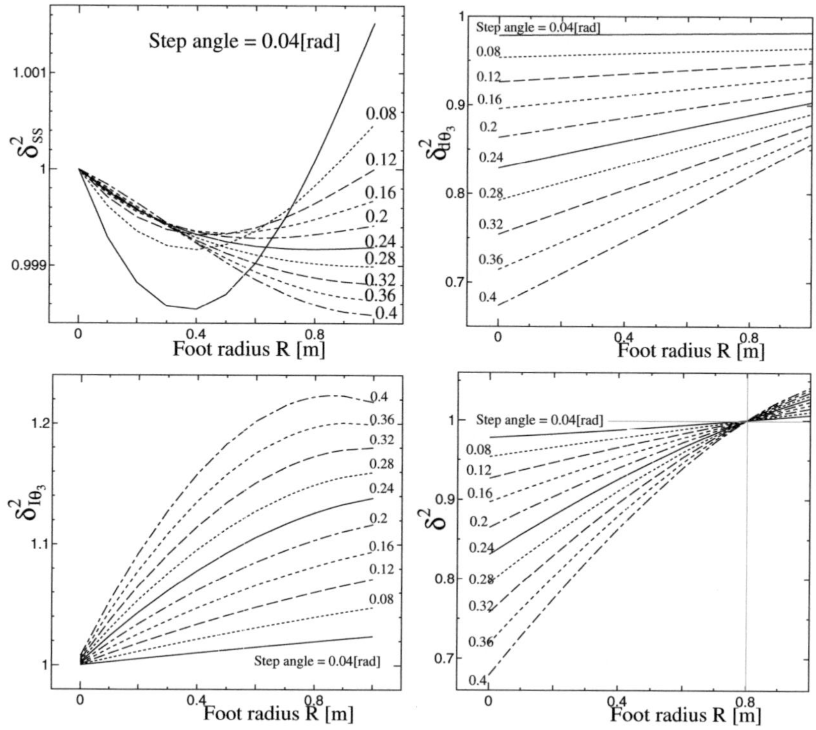

Fig. 8. Slope of the Poincaré return map. Step angle which means angles between two legs at the impact varies from 0.04 [rad] to 0.40 [rad]. The figures show δ^2_{SS} (above left), $\delta^2_{\dot{\theta}_3}$ (above right), $\delta^2_{I_{\theta_3}}$ (bellow left) and δ^2 with respect to the foot radii $R = 0 \sim 1.0$ [m] (bellow right). $R = 0.8$ [m] means that the radius is the same as the leg length at the impact for the analytic solution. For $R = 0.8$ [m], the cyclic motion is not stable.

'Totter mode'. However the crossing point changes with respect to slope angle and physical parameters of bipeds. The 14% of leg length is not the best radius, generally speaking. In our controlled system, it is predictable that the 'Totter mode' is close to zero or much smaller than the 'Speed mode', since the 'Speed mode' is expressed by the zero dynamics of the controlled system and the 'Totter mode' is depending on the controller gains. Term δ^2 has the same property of the 'Speed mode', and thus is increasing with respect to R. In our case we are not interested in the best solution but in the limit where stability exists, thus there are no contradiction with the results of Wisse Wisse et al. (2006).

6.3 Basin of attraction

Basins of attraction determined by numerical computations are shown in Fig.10. The larger the foot radii are in the stable domain, the wider the basin of attraction is but the slower the speed of convergence is. If the foot radius is the same as the leg length, the motion is neutral, that is, all motions are cyclic.

In Fig.10, the area between the line of $\xi^-_{\min}$ and $\xi^-_{\max}$ is the basin of attraction. The variable ξ just before the impact is used for expressing the basin of attraction. The line ξ_c represents the cyclic motions. Fig.11 presents time evolutions of θ_3, $\dot{\theta}_3$ for 100 steps. The following foot radii are considered: $R = 0$ [m], 0.5 [m] , 0.8 [m] and 1.3 [m]. The first two cases are clearly

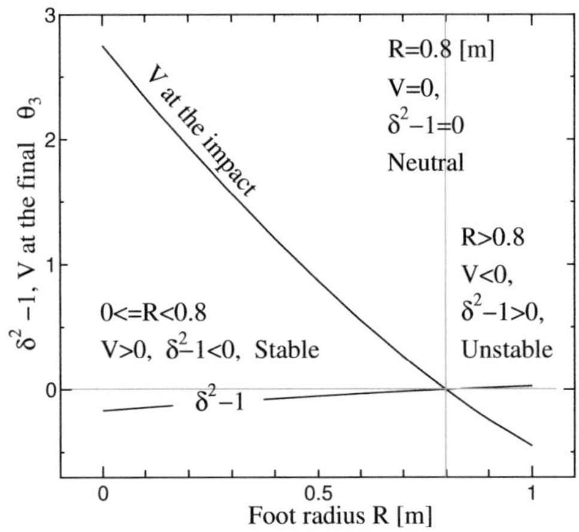

Fig. 9. The property of stability with respect to the foot radii R. Two black rigid lines show V and $\delta^2 - 1$. V and $\delta^2 - 1$ have opposite sign thus a cyclic motion may exist such that (45) is satisfied. For $R < 0.8$ [m] the motion is stable. For $R > 0.8$ [m] the motion is unstable. For $R = 0.8$ [m] the motion is neutral, that is all of ξ_c are cyclic motions.

stable, the case, $R = 0.8$, is neutral, and the case, $R = 1.3$, is unstable. Simulations confirm the existence of the neutral condition.

The property of the basin of attraction with respect to the radius is also analogous to the results of passive dynamic walking by Wisse Wisse & van Frankenhuyzen (2003). As depicted in Fig.10, the bottom line shows minimal ξ corresponding to ξ_{min}. It means a required minimal angular momentum to overcome a gap from a minimum of a vertical position of CoM to a maximum. If the momentum is smaller than the minimum, the complete step is not achieved, the step begins and then the robot goes backward to return to its initial configuration for the step. After that, the robot stops, but it does not fall down contrarily to a passive dynamic walker Wisse et al. (2005) that falls down backward.

From Fig.5, the smaller the radius is, the larger the gaps of the vertical positions of CoM and the minimal ξ_{min}^- are. Thus the circular arc feet broaden the minimal bounds. The variation of the maximal bounds is caused by limits on the vertical reaction forces to avoid taking-off. The reaction force vector R_1 at the point N_1 is given by the following equation:

$$R_1 = \begin{bmatrix} R_{x_1} \\ R_{y_1} \end{bmatrix} = \begin{bmatrix} M\ddot{x}_G \\ M(\ddot{y}_G + g) \end{bmatrix}. \tag{60}$$

The vertical acceleration $\ddot{y}_G$ is decided by the the centrifugal force caused by the angular velocity of the stance leg $\dot{\theta}_3$ and an acceleration of the leg variation $\ddot{l}_i(t)$. The radius smoothens the variation of CoM, and consequently the centrifugal force is reduced. We observe that the acceleration of the leg is smaller when the radii increase. Thus, the maximal ξ_{max}^- is extended when the radius increases. Namely, the basin of attraction is broaden by physical properties such as the feet radii. Globally, our controlled system has similar properties for stability and basin of attraction to the passive dynamic walking.

6.4 Consumed energy

Consumed energies and specific resistance for one cyclic step with respect to the foot radii R are described in Fig.12. The following formula is used for computing the consumed energy

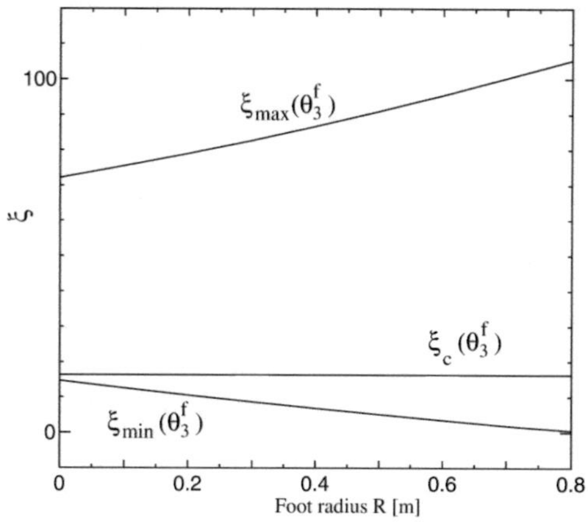

Fig. 10. Basin of attraction of ξ w.r.t. the foot radii R. The area between the line of $\xi_{\min}(\theta_3^f)$ and $\xi_{\max}(\theta_3^f)$ is the basin of attraction by the numerical method. The line ξ_c means the cyclic motions. In the upper area of $\xi_{\max}(\theta_3^f)$, vertical reaction forces are negative. There would be a flight phase. In the lower area of $\xi_{\min}(\theta_3^f)$, the velocity of the monotonic variable after the impact is not large enough to produce a step, $\xi_{\min}(\theta_3^f)$ is given by (45). After the beginning of the step, the biped goes backward or stands still eventually.

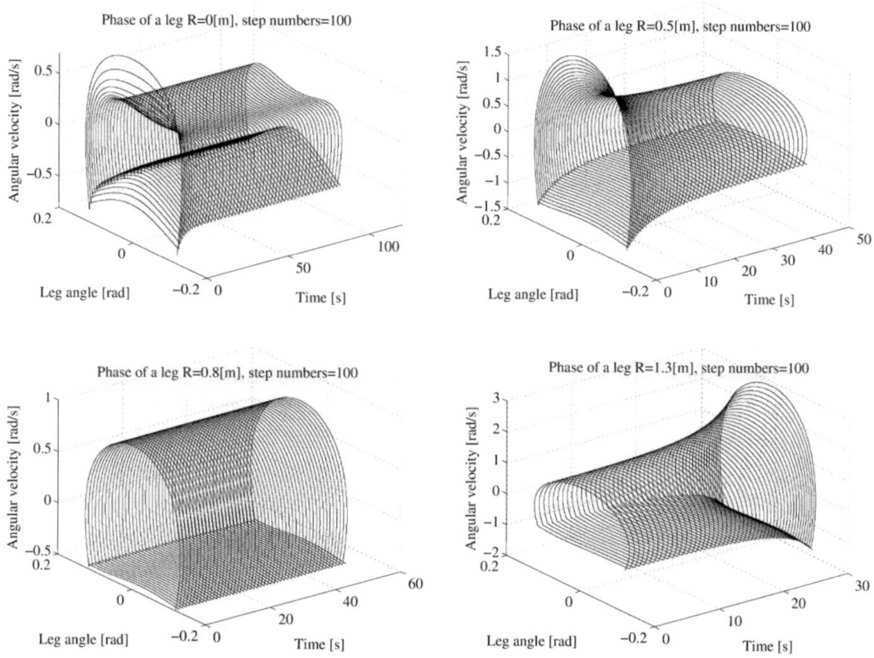

Fig. 11. Time evolutions of phases for the first leg at the foot radii $R = 0$ [m] (stable, above left), 0.5 [m] (stable, above right), 0.8 [m] (neutral, bellow left) and 1.3 [m] (unstable, bellow right).

E_c:

$$E_c = \int_0^T |\dot{\theta}' \cdot B \cdot \Gamma| dt. \tag{61}$$

The specific resistance SR is computed by the following fomula:

$$SR = \frac{E_c}{Mgd_{x_G}} \tag{62}$$

d_{x_G} indicates distance of total CoG for one step in horizontal direction. The larger the foot radius is, the smaller the consumed energy as well as the specific resistance is for the cyclic motion, even if the motion becomes unstable. Thus, the circular arc feet are effective in reducing the consumed energy.

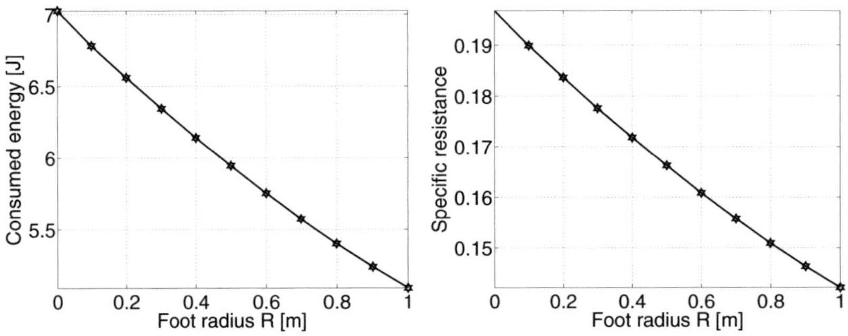

Fig. 12. Consumed energy (left) and specific resistance (right) for one cyclic step w.r.t. the foot radii R by the numerical simulation. The torso angle is chosen so that $\xi = 16.27$ by the analytic solution for all R.

6.5 Optimal radius

There is a trade-off property between the convergence speed, the basin of attraction and the energy consumption. What we can say is that the nearer the radius is to the leg length, the slower the speed of convergence is and the larger the basin is. 'Foot clearance problem' does not appear because of the variable length legs in our case. In the cases of 'Anthropomorphic Model' and 'Simplest Model' of Adamczyk's result Adamczyk et al. (2006), the CoM mechanical work property with respect to feet radii is similar to our result of consumed energy. However, in their cases of 'Forward-foot Model' and 'Kneed Model', the work had a minimum.

The suggestion of McGeer's to choose a foot radius of 1/3 of leg lengths can also be considered in our discussion. It might be better to choose a larger radius (e.g. between a half and three quarters) to have a large basin of attraction even if the speed of convergence is worth.

6.6 Unstable walking with radii greater than the leg length

Kuo's analysis Kuo (2001) of the CoM velocity contradicts our study because he considers a simple model with rigid legs and circular arc feet and the CoM is located at hip position, and we consider prismatic knees. The right of Fig.5 presents the evolution of the CoM relative to the simple model Kuo (2001). Tangent vectors of right ends of lines are expressing the pre-impact velocity of CoM, and tangent vectors of left ends of lines are expressing the post-impact velocity of CoM. When R > 0.8 [m], the change of CoM velocities are upward, which means the impulsive force at the impact is negative. It actually would be a flight phase. Left part of Fig.5 gives the CoM evolution in the case of our biped shown in Fig.2. Since all of the ranges of velocities of CoM at the impact are downward, it never fails to flight phase for any radius. In fact, our biped has prismatic knees and CoM is mainly distributed on the torso which is swinging a little. A lot of paths can be chosen for the CoM position differently from the simple model.

7. Conclusion

In the paper, some effects of circular arc feet for a planar biped via a geometric tracking were taken into account. An analytic solution of Poincaré map was given for the controlled system. Stability of walking was analyzed by the Poincaré map and the following results are obtained:

- Radii of the circular arc feet affect the stability of walking, and the speed of convergence decreases when the radii approaches to a leg length.
- A basin of attraction is broadened by choosing larger radii and the controller can stabilize the biped walking in the largest basin of attraction for the radii less than the leg length.

The leg length and the radius smoothen the variation and reduce the impact velocity. From the properties of the reference paths, The radius of the foot has a significant effect for the stability and the basin of attraction. The results are analogous to those McGeer (1990); Wisse & van Frankenhuyzen (2003) and the prospect Wisse et al. (2005) on passive dynamic walking. The geometric tracking method does not change the general effect of the circular arc feet. A reduction of the vertical CoM variation by the foot radius is functional not only for the geometric tracking method but for general biped walking. However the motion of CoM and the consumed energy are different from some very simple models because our model has variable length of legs and a torso.

8. References

Aoustin, Y. & Formalsky, A. M. (1999). Design of Reference Trajectory to Stabilize Desired Nominal Cyclic Gait of a Biped, *Proceedings of the International Workshop on Robot Motion and Control, ROMOCO'99*, IEEE, Kiekrz , Poland, 1999, pp.159-165.

Grizzle, J.W.; Abba, G. & Plestan, F. (2001). Asymptotically stable walking for biped robots: Analysis via systems with impulse effects, *IEEE transaction on Automatic Control*, Vol. 46, No, 1, (Jan. 2001), 51-64, ISSN 0018-9286.

Aoustin, Y. & Formalsky, A. M. (2003). Control Design for a Biped Reference Trajectory Based on Driven Angles as Functions of the Undriven Angle, *Journal of Computer and System Sciences International*, Vol.42, No.4, (Apr. 2003), pp.159-176, ISSN 1064-2307.

Chevallereau, C.; Abba, G.; Aoustin, Y.; Plestan, Y.; Westervelt, E.R.; Canudas-De-Wit, C. & Grizzle, J.W. (2003). Rabbit: A test bed for advanced control theory, *IEEE Control System Magazine*, Vol.23, No.5, (Oct. 2003), pp.57-79, ISSN 0272-1708.

Furusho, J.; Moritsuka, H. & Masubuchi, M. (1981) Low Order Modeling of Biped Locomotion System Using Local Feedback, *Transactions of the Society of Instrument and Control Engineers*, Vol.17, No.5, (May 1981), pp.596-601, ISSN 0453-4654

Kajita, S. & Tani, K. (1991) Study of Dynamic Biped Locomotion on Rugged Terrain – Derivation and Application of the Linear Inverted Pendulum Mode –, *Proceedings of IEEE International Conference on Robotics and Automation, ICRA1991*, IEEE, California, USA, Apr-1999, pp.1405–1411.

Westervelt, E.R.; Buche, G. & Grizzle, J.W. (2004) Experimental Validation of a Framework for the Design of Controllers that Induce Stable Walking in Planar Bipeds, *International Journal of Robotics Research*, Vol.23, No.6, (Jun. 2004), pp.559-582, ISSN 0278-3649.

Kinugasa, T; Hashimoto, Y & Fushimi, H. (2003) Passive Walking of Biped Emu with Attitude Control of Body, *Proceedings of IEEE/RSJ International Conference on Intelligent Robots and Systems, IROS 2003*, IEEE, Las Vegas, USA, Oct-2003, pp.346–359.

Kinugasa, T.; Osuka, K. & Miwa, S. (2007) Biped Walking by Variations of Knee Lengths and Attitude Control of a Body and its Frequency Analysis, *Journal of the Robotics Society of Japan*, Vol. 25, No. 3, (Apr. 2007), pp.116–123, ISSN 0289-1824

Chevallereau, C. & Djoudi, D. (2006) Feet can improve the stability property of a control law for a walking robot, *Proceedings of IEEE International Conference on Robotics and Automation, ICRA2006*, IEEE, Florida, USA, May-2006, pp.1206–1212.

McGeer, T. (1990) Passive Dynamic Walking, *International Journal of Robotics Research*, Vol.9, No.2, (Feb. 1990), pp.62–82, ISSN 0278-3649.

Wisse, M. & van Frankenhuyzen, J., (2003) Design and construction of mike; a 2d autonomous biped based on passive dynamic walking, Proceedings of Conference on Adaptive Motion of Animals and Machines, Kyoto, Japan, Mar-2003, WeP-I-1.

Wisse, M.; Schwab, A.L.; van der Linde, R.Q. & van der Helm, F.C.T., (2005) How to Keep From Falling Forward: Elementary Swing Leg Action for Passive Dynamic Walkers, IEEE Transactions on robotics, Vol.21, No.3, (May 2005), pp.394–401, ISSN 1552-3098.

Wisse, M.; obbelen, D.G.E.; Rotteveel, R.J.J.; Anderson, S.O. & Zeglin, G.J.,(2006) Ankle Springs Instead of Arc-shaped Feet for Passive Dynamic Walkers, Proceedings of IEEE-RAS International Conference on Humanoid Robots 2006, Genova, Italy, Dec-2006, pp.110–116.

Asano, F. & Zhi-Wei Luo, (2007) The Effect of Semicircular Feet on Energy Dissipation by Heel-strike in Dynamic Biped Locomotion, Proceedings of IEEE International Conference on Robotics and Automation 2007, Rome, Italy, Apr-2007, pp.3976–3981.

Adamczyk, P. G.; Collins, S. H. & Kuo, A. D., (2006) The advantages of a rolling foot in human walking, The Journal of Experimental Biology 209, (Oct. 2006), pp.3953–3963, ISSN 0022-0949.

Morris, B. & Grizzle, J. W., (2005) A Restricted Poincare Map for Determining Exponentially Stable Periodic Orbits in Systems with Impulse Effects: Application to Bipedal Robots, Proceedings of IEEE Conference on Decision and Control 2005, Seville, Spain, Dec-2005, pp.4199–4206.

Kuo, A. D., (2001) Energetics of Actively Powered Locomotion Using the Simplest Walking Model, Journal of Biomechanical Engineering, 124, (2001), pp.113–120, ISSN 0148-0731.

Chevallereau, C.; Formal'sky, A. & Djoudi, D., (2004) Tracking of a Joint Path for the Walking of an Underactuated Biped, Robotica, vol.22, (2004), pp.15-28, ISSN 0263-5747.

Farrell, K. D.; Chevallereau, C. & Westervelt, E. R. (2007) Energetic Effects of Adding Springs at the Passive Ankles of a Walking Biped Robot, Proceedings of IEEE International Conference on Robotics and Automation, Roma, Italy, Apr-2007, pp.3591–3596, ISSN 1050-4729.

5

SVR Controller for a Biped Robot with a Human-like Gait Subjected to External Sagittal Forces

João P. Ferreira[2,1], Manuel Crisóstomo[1],
A. Paulo Coimbra[1] and Bernardete Ribeiro[3]
[1]*Institute of Systems and Robotics, Dept. of Electrical and Computer Engineering, Univ. of Coimbra,*
[2]*Dept. of Electrical Engineering, Superior Institute of Engineering of Coimbra, Coimbra,*
[3]*CISUC, Dept. of Informatics Engineering, Univ. of Coimbra, Coimbra,*
Portugal

1. Introduction

This paper describes the control of a biped robot that uses an SVR (Support Vector Regression) for its balance. The control system was tested subjected to external sagittal pulling and pushing forces. Biped robots have leg link structures similar to the human's anatomy. To be able to maintain its stability under dynamic situations such robotic systems require good mechanical designs and force sensors to acquire the zero moment point (ZMP). Research in biped robotics has recently had a great surge due to the challenges of the subject and the media impact of famous biped robots like Honda's.

(Vukobratovic, 1990) developed a mathematical model of a biped robot and its method of control. Some research works (Zarrugh & Radcliffe, 1979), (Nakamura et al., 2004), (Jang et al., 2002) have reported the gait of biped robots based on human kinematics data, and a very good study of human body kinematics was done by Winter (Winter, 1990). Because a biped robot is easily knocked down, its stability must be taken into account in its gait design. Zheng (Zheng & Shen, 1990) proposed a method of gait synthesis taking into account the static stability. Chevallereau (Chevallereau et al., 1998) discussed dynamic stability through the analysis of the reaction force between the base of the foot and the ground. Unfortunately the defined trajectory does not assure the satisfaction of the stability restriction.

To assure the dynamic stability of a biped robot, Shin (Shin et al., 1990) and Hirai (Hirai et al., 1998) proposed standard methods for gait synthesis based on the zero moment point (ZMP). Basically this method consists of designing a desired ZMP trajectory, duly correcting the movement of the torso to maintain the ZMP trajectory as designed. However, because the change of the ZMP to accommodate the movement of the torso is limited, not all desired ZMP trajectories are possible (Park & Kim, 1998). The ZMP position can be obtained computationally using a model of the robot. However there might be a significant difference between the real and the calculated ZMP due to the difference between the real robot's physical parameters and its approximated mathematical model. To avoid this error, four force sensors are usually used on each foot to obtain an estimate for the real ZMP.

In the simplified 2-link control model the biped robot is divided into two masses, the torso and the legs, and can be used online. In the method described in this work the robot is divided into 8 links, which increases the accuracy, but also increases the computation time making it difficult to be used online. To overcome this limitation, some researchers propose computational intelligence techniques such as neuro-fuzzy nets and fuzzy systems for the control of biped robots (Choi et al., 2006), (Behnke, 2006), (Ferreira et al., 2004). These techniques have been surveyed by Katić (Katić & Vukobratović, 2005). As the ZMP control is non-linear an SVR (Vapnik, 1998) is appropriate. SVR calculates the optimal hyper plane for the training data and works faster than several other computational intelligence techniques (Ferreira et al., 2007a).
In this work the training of the SVR uses values obtained by simulating the full dynamic model of the biped robot. Some walking experiments where conducted to test the implemented control algorithm. The robot was subjected to external sagittal pushing forces and the dragging of masses. The experiments were performed with and without the SVR controller active and the results show the effectiveness of the SVR controller together with the presented gait, based on the human locomotion.

2. Implemented robot and software

A biped robot was designed and built at the Institute of System and Robotics of the Department of Electrical and Computer Engineering of the University of Coimbra, in Portugal. The mechanical structure of the robot shown in Fig. 1 has the main joints of hip, knee, and ankle, for each leg. There is another joint, an active inverted pendulum that is

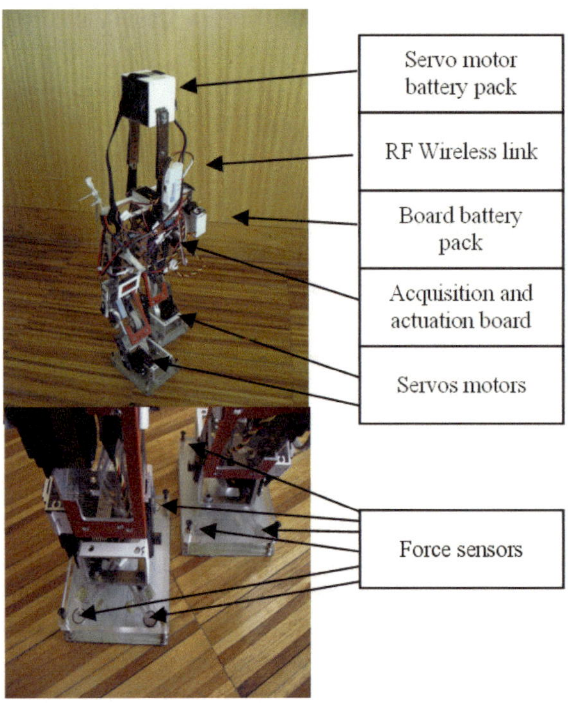

Fig. 1. Implemented robot.

used for the lateral balance of the structure. The robot carries its battery pack on this inverted pendulum. The robot is actuated by seven servo motors and the structure is made of acrylic and aluminium. It weighs 2.3 kg and is 0.5 m tall.
The robot was designed to move in both horizontal and inclined planes, to go up and down stairs, and has a speed of approximately 0.05 m/s. A 9600 bit/s RS232 wireless transmission link binds the control software, that is running on a PC, to the robot. The robot board has two PIC microprocessors, one to acquire the digital values of the force sensors and the other to actuate the servo motors.
An integrated software platform was developed which allows both the simulation and the real time control of the biped robot. The software screen layout is illustrated in Fig. 2.

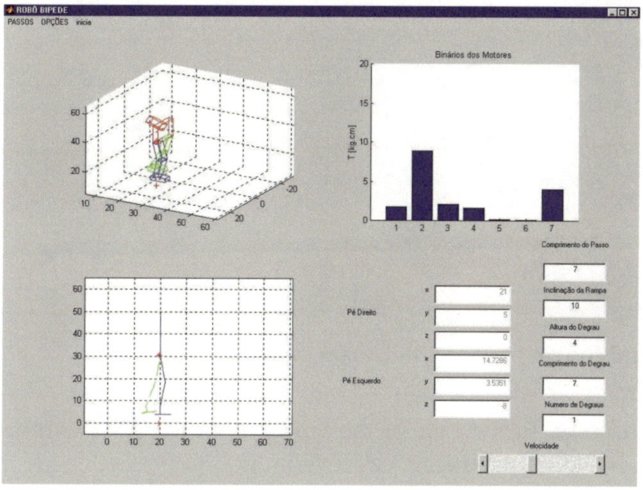

Fig. 2. Simulation and control software window.

Because it is very important to have very good mechanical contact between the force sensors (strip type) and the robot's feet, two different mechanical configurations were tested, as shown in Fig. 3. One of them uses a 9 mm of diameter cylinder (size of the active area of the sensor), and the other uses a semi-sphere to contact each sensor. Initially the cylindrical configuration was used and the signal was found to have a poor precision, due to the distribution of the force over the surface of the sensor. With the spherical configuration this problem was solved because the force is applied at a single point.

Fig. 3. Detail of the mechanical configurations used to improve the contact of the robot's feet with the force sensors.

3. Robot's 3D model and dynamics

Kinematics allows the movement of a robot to be described in terms of its position, speed and acceleration. Kinematics ignores the concepts of force, mass and inertia. The most widely-used representation is the Denavit-Hartenberg formalism. The systems of axes adopted to model the biped robot are shown in Fig. 4, together with a 3D model of the robot.

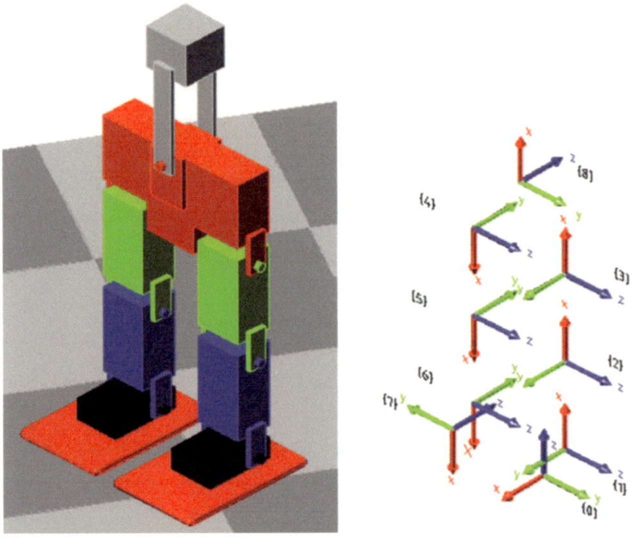

Fig. 4. 3D model and the respective coordinate systems.

Dynamics is the field of physics that studies the application of systems of forces and moments to a rigid body. In the case of the biped robot the ZMP formulation can be used (Hang et al., 1999), based on a dynamic calculation.

For a robot with four or more legs it is possible to consider the static stability that uses the center of gravity, but for a biped robot the dynamic stability may have to be taken into account, and the calculation of the ZMP makes this possible. The ZMP is defined as the point on the ground where the sum of all the active moments of force is null. In Fig. 5, the minimum distance between the ZMP and the border of the stable region is called the stability margin, and this can be considered as an indicator of the quality of the robot's stability.

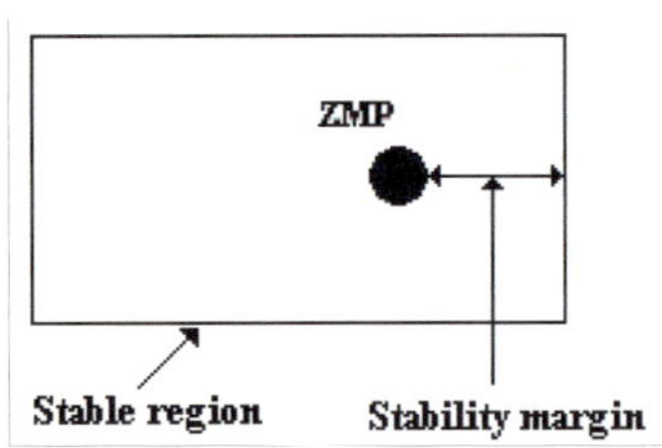

Fig. 5. Definition of the stability margin.

If the ZMP is inside the polygon of contact (stable region of the foot) between the foot and the ground, it can be said that the biped robot is stable. When the distance between the ZMP and the border of the stable region of the foot is the greatest, that is, when the coordinates of the ZMP are next to the center of the stable region, it can be said that the biped robot exhibits a high stability. For the robot model in Fig. 6, with the physical characteristics presented in Table I, the ZMP location is calculated by

$$X_{zmp} = \frac{\sum_{i=0}^{7} m_i(\ddot{z}_i + g)x_i - \sum_{i=0}^{7} m_i\ddot{x}_i z_i - \sum_{i=0}^{7} I_{iy}\alpha_{iy}}{\sum_{i=0}^{7} m_i(\ddot{z}_i + g)} \quad (1)$$

$$Y_{zmp} = \frac{\sum_{i=0}^{7} m_i(\ddot{z}_i + g)y_i - \sum_{i=0}^{7} m_i\ddot{y}_i z_i + \sum_{i=0}^{7} I_{ix}\alpha_{ix}}{\sum_{i=0}^{7} m_i(\ddot{z}_i + g)} \quad (2)$$

where $\ddot{x}$, $\ddot{y}$ and $\ddot{z}$ are linear accelerations, I_{ix} and I_{iy}, are inertia coefficients, α_{ix} and α_{iy} are angular accelerations, m_i is the mass of the link i, and g is the gravitational acceleration. The aim of this work is the control of the robot in the sagittal plane. The control variable is the X_{zmp}. The lateral control is assured by the pendulum movement and is not studied here.

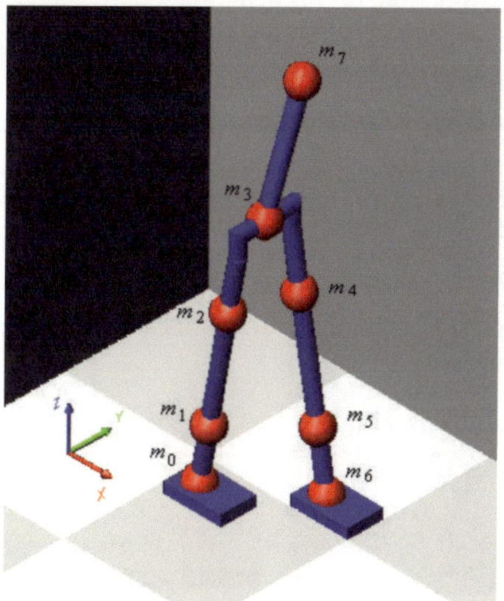

Fig. 6. Biped lumped mass model.

Biped	Mass (kg)	Length (m)	I_x (×10^{-4} kg.m^2)	I_y (×10^{-4} kg.m^2)
Foot	0.28x2	0.035	2.64	3.51
Shank	0.15x2	0.115	7.23	6.81
Thigh	0.15x2	0.115	7.23	6.81
Haunches	0.60	0.065	21.25	9.70
Pendulum	0.54	0.170	159.58	160.17

Table 1. Physical characteristics of the biped robot.

4. Designed gaits

A gait has been chosen that was conceived with the goal of being similar to human locomotion in horizontal planes. To describe a robotic human-like gait only the hip and ankle trajectories are needed. The knee trajectory depends on these two trajectories.

Humans are among the best biped walkers, which is a good reason for obtaining their joint trajectories when they walk, and then applying this information to a biped robot, even though its physical characteristics differ from those of a human being.
The joint trajectories of hip, knee and ankle in the sagittal plane during a single gait cycle of a normal human being have been presented by Winter (Winter, 1990).
To acquire human trajectories an acquisition system was developed (Ferreira et al., 2007b). Images were captured with a color webcam that has the following characteristics: CMOS 640 x 480 (VGA) sensor, maximum of 30 frames per second, USB 2 interface. Trajectory data were obtained with a 26 year-old male, 1.85 m tall. Several light emitting diodes (LED) were placed on strategic points on the male (Fig. 7a). Fig. 7b shows the model used to calculate the joint angles, where H is the point on the heel, T is the tip of the foot, K is the point on the knee, HI is the point on the hip and SH is the point on the shoulder. These reference points were captured by placing the camera perpendicular to the background, 3 m away from it and 0.75 m from the floor. This latter distance is half the distance from the floor to the highest reference point, which is the shoulder mark.

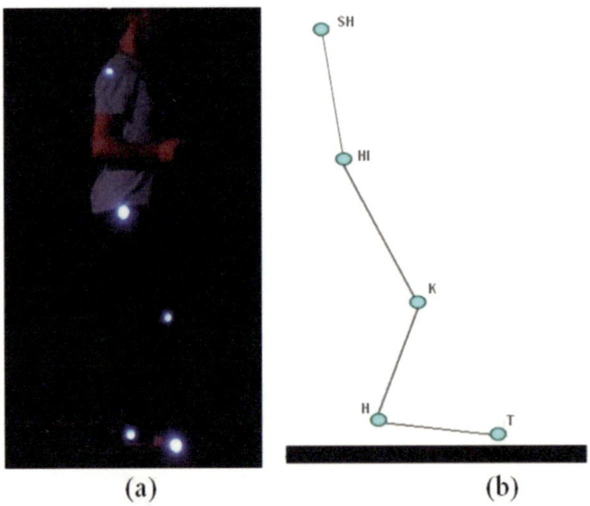

Fig. 7. (a) White light LED used as reference points on the person and (b) reference points on the person's model.

After the image acquisition of the reference points, the coordinates of the mass centers of these reference points were calculated for each frame and their trajectories obtained. The next step was to normalize the points of the trajectories and obtain a polynomial regression (which has square coefficient relation 0.996). This can be applied to any robot, using the height of the leg (Z_L) and the step length (X_S) as scale factors. So the hip and ankle trajectories are calculated by (3) and (4).

$$\begin{bmatrix} X_{hip} \\ Z_{hip} \\ X_{ankle} \\ Z_{ankle} \end{bmatrix}_{mov}^{T} = \left(A \cdot G_{mov}\right)^{T} \cdot XZ_{mov} \tag{3}$$

$$\begin{bmatrix} X_{hip} \\ Z_{hip} \\ X_{ankle} \\ Z_{ankle} \end{bmatrix}_{gnd}^{T} = \left(A \cdot G_{gnd}\right)^{T} \cdot XZ_{gnd} \quad (4)$$

where

$$A = \begin{bmatrix} 0 & 0 & 0 & 0 & 10.263\text{E-}03 & 249.033\text{E-}03 \\ 0 & 0 & 0 & -158.826\text{E-}06 & 78.736\text{E-}04 & 897.685\text{E-}03 \\ 0 & 0 & -15.67\text{E-}06 & 121.278\text{E-}05 & -14.623\text{E-}04 & 26.52\text{E-}04 \\ -180.869\text{E-}10 & 208.426\text{E-}08 & -745.788\text{E-}07 & 585.999\text{E-}06 & 954.539\text{E-}05 & -250.062\text{E-}06 \end{bmatrix}$$

is a coefficient polynomial matrix,

$$G_{mov} = \begin{bmatrix} (S_{\%}-50)^5 \\ (S_{\%}-50)^4 \\ (S_{\%}-50)^3 \\ (S_{\%}-50)^2 \\ (S_{\%}-50) \\ 1 \end{bmatrix} \quad \text{and} \quad XZ_{mov} = \begin{bmatrix} X_S & 0 & 0 & 0 \\ 0 & Z_L & 0 & 0 \\ 0 & 0 & X_S & 0 \\ 0 & 0 & 0 & Z_L \end{bmatrix}$$

are the percentage of the stride cycle for the moving foot and a scale factor, respectively,

$$G_{gnd} = \begin{bmatrix} S_{\%}{}^5 \\ S_{\%}{}^4 \\ S_{\%}{}^3 \\ S_{\%}{}^2 \\ S_{\%} \\ 1 \end{bmatrix} \quad \text{and} \quad XZ_{gnd} = \begin{bmatrix} X_S & 0 & 0 & 0 \\ 0 & Z_L & 0 & 0 \\ 0 & 0 & 0 & 0 \\ 0 & 0 & 0 & 0 \end{bmatrix}$$

are the percentage of the stride cycle for the grounded foot and a scale factor, respectively. The values of the parameters used in the experiment were $X_S = 0.23$ m and $Z_L = 0.07$ m. The trajectories obtained for the hip and the ankle are shown in Fig. 8 for the grounded foot, and in Fig. 9 for the moving foot. The torso angle in the human walk is similar to a cosine function.

5. Trajectory planning algorithm

The control method used in this work to achieve the sagittal equilibrium of the robot consists of correcting the hip and pendulum angle (torso angle) in order to keep the ZMP in the center of the grounded foot (with a tolerance of 4 mm).

We describe next the torso trajectory planning algorithm. It was executed off-line, and the result was used for the SVR training and for the amplitude setting of the initial predictive trajectory of the torso. First, taking into account the gait's characteristics, several via points

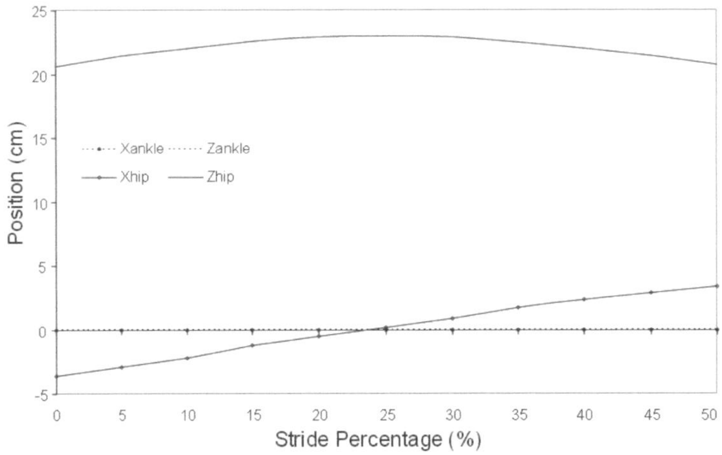

Fig. 8. Trajectories of the hip and ankle for the grounded foot.

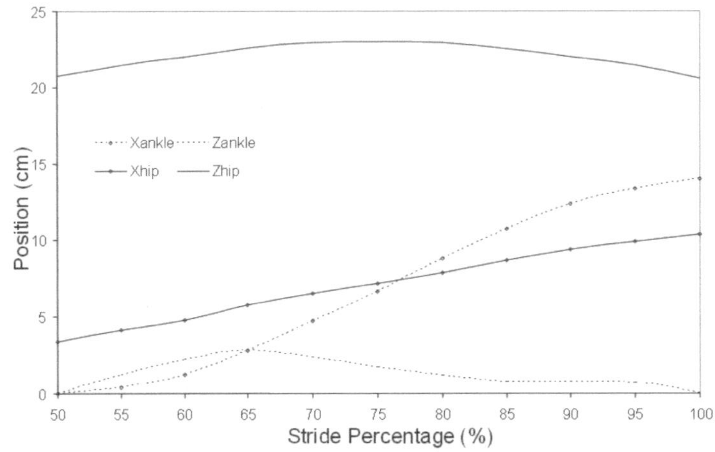

Fig. 9. Trajectories of the hip and ankle for the moving foot.

of the trajectory are obtained, and the trajectory of the foot is generated using a cubic spline. After that, the trajectories of the joints are calculated using direct and inverse kinematics; finally, the ZMP is calculated and the stability margin determined. This stability margin is then used to make iterative corrections to the torso angle, using the desired ZMP (center of the grounded foot) as reference. This procedure is repeated for all frames of the step.

It must be noticed that some gaits may not be stable because the working range of the servo motors used is limited. Fig. 10 shows the algorithm explained before.

This algorithm is used for the static and dynamic models, but the spline is not needed in the static model, which greatly reduces computational effort. The dynamic model requires four splines for each link, resulting in a great computational effort and making real-time computation difficult.

Fig. 11 shows the ZMP trajectory obtained with the previous algorithm for the "walking in horizontal planes" gait with a step length of 0.11 m. The figure shows ZMP points marked with crosses.

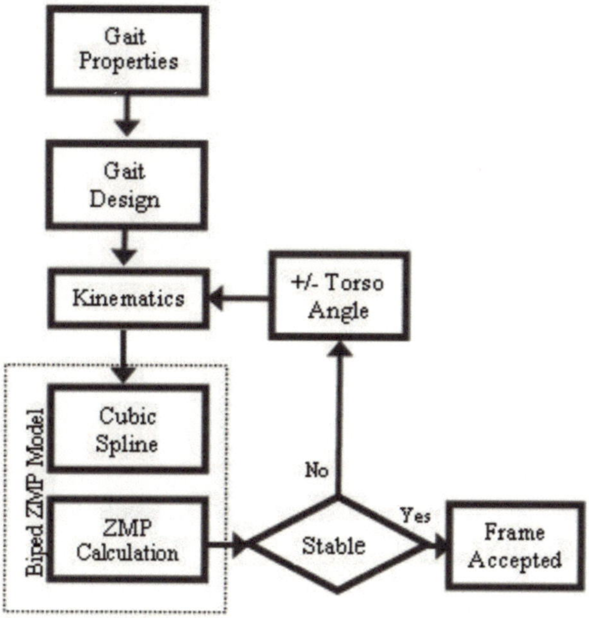

Fig. 10. Trajectory planning algorithm.

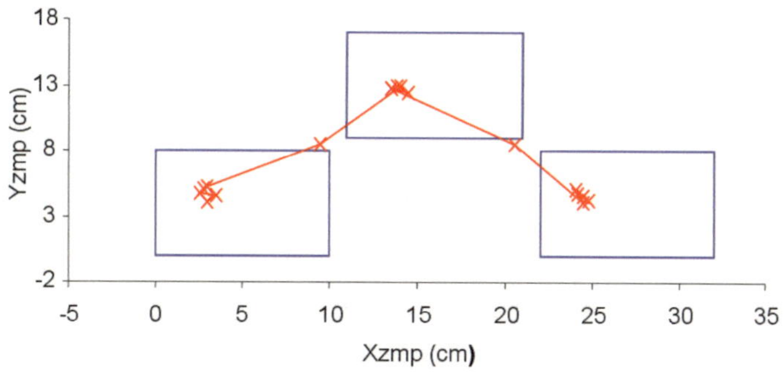

Fig. 11. ZMP Trajectory.

The SVR training data are the $\Delta\theta_{Torso}$ (torso angle correction), the X_{ZMP} and DX_{ZMP} (X_{ZMP} discrete time derivative). The $\Delta\theta_{Torso}$ versus DX_{ZMP} data were obtained by expert knowledge having a PD controller in mind. The $\Delta\theta_{Torso}$ versus X_{ZMP} training data were obtained by simulation using the trajectory planning algorithm, described above, for several step lengths and times. The result of the simulation was the relation K(t) between the required torso angle correction ($\Delta\theta_{Torso}$) and the difference between the actual X_{ZMP} and the desired X_{ZMP} (X_{ZMPD}, which is zero because it is at the origin of the coordinate system), to maximize the biped robot stability margin, as a function of the step time T_{step} (see Fig. 12).

Analyzing Fig. 12, it can be considered that K does not depend on the length of the gait, but only on the step time. Using a curve fitting algorithm, the expression

$$K(t) = 0.002 \cdot t^4 - 0.0637 \cdot t^3 + 0.7795 \cdot t^2 - 4.4803 \cdot t + 2.7866 \quad (5)$$

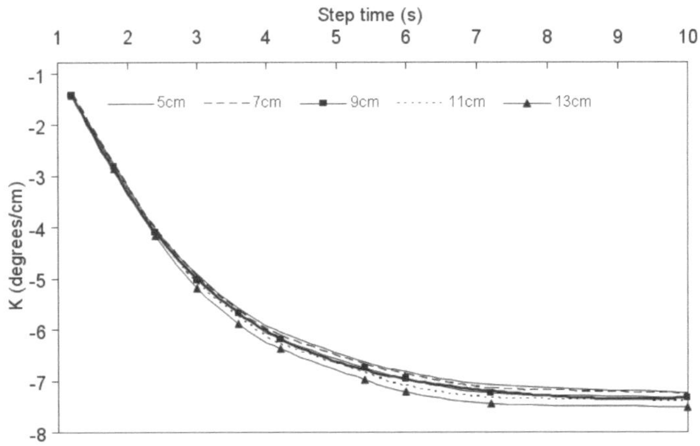

Fig. 12. The relation $K = \Delta\theta_{Torso} / (X_{ZMP} - X_{ZMPD})$.

was obtained. With this relationship the static model can be used to generate a first estimate for θ_{torso}, using the center of gravity (CoG), instead of the X_{ZMP}.
It was found that the human torso angle trajectory is similar to a cosine function. The predicted torso angle is then given by

$$\theta_{torso} = A \cdot \cos\left(\frac{2\pi \cdot S_{\%}}{100}\right) \tag{6}$$

where $A = X_{CoG} \cdot K(T_{Step})$ and S% is the gait time percentage. The value of X_{CoG} (X coordinate of the center of gravity of the robot) is calculated in the transition from the double phase to the single phase using

$$X_{CoG} = \frac{\sum_{i=0}^{7} m_i \cdot g \cdot x_i}{\sum_{i=0}^{7} m_i \cdot g} \tag{7}$$

where m_i is the mass of the link i and g is the gravitational acceleration.

6. Control strategy

The proposed biped control model is shown in Fig. 13. The joint angles of the robot, except the torso angle, are determined by inverse kinematics using the designed gait (3) and (4) for the entire step. The values of the torso angle are predicted by (6). After that, for each frame, all joint angles values are sent to the biped robot and the real X_{ZMP} (X_{ZMPR}) is determined reading the force sensors placed under its feet. The X_{ZMPR} value is then used by the SVR to correct the torso angle in real time. The output of the SVR is an increment to be added to the torso angle given by (6).
This control strategy assures the stability of the robot, even if external disturbances occur.
In order to allow real time control, the actual (real) value of the ZMP (ZMPR) is needed. When the ZMP is within the stable area, the ZMP and the centre of pressure (CoP) are the same. To determine the CoP, four force sensors are used in each foot of the robot, as Fig. 14

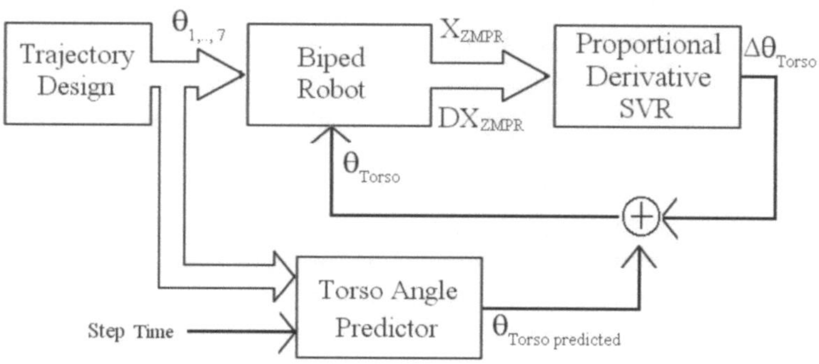

Fig. 13. SVR control of the biped robot.

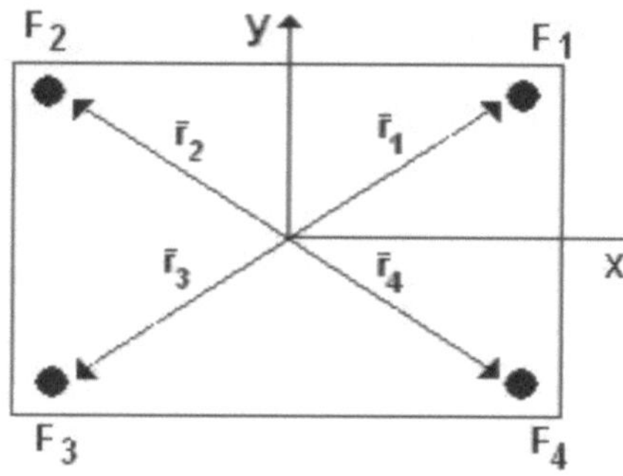

Fig. 14. Top view diagram of the foot, showing the location of the force sensors.

shows. These sensors are used to detect the force intensity and where the force is exerted, which is determined by

$$CoP = \frac{\sum_{i=1}^{8} F_i \cdot \overline{r_i}}{\sum_{i=1}^{8} F_i} \quad (8)$$

where F_i is the force measured in sensor i, and $\overline{r_i}$ is the sensor i position vector.
Force sensor signals are acquired by an analog to digital converter (ADC) with 10-bit resolution and with a maximum of 30 Hz sampling rate. The force measurements are noisy and the force sensor is sensitive to vibrations during the motion, so a second order Butterworth low pass filter is used to remove noise and high frequency vibrations from the force sensor signal. The difference equation for a second-order low-pass Butterworth digital filter has the form

$$y_k = b_1 x_k + b_2 x_{k-1} + b_3 x_{k-2} - a_2 y_{k-1} - a_3 y_{k-2} \quad (9)$$

where y is the filtered variable, x is the unfiltered variable, x_k is the value of x at time t_k, y_k is the value of y at time t_k, $t_k = k \cdot T$ is the current time, $T = t_k - t_{k-1}$ is the constant sampling interval, and k is an integer.

7. Support vector machines

Support Vector Machines (SVM) were developed by Vapnik (Vapnik, 1998) first to solve classification problems, and then they were successfully extended to regression and density estimation problems (Mohamed & Farag, 2003). SVM have gained popularity due to their many attractive features and promising empirical performance. The formulation of SVM employs the Structural Risk Minimization (SRM) principle, which has been shown to be superior to the traditional Empirical Risk Minimization (ERM) principle employed in the other non-parametric learning algorithms (e.g. neural networks) (Vapnik et al., 1999). SRM minimizes an upper bound on the generalization error as opposed to ERM, which minimizes the error on the training data. This difference makes SVM more attractive in statistical learning applications. SVM are used for classification and regression. In this work an SVM is used for regression, being usually designated by SVR.

7.1 Support Vector Regression (SVR)

Given a set of data points, $\{(x_1, z_1), \ldots, (x_k, z_k)\}$, such that $x_i \in R^n$ is an input and $z_i \in R$ is a target output, the standard form of support vector regression (Vapnik, 1998) is:

$$\min_{w,b,\xi,\xi^*} \frac{1}{2}w^T w + C\sum_{i=1}^{l}\xi_i + C\sum_{i=1}^{l}\xi_i^* \tag{10}$$

$$\text{subject to} \quad w^T\phi(x_i) + b - z_i \le \varepsilon + \xi_i,$$

$$z_i - w^T\phi(x_i) - b \le \varepsilon + \xi_i^*,$$

$$\xi_i, \xi_i^* \ge 0, \quad i = 1,\ldots,l.$$

For practical reasons, instead of solving this minimization problem its dual form, more manageable, is used:

$$\min_{\alpha,\alpha^*} \frac{1}{2}(\alpha - \alpha^*)^T Q(\alpha - \alpha^*) + \varepsilon\sum_{i=1}^{l}(\alpha_i + \alpha_i^*) + \sum_{i=1}^{l} z_i(\alpha_i - \alpha_i^*) \tag{11}$$

$$\text{subject to} \quad \sum_{i=1}^{l} z_i(\alpha_i - \alpha_i^*) = 0, \quad 0 \le \alpha_i, \alpha_i^* \le C, \quad i = 1,\ldots,l,$$

where $Q_{ij} = k(x_i, x_j) \equiv \phi(x_i)^T\phi(x_j)$.

The approximate function is

$$f(x) = \sum_{i=1}^{l}(-\alpha_i + \alpha_i^*)k(x_i, x) + b \tag{12}$$

For the biped robot controller we used the Gaussian kernel

$$k(x_i, x_j) = \exp(-\gamma\,||x_i - x_j||^2) = \exp\left(-\frac{(x_i - x_j)^2}{2\sigma^2}\right) \tag{13}$$

since it demonstrated to perform well on this problem.

Fig. 15 shows the insensitive band (tube) of a typical non-linear regression function when the SVR method is used. To solve the SVR with this kernel, the LIBSVM (Chang & Lin, 2007) was used, where the C, ε and γ parameters of (10) and (13) must be chosen to generate the best final model, for this training data. Parameter C represents the importance of the values outside the regression tube, ε corresponds to the *radius* of the regression function tube and γ represents the Gaussian kernel width. Therefore, the number of support vectors is a decreasing function of ε and a nonzero value of ε is required to avoid overfitting. On the other hand, a too large value of ε could result in an underfitting.

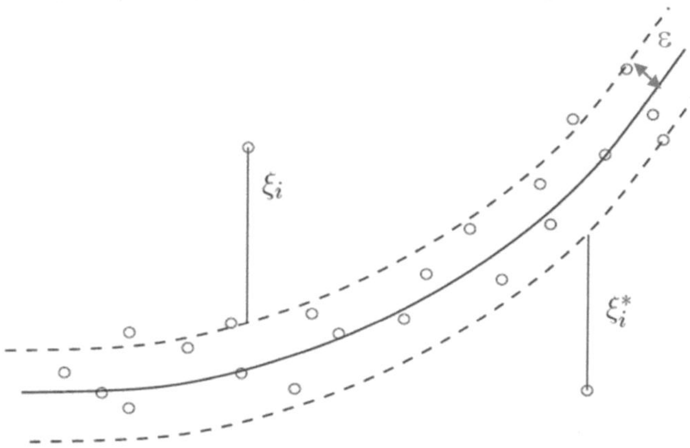

Fig. 15. The insensitive band for a non-linear regression function.

7.2 Support Vector Regression simulation

In this work the SVR was trained with 239 normalized data points and tested with another 68 data point, as defined in Section 5. To choose the above parameters, ε was varied from 0.01 to 0.2 with increments of 0.01, C was varied from 0.1 to 2 with increments of 0.1 and γ was varied from 1 to 4 with increments of 0.1. Fig. 16 plots the minimum MSE (mean square

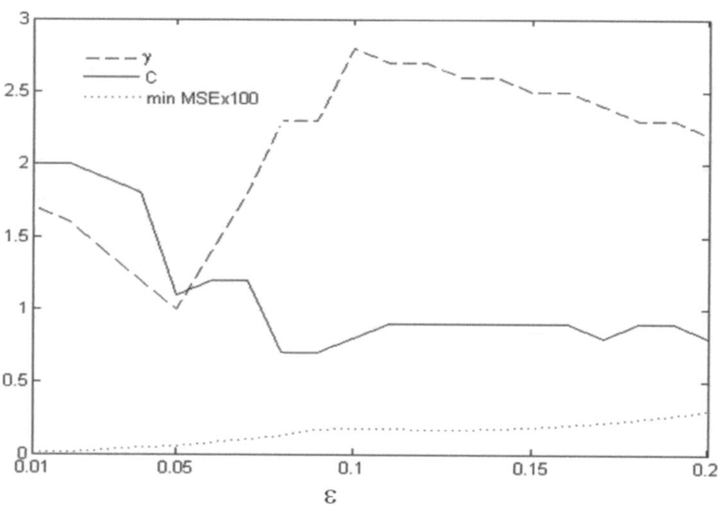

Fig. 16. γ, C and MSE versus ε.

error) for the combinations defined above, as well as the corresponding C and γ parameters, as functions of ε.

Analyzing Fig. 16, the value of ε = 0.13 was chosen because it corresponds to the highest point of the region where the MSE is constant and small. For this value of ε the other parameters are γ = 2.6 and C = 0.9. To complement the performances of the SVR, Fig. 17 and 18 show the variation of the mean square error (MSE) and the mean absolute error (MAE) versus C and γ respectively. The average time to calculate the SVR was 0.2 ms, which proved to be adequate for our problem. Fig. 19 shows the behavior of the final model.

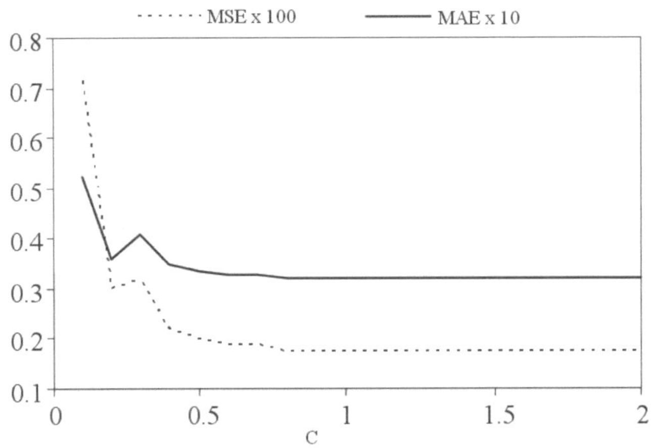

Fig. 17. MSE and MAE against C.

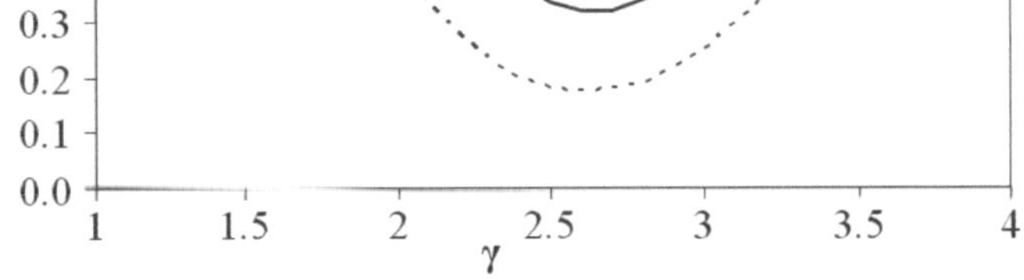

Fig. 18. MSE and MAE against γ.

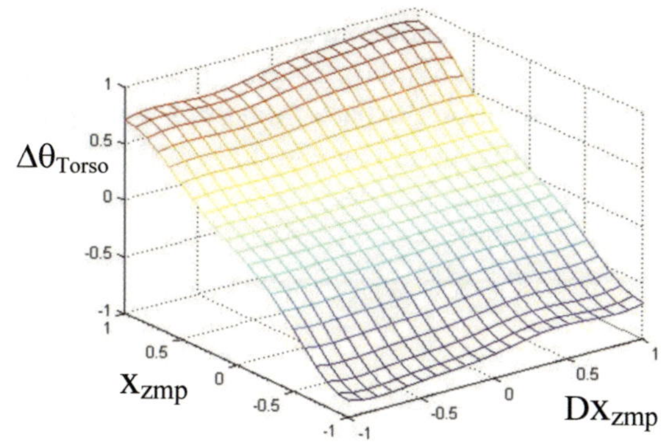

Fig. 19. Behavior of the final model.

8. Experimental results

To determine the functionality of the balance control system based on an SVR, several experiments were performed. With this control technique, the stability of the robot is assured, even in the event of external disturbances.

The values presented in the next figures were normalized such that unit values correspond to 25 degrees for θ_{torso}, 10 degrees for θ_{ankle}, 55 degrees for the pendulum lateral angle ($\theta_{lateral}$), 0.047 m for X_{ZMP} and by 9.8 N for the external force.

In the first experiment the biped robot was walking on a flat surface, playing back the human leg trajectories obtained with the video camera, without the SVR controller active. The torso was kept vertical. The X_{zmp} behavior is shown in Fig. 20 and it can be seen that it is very irregular, and that after some steps the robot falls. The origin of the Xzmp axes corresponds to the center of the support area.

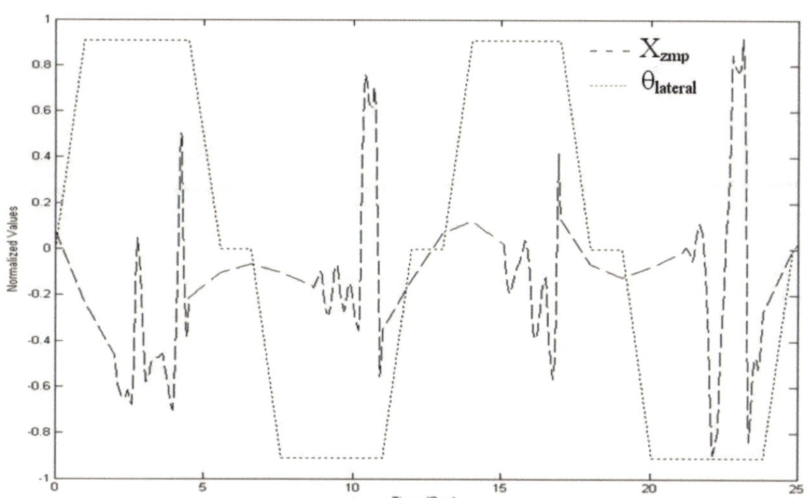

Fig. 20. $\theta_{Lateral}$ and X_{zmp} obtained with the robot walking on a flat surface and the torso always vertical, without the SVR controller active.

In the second experiment the robot was walking on a flat surface with the SVR controller active. In this case both the X_{zmp} and the torso angle exhibit regular behavior (see Fig. 21). A walking snapshots of one step is shown in Fig. 22. At the end of this experiment an external force was applied, pushing the robot on the back. The force was applied at a height of 0.3 m, and the robot maintained its stability, as can be seen in Fig. 23.

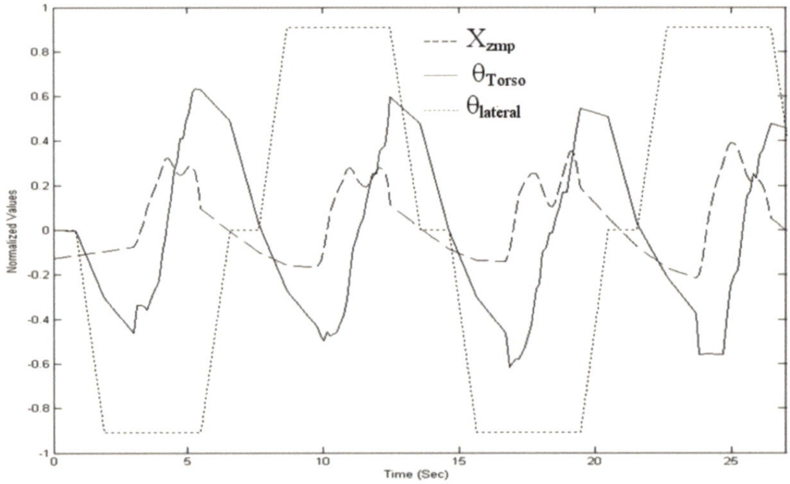

Fig. 21. $\theta_{Lateral}$, θ_{Torso} and X_{zmp} obtained with the robot walking on a flat surface and the torso control active.

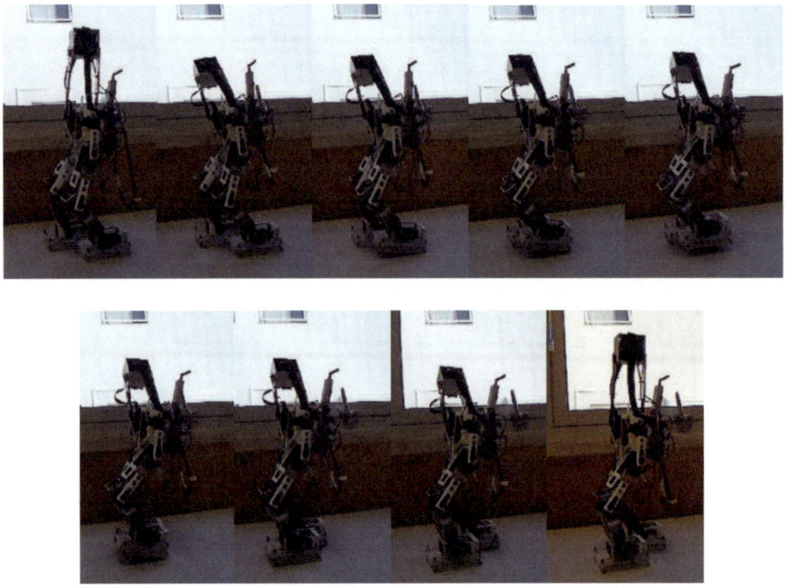

Fig. 22. Walking snapshots on a flat surface and the torso control active.

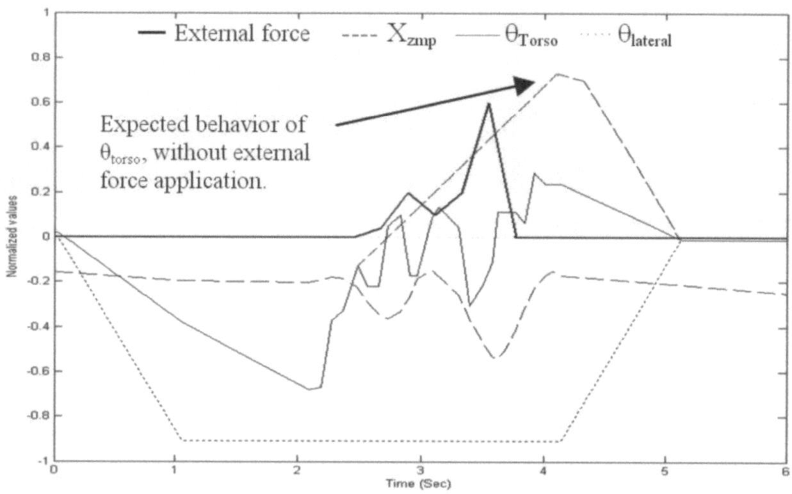

Fig. 23. $\theta_{Lateral}$, θ_{Torso} and X_{zmp} obtained with the robot walking on a flat surface and the torso control active, and under an external force.

In the third experiment an external pushing force was applied on the rear side of the robot when it was standing with only one foot on the ground. As can be seen in Fig. 24 and in the snapshots shown in Fig. 25, when the force is applied the controller maintains the stability. In the final experiment the robot suffers another external pushing force (greater than the force applied in the third experiment and for a little longer) applied on its front side. As shown in Fig. 26 and Fig. 27 the controller is able to stabilize the robot.

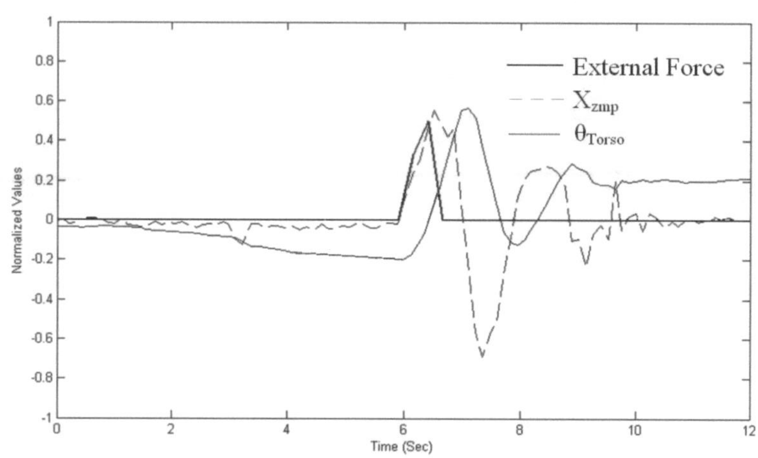

Fig. 24. θ_{Torso} and X_{zmp} obtained when the robot is pushed from behind, standing with only one foot on the ground.

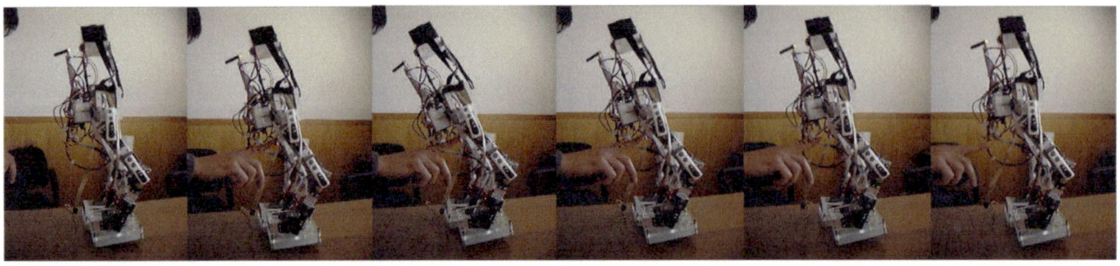

Fig. 25. Snapshots of a rear push.

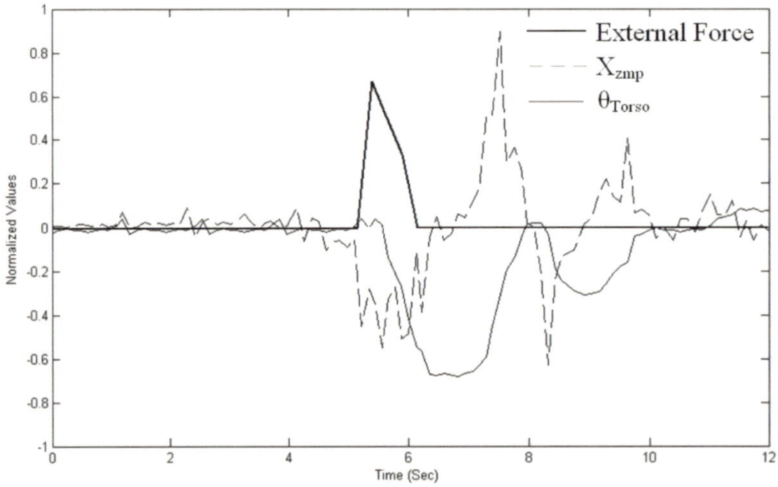

Fig. 26. θ_{Torso} and X_{zmp} obtained when the robot is pushed from the front, standing with only one foot on the ground.

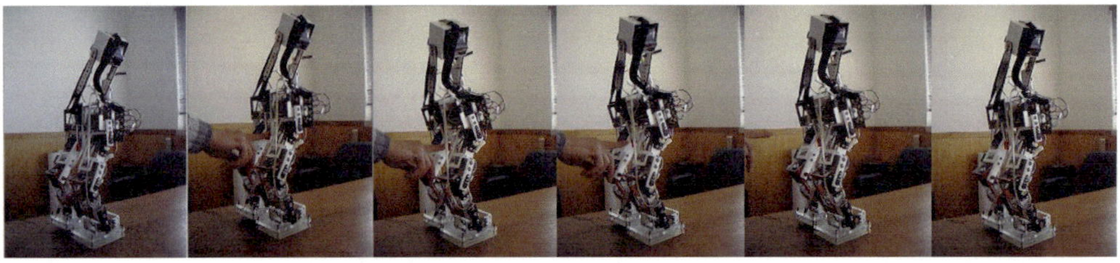

Fig. 27. Snapshots of a front push.

In these two next experiments the robot was walking on a flat surface, with a step length of 0.07 m, using the trajectories of the human gait, dragging a mass of 1.5 kg (that provides a pulling force of about 5 N), with (Fig. 28 and 29) and without (Fig. 30 and 31) the SVR controller active.

It is visible in Fig. 29 (with the SVR balance controller active) that the robot is able to pull the mass along the step, i.e., the mass moves forward 0.07 m (the step length), while in Fig. 31 (without the SVR balance controller active) the mass moves forward only 0.035 m and the robot falls in the next step. In Fig. 30 it is possible to see that the value of X_{ZMP} is in the limit of the stable area.

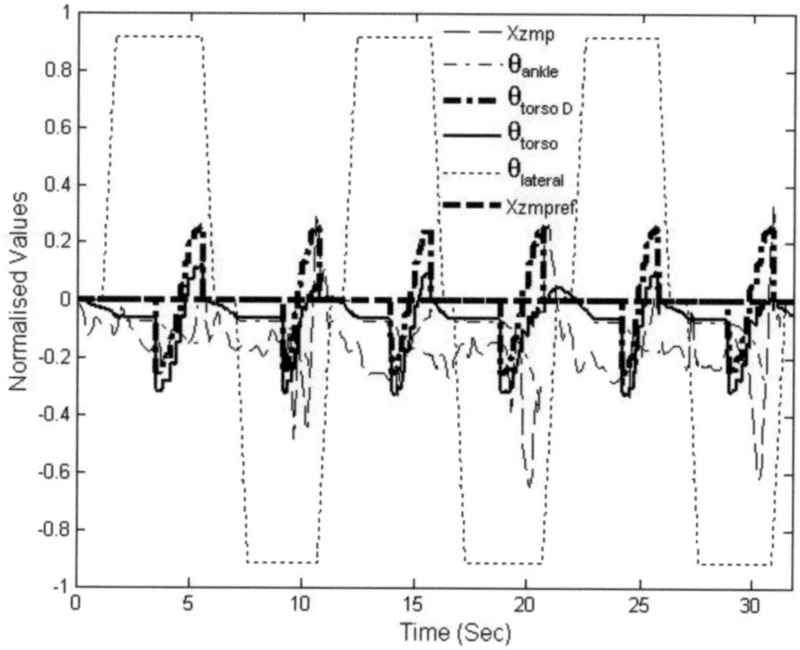

Fig. 28. X_{ZMP}, X_{ZMPref}, ankle, designed torso (θ_{torsoD}), torso and lateral angles on a horizontal flat surface pulling a mass with the SVR controller active.

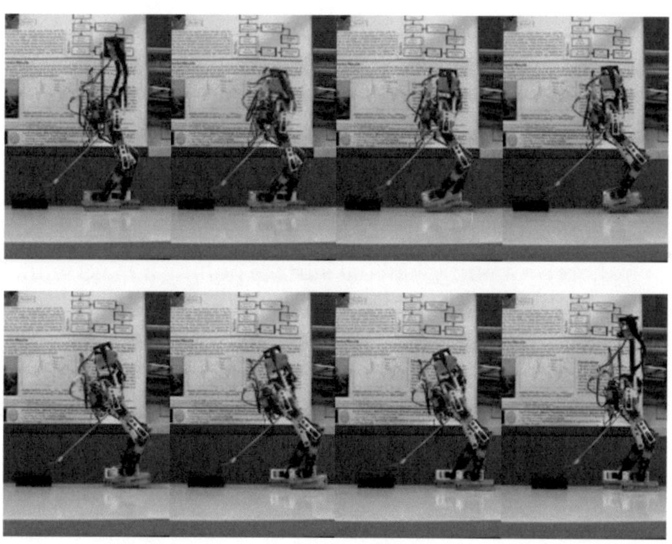

Fig. 29. Walking snapshots of one step on a horizontal flat surface pulling a mass with the SVR controller active.

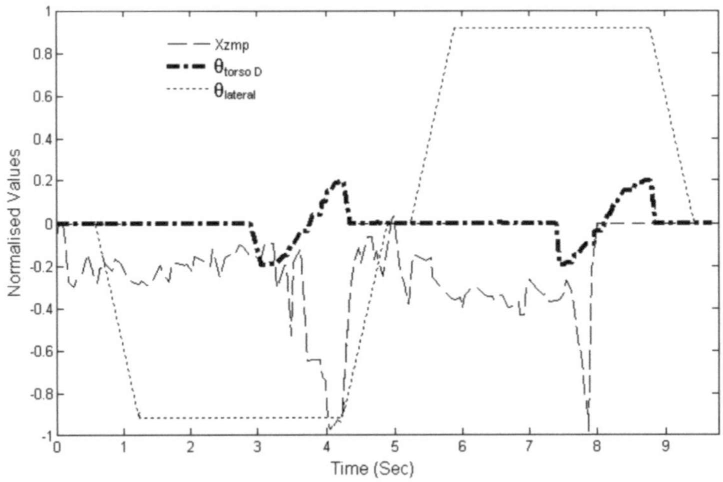

Fig. 30. X_{ZMP}, designed torso (θ_{torsoD}) and lateral angles on a horizontal flat surface pulling a mass without the SVR controller active.

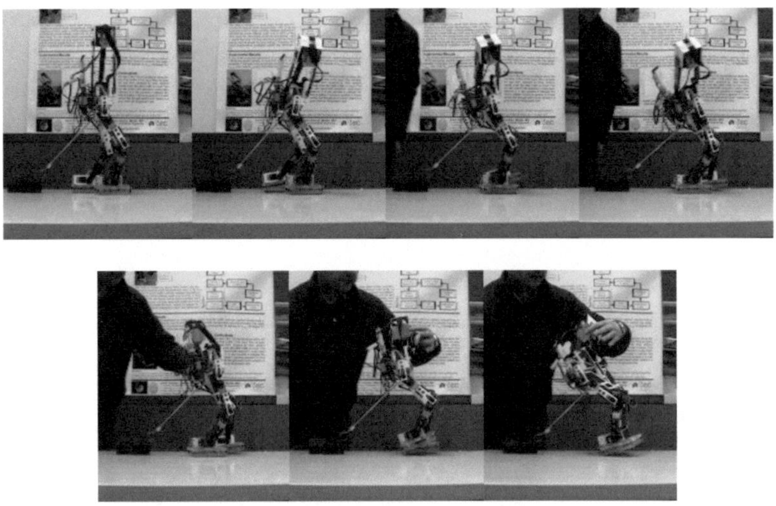

Fig. 31. Walking snapshots on a horizontal flat surface pulling a mass without the SVR controller active. The robot falls down at about t = 8s.

The effectiveness of the SVR controller is illustrated in Fig. 28 where the robot presents a good stability margin compared to Fig. 30 where the X_{ZMP} profile is irregular and the stability margin is close to zero at time about 4 seconds and zero at about 8 seconds, when the robot falls down.

9. Conclusions

The real time control of an 8-link biped robot using the dynamic model of the ZMP is difficult, due to the processing time of the corresponding equations. A simulator of an 8-link biped robot model was developed that allows training data to be obtained for an SVR. The

input variable DX_{zmp} was used to include the behavior of a PD controller in the SVR.The application of the SVR method requires a good trade off between over-fitting and under-fitting to obtain a smooth behavior for the final model.
The major advantage of the SVR is running in 0.2 ms which is 50 times faster than a neuro-fuzzy controller. The use of the SVR allows the real time control of the robot. This SVR uses the real CoP, acquired through force sensors placed under the robot's feet, as input. The SVR was tested and satisfactory results were obtained when the robot pulls a mass that is 65% of the robot's mass.
To obtain a good stable step it is very important to design good leg trajectories so that only a small variation of torso movement is needed to maintain balance. It was verified that the human-like gait is a good choice to use in this biped robot.

10. Acknowledgments

The authors would like to thank the Portuguese Fundação para a Ciência e a Tecnologia for financial support.

11. References

Behnke, S. (2006). Online trajectory Generation for Omnidirectional Biped Walking, *Proc. of the 2006 IEEE International Conference on Robotics and Automation*, Orlando, Florida

Chang, C. & Lin, C. (2007). LIBSVM: a Library for Support Vector Machines, January 2, 2007

Chevallereau, C., Formal'sky, A. & Perrin, B. (1998). Low Energy Cost Reference Trajectories for a Biped Robot, *Proc. IEEE Int. Conf. Robotics and Automation*, pp. 1398-1404

Choi, K. C., Lee, M. C. & Lee, J. M. (2006). Fuzzy Posture Control for a Biped Walking Robot Based on Force Sensor for ZMP, *The Eleventh International Symposium on Artificial Life and robotics*, Oita, Japan, January 23-25

Ferreira, J. P., Amaral, T. G., Pires, V. F., Crisóstomo, M. M. & Coimbra A. P. (2004). A Neural-Fuzzy Walking Control of An Autonomous Biped Robot, *WAC - IEEE Conference*, Sevilha, 21-23

Ferreira, J. P., Crisóstomo, M. M., Coimbra A. P. & Ribeiro, B. (2007a). Simulation control of a biped robot with Support Vector Regression, *IEEE International Symposium on Intelligent Signal Processing*, Madrid, Spain, 3-5 October

Ferreira, J. P., Crisóstomo,M. M., Coimbra, A. P., Carnide, D. & Marto, A. (2007b). A Human Gait Analyzer, *IEEE International Symposium on Intelligent Signal Processing-WISP'2007*, Madrid, Spain, 3-5 October

Jang, Y., Mark, S., Nixon & Chris, J. H. (2002). Extracting Human Gait Signatures by Body Segment Properties, *Fifth IEEE Southwest Symposium on Image Analysis and Interpretation*

Hirai, K. M., Haikawa, Y. & Takenaka, T. (1998). The Development of Honda Humanoid Robot, *Proc. Int. Conf. Robotics and Automation*, pp. 1321-1326.

Huang, Q., Kajita, S., Kaneko, K., Yokoi, K., Komoriya, K. & Tanie, K. (1999). A High Stability, Smooth Walking Pattern for a Biped Robot, *Proc. Int. Conf. Robotics and Automation*, pp. 65-71

Katić, D. & Vukobratović, M. (2005). Survey of Intelligent Control Algorithms For Humanoid Robots, Proceedings of the 16th IFAC World Congress, July

Mohamed, R. M. & Farag, A. A. (2003). Classification of Multispectral Data Using Support Vector Machines Approach for Density Estimation, *IEEE Seventh International Conference on Intelligent Engineering Systems*, INES03, Assiut, Egypt, March 2003

Nakamura, M., Mori, M. & Nishii, J. (2004). Trajectory planning for a leg swing during human walking, *IEEE International Conference on Systems, Man and Cybernetics*

Park, J. H. & Kim, K. D. (1998). Biped Robot Walking Using Gravity-Compensated Inverted Pendulum Mode and Computed Torque Control., *International Conference on Robotics and Automation*, volume 4, pp. 3528-3533

Shin, C. L., Li, Y. Z., Churng, S., Lee, T. T. & Cruver, W. A. (1990). Trajectory Synthesis and Physical Admissibility for a Biped Robot During the Single-Support Phase, *Proc. Int. Conf. Robotics and Automation*, pp. 1646-1652

Vapnik, V. (1998). The Nature of Statistical Learning Theory, *Springer*, New York

Vapnik, V., Golowich, S. & Smola, A. (1999). Support Vector Method for Multivariate Density Estimation, *Advances in Neural Information Processing Systems*, Vol. 12, pp. 659-665, MIT Press, April 1999

Vukobratovic, M. (1990). Biped locomotion: Dynamics, Stability, Control and Aplication, *Berlin: Springer-Verlag*

Winter, D. A. (1990). The Biomechanics and Motor Control of Human Movement, 2nd Eds., *John Wiley & Sons*

Zarrugh, M. Y. & Radcliffe, C. W. (1979). Computer Generation of Human Gait Kinematics, *J. of Biomechanics*, Vol. 12, pp. 99 - 111

Zheng, Y. F. & Shen, J. (1990). Gait Synthesis for the SD-2 Biped Robot to Climb Sloping Surface, *IEEE Trans. on Robotics and Automation*, Vol. 6, No. 1, pp. 86-96

6

Semi-Passive Dynamic Walking Approach for Bipedal Humanoid Robot Based on Dynamic Simulation

Aiman Omer[1], Reza Ghorbani[2], Hun-ok Lim[3], and Atsuo Takanishi[1]
[1]*Waseda University*
[2]*University of Hawaii at Manoa*
[3]*Kanagawa University*
[1,3]*Japan*
[2]*USA*

1. Introduction

The research on the principles of legged locomotion is an interdisciplinary endeavor. Such principles are coming together from research in biomechanics, neuroscience, control theory, mechanical design, and artificial intelligence. Such research can help us understand human and animal locomotion in implementing useful legged vehicles. There are three main reasons for exploring the legged locomotion. The first reason is to develop vehicles that can move on uneven and rough terrain. Vehicles with wheels can only move on prepared surfaces such as roads and rails; however, most surfaces are not paved. The second reason is to understand human and animal locomotion mechanics. The study of the mechanisms and principles of control found in nature can help us develop better legged vehicles. The third reason which motivated the study of legged locomotion is the need to build artificial legs for amputees. Although some effective artificial legs have been built to date, more in-depth research is required to fully understand the mechanisms and movements necessary to substitute the actual limbs.

The research in this paper concerns a group of legged robots known as bipedal walking robots. Research on this subject has a long history; however, it is only in the last two decades that successful experimental prototypes have been developed. The vast majority of humanoid and bipedal robots control the joint angle profiles to carry out the locomotion. Active walking robots (robots with actuators) can do the above task with reasonable speed and position accuracy at the cost of high control efforts, low efficiencies, and most of the time unnatural gaits. WABIAN-2R is among the most successful bipedal walking humanoid robots. In spite of the extensive research on humanoid robots, the actions of walking, running, jumping and manipulation are still difficult for robots.

Passive-dynamic walking robots have been developed by researchers to mimic human walking. The main goal of building passive-dynamic walking robots is to study the role of natural dynamics in bipedal walking. Passive-dynamic walkers use gravitational energy to walk down a ramp without any actuators. They are energy efficient but have weak stability in the gait. In addition, the major cause of the energy loss in the current passive-dynamic walking robots is the instantaneous change in the velocity of the mass centre during each leg transition.

Recently, to overcome the limitations and disadvantages of the above walking robots (active and passive), researchers have proposed energy-efficient walking robots which can be divided into two major research areas. The first research area is the walking robots with actuators which track the optimized joints angle trajectories. The trajectories are determined from an optimization procedure used to minimize an objective function. The second research area is the passive–dynamic robots with direct drive or elastic actuators installed at some of the joints of the biped. Three successful dynamic walking robots are the Cornell Robot, Denise and Toddler. The main goal of developing dynamic walking robots is to increase the efficiency of locomotion.

The bipedal humanoid robot WABIAN-2R was developed in Takanishi Laboratory at Waseda University to simulate human motion. Compared to most bipedal humanoid robots, WABIAN-2R is able to perform a human-like walking with stretched knees during the stance phase while other robots walk with bent knees (Fig. 1). However, its walking performance requires a large torque and a rapid change in velocity. This requires a harmonic drive gear with high ratio to increase the torque as well as a fast rotating motor (Fig. 2). Therefore, WABIAN-2R needs a lot of energy in each walking step with heavy foot and respectively oversized actuators. This is a problem that can be seen in most of the advanced humanoid robots developed for various tasks. However, the energy loss could be prevented by modifying the design of the ankle joint. A spring mechanism could be added at the ankle joint in order to store part of the energy of the robot during the collision phase and to release it by continuing the motion passively. By combining the passive motion and the actively controlled joints, the humanoid robot can realize walking with more similarity to human motion. This paper investigates the idea through simulation of WABIAN-2R with passive ankle joints that has a back actuator attached in series with springs. This study is currently focusing on dynamic motion on the sagittal plane while the lateral plane is fully active.

2. Dynamic simulation

Dynamic simulation could be used the purpose of testing and checking the dynamic motion of a mechanical structured model. It has the advantages of saving cost and risk which are highly needed in a development of a mechanical structure. There are many simulation software have been developed for robotics application, mainly for the industrial robot applications. However, there are some software packages used for mobile robot simulation. For examples, RoboWorks, SD/FAST, OpenHRP, Webots, and Yobotics are used for mobile and legged robot simulation. Webots is high and advanced simulation software used in Robotics simulation. It is use for prototyping and simulation of mobile robots. It has many advanced functions and techniques. Webots is very easy to use and implement. Therefore, we choose it as simulation software for our research.

2.1 Modeling

In order to develop a dynamic simulation, we need to go through several steps. First is modeling where we set up the simulation environment and initial parameters. We set up a full structure of WABIAN-2, based on the specifications (size, shape, mass distribution, friction, .etc) of components of WABIAN-2 (Fig. 3).

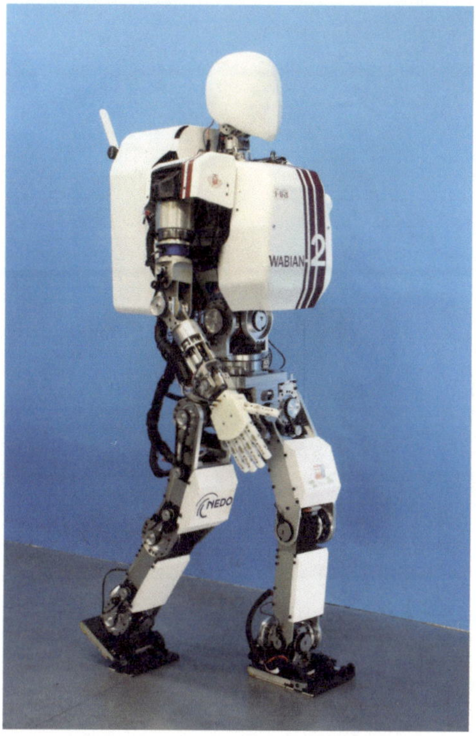

Fig. 1. WABIAN-2R

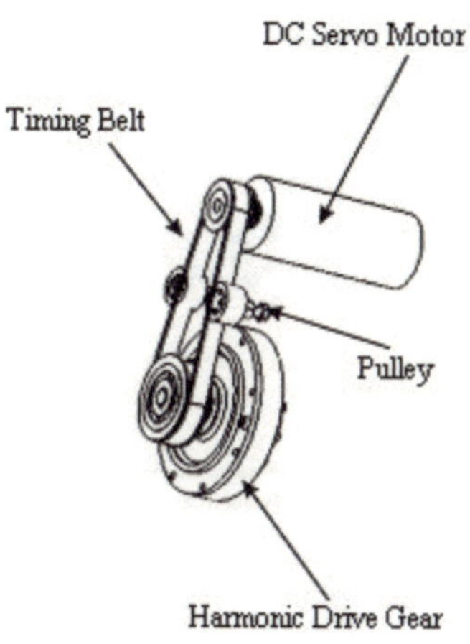

Fig. 2. The Joint Gear System

2.2 Controlling

Second is controlling, which identifies simulation objects and controls the simulation procedures. The controller is some how similar to the WABIAN-2R control. It gets the input data from the CSV pattern file, and sets the position angle of each joint through inverse kinematics techniques. Moreover, the controller sets the simulate time step and the measurement of data.

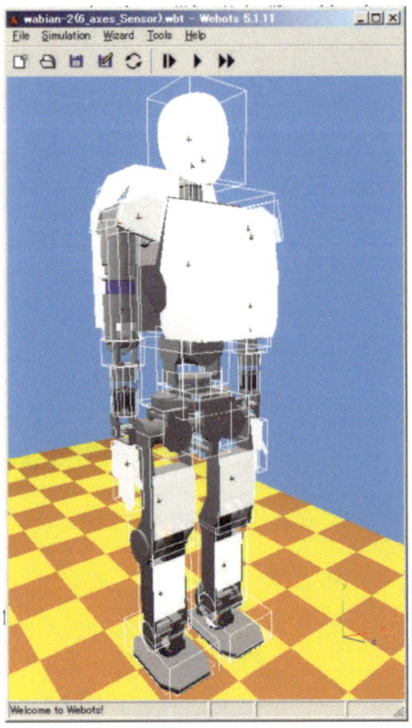

Fig. 3. Modeled WABIAN-2R in the simulation world

2.3 Running

The program in the controller section of the simulator will run by going through the main function. There are several steps the controller will go through. First, check the pattern file and prepare to read through the lines. Then read the data from one line. The data is in terms of position and orientation of foots and hands. Using these data we calculate each joint position through inverse kinematics techniques. After that it will set all positions to its joint. The controller runs one control step of 30ms which is similar to the real robot. The controller goes through all the lines in the pattern file until it is completed in the last line. When the simulation runs it can be viewed the simulation from different view sides. This can gives us a clear idea about the simulation performance. Moreover, most of the needed data could be measured through several functions.

3. Robot model

WABIAN-2R is developed to simulate human motion. Thus the DOF configuration and design structure of the robot is made according to the human body. The design of the robot waist with a 2 DOF helps the robot to perform stretched knee walking, which is similar to the human's. The leg model and ankle joint is detailed in this section.

3.1 Leg model

In most of the passive dynamic walking robots, the legs consist of a hip joint and a knee joint. The knee joint is useful when lifting the leg above the ground during the swing phase. The ankle joint could be eliminated in case that semicircular feet are used. Otherwise, it would be necessary to use an ankle joint in addition to the knee and hip joints.
WABIAN-2R is able to perform a fully stretched walking. This walking is made using only the hip and ankle joints without the use of the knee joint. In this case, the robot leg is

simplified from 6 DOF to 5 DOF (3 DOF at the hip and 2 DOF at the ankle). However, the joints movements in the sagittal plane will be made only through 2 DOF (the ankle pitch joint and the hip pitch joint) (Fig. 4).

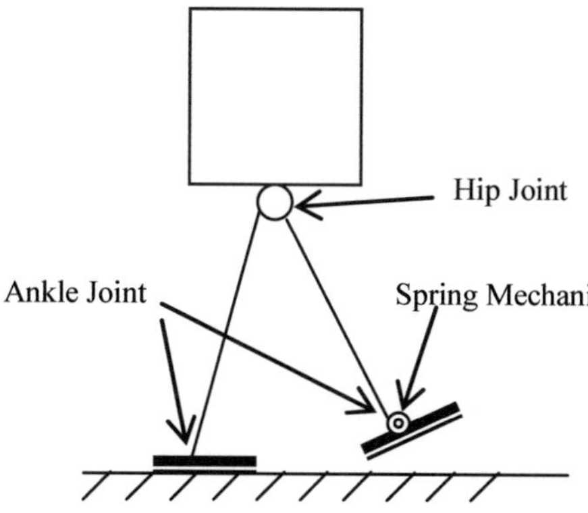

Fig. 4. Simple Robot Model

3.2 Ankle joint

In order to have partly passive walking motion, named by semi-passive motion in this paper, the actuator of the ankle joint in the robot is redesigned in a way to be switched between passive and active mode. Moreover, the ankle should have a level of controllable mechanical compliance for better stability. Therefore, the ankle joint is modified by adding a mechanism called "Rotary Adjustable Stiffness Artificial Tendon" (RASAT). RASAT provides both active and passive rotational motion at the ankle joint. Here we proposed a new design of the ankle, which makes the RASAT bi-directional performing in both active and passive modes. Moreover, it keeps the compliancy of the joint. The elasticity of the joint could overcome the stability difficulties disturbances that might occur in case the foot landing with an orientation (Fig. 5).

The rotary adjustable stiffness artificial tendon (RASAT) is specially designed to provide a wide range of the stiffness (Fig. 6). In RASAT, a pair of compression springs is intentionally inserted between the two concentric input and output links. Each spring pair consists of a low stiffness spring with a stiffness of K1 and a high stiffness spring with a stiffness of K2. The offset between the low and high stiffness springs with value l, is adjustable. Distance d, of the spring pairs with respect to the center of rotation of the links. In this case, the internal torque T, between the concentric input and output links is calculated from:

$$T = \begin{cases} K_1 dx = K_1 d^2 \tan\theta & \dfrac{l}{d} \geq \tan\theta \\ K_1 dl + (K_1 + K_2)d(d\tan\theta - l) & \dfrac{l}{d} < \tan\theta \end{cases} \tag{1}$$

The rotation around the ankle joint is very small, ranging between -0.25 to 0.25. Therefore, the rotation angle equal to:

$$\theta \approx \tan\,\theta \tag{2}$$

From equation (2) the torque provided by the mechanism could be calculated from:

$$T = \begin{cases} K_1 d^2\theta & \frac{l}{d} \geq \theta \\ K_1 dl + (K_1 + K_2) d(d\theta - l) & \frac{l}{d} < \theta \end{cases} \quad (3)$$

In addition to RASAT, a back actuator is attached in serial (Fig. 7). The purpose of the Back Actuator is not only to adjust the offset between two springs, but also to provide the required torque at the ankle joint in order for the robot to move forward(Fig. 8).

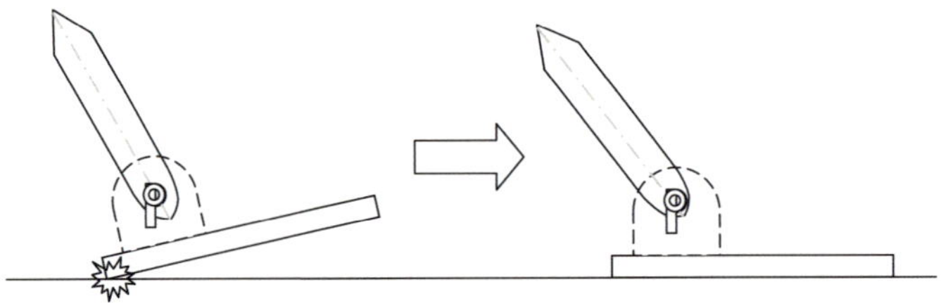

Fig. 5. Foot Landing

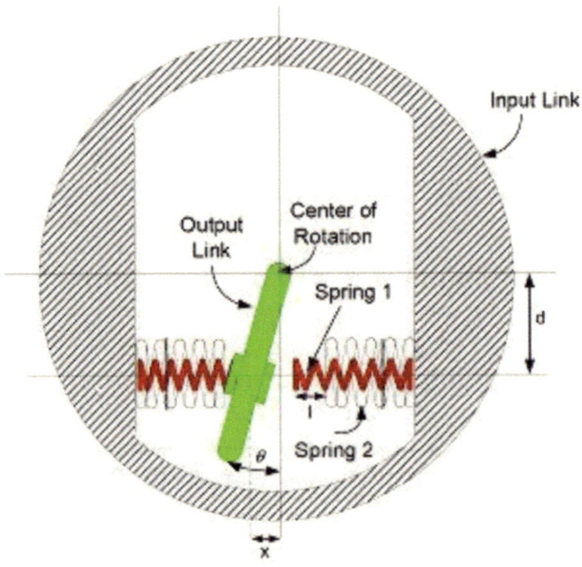

Fig. 6. General schematic of RASAT

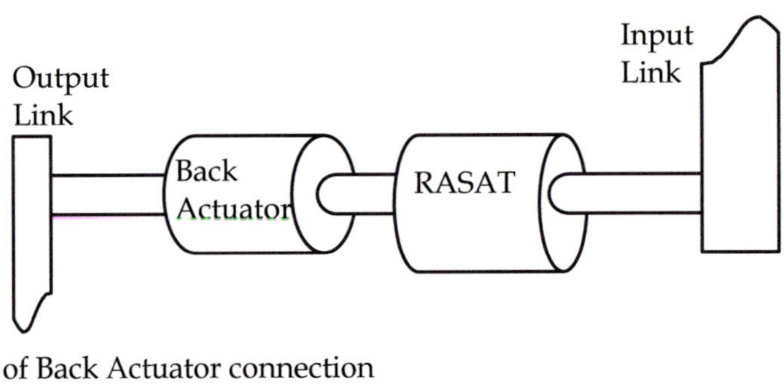

Fig. 7. Structure of Back Actuator connection

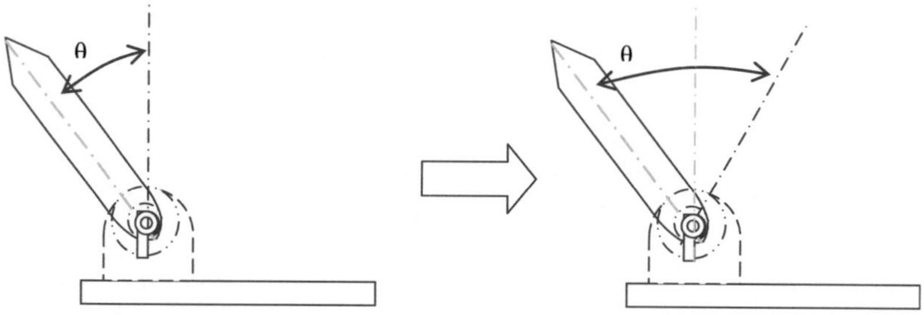

Fig. 8. Changing the Twisting Angle to increase the Joint Torque

4. Walking control

WABIAN-2R is based on joint position control according to the trajectory planning of the foot. The step length and height are calculated from the robot's pattern generator, which provides the robot trajectory planning in joint or Cartesian space.
The stability of passive dynamic walking relies on the energy consumption of the robot. Therefore, the trajectory based joint control method of WABIAN-2R should be partially switched to the torque controlled method instead of position control. Since the ankle joint is set to be passive in this research, the hip joint is the only leg joint that is being controlled. However, in this paper, only the joints on the sagittal plane of the leg are torque controlled while the other joints on the robot are position controlled.

4.1 Hip joint control

The hip joint is controlled using a PD controller to provide the required torque to perform the motion (Fig. 9). The equation for the torque is given below:

$$\tau = K_p(\theta_d - \theta_c) - K_v\omega \qquad (4)$$

where K_p is the spring constant gain and K_v is the damper constant gain. θ_d is the reference position set in the pattern, θ_c is the current position, and ω is the joint velocity.

Fig. 9. Hip joint controlled using torque control

4.2 Ankle joint control

The ankle joint provide the required torque in order for the robot to move forward. The design of the joint help to store some the energy in terms of elastic potential energy and release it in form of a joint torque. The control method depends on the mode of action; RASAT mechanism as passive and Back Actuator as active mode.

To have a fully passive mode the offset l in the RASAT is set to maximum value. The makes the low stiffness spring is the only acting spring around the ankle joint. On the other hand, setting the offset to zero will limit the motion making the RASAT unmovable (Fig. 10). In this way the rotation around the ankle pitch joint is only provide the back actuator.

In the case of setting the mode to semi passive, both the RASAT and the back actuator is used. The offset value of l is initially adjusted according to the required torque need. The back actuator controller set the twisting angle according the velocity feedback. The controller the measure the velocity of the robot body and compare it with the reference velocity which set for the robot. The twisting angle increases and decreases according to the amount of differences of the robot velocity. The difference is set in the equation below:

$$\frac{1}{2}m\Delta v^2 = \frac{1}{2}k\Delta\theta^2 \tag{5}$$

From equation (5) the desired spring deflection can be obtained using

$$\Delta\theta = \sqrt{\frac{m}{k}} \cdot (v_d - v_c) \tag{6}$$

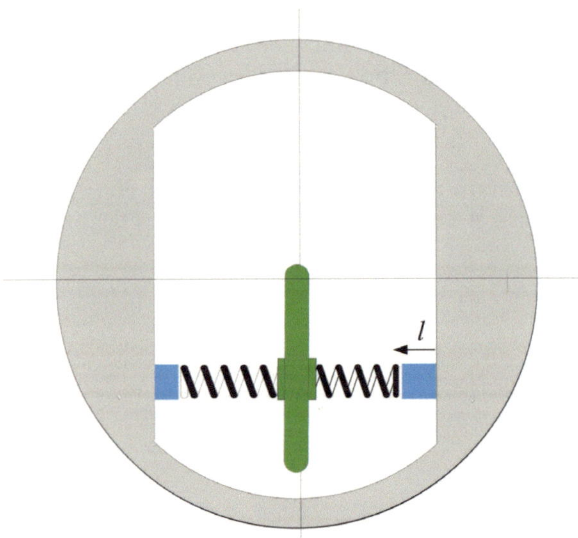

Fig. 10. RASAT mechanism set to active mode

The twisting angle controller is given below:

$$\theta_t(n+1) = \theta_t(n) + G \cdot (v_d - v_c) \tag{7}$$

Where θ_t is the twisting angle, v_d is the desired velocity, v_c is the measured velocity, and G is the control gain. The variable n is the control time step number.

In case that the offset l is set to small value, this can help to provide mechanical compliance. This will support the foot landing in case the foot plane is not in parallel to the surface.

4.3. Robot walking

In order for the robot to make the passive move, it goes through several stages, as shown in Figure 14. First the robot takes the passive leg forward to have a step. Second, the heel of the foot touches the ground (until this stage, there is no energy stored in the passive joint). Third, the joint starts twisting while the leg is forced downward, making the joint store the energy. After the foot landing has been completed, the forward passive motion starts by releasing the stored energy in the passive joint. Finally, the step ends by landing the other foot on the ground. After the step is complete it either stops the motion or takes another step (Fig. 11).

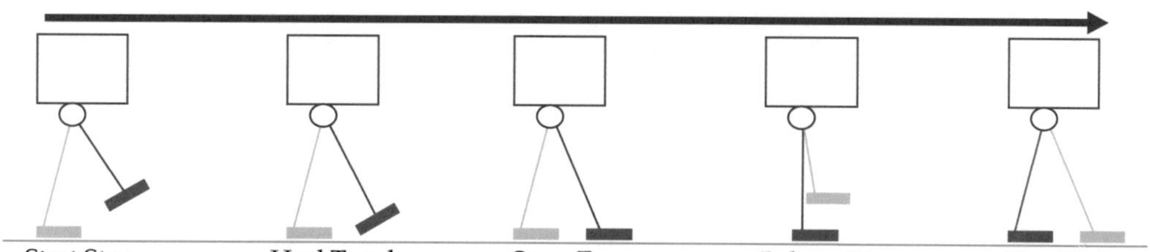

Fig. 11. Passive Step Taking Stages

5. Experiments

Many experiments are conducted in a simulation to check the performance of the robot. WABIAN-2R is simulated with all of its 39 DOF in the simulation package. Different ways were check to achieve the semi passive dynamic walking.

5.1 Natural mode

We conducted many simulations to achieve the semi passive walking using only the spring mechanism on the right ankle pitch. The spring mechanism in the ankle joint is a torsion spring with a stiffness of 50 N.m /rad. The right leg only has the passive joint while the other leg is full active. This helps the robot to stand before it starts to walk. The controller gains for the hip joint are set to Kp = 3000 and Kv = 25.

In the simulation the robot starts by lifting the right leg and pushing it forward using the hip joint. Then the left leg stands on the ground, lifting the whole body of the robot forward using the hip and ankle joints. The right foot, with a passive spring touches the ground and the ankle joint stores energy in the spring during the collision phase. The stored energy provided enough torque to push the body of the robot forward. A semi-passive walking for one step or two steps can be easily achieved (Fig. 12 & Fig. 15). Several other experiments were conducted with different parameters for the controller. Moreover, the spring stiffness was also adjusted and the effects on the walking performance were checked. We realised that the higher the spring stiffness the first steps were difficult for the robot to complete. On the other hand, when the spring stiffness is low, the walking performance goes smoothly but further on the robot velocity becomes slower which makes the robot unable to complete the semi-passive walking.

5.2 Using RASAT mechanism

Many simulations were conducted to achieve the semi-passive walking. Some experiments were successful by setting the ankle joint spring stiffness to 100 N.m/rad which is equal to 160kN/m for the compression spring in the ASAT mechanism. For the hip joint torque control the spring constant (K_p) is set to 5000 and the damper constant (K_v) is set to 25. Both ankle pitch joints are set to active mode in order to keep the robot standing before start walking. Whence the robot gain some velocity for it motion both joint are set to passive mode to be passive dynamic walking motion. A simulation was conducted and the robot was able to walking for 8 steps (Fig. 18).

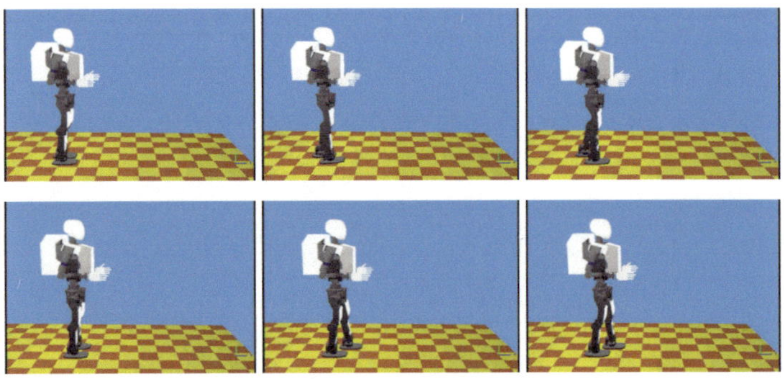

Fig. 12. Simulation of A Semi-Passive Walking for One Step

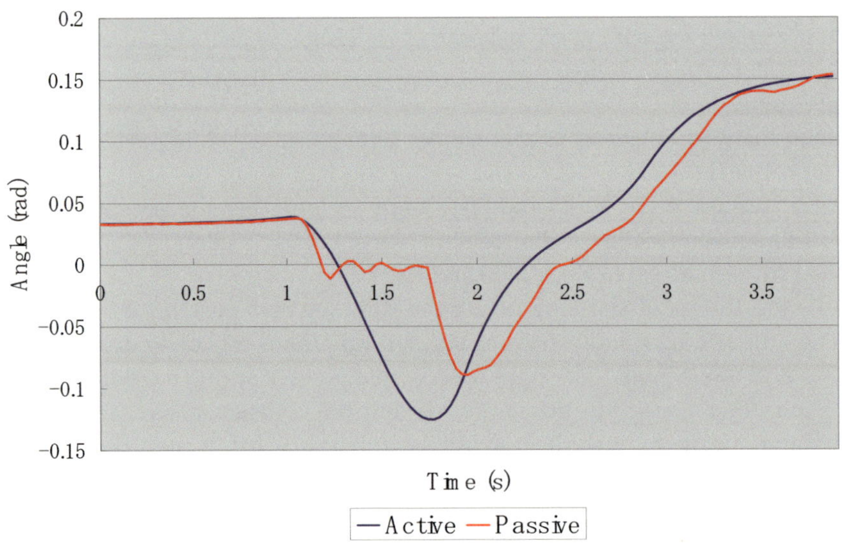

Fig. 13. Passive Joint Angle Measurement

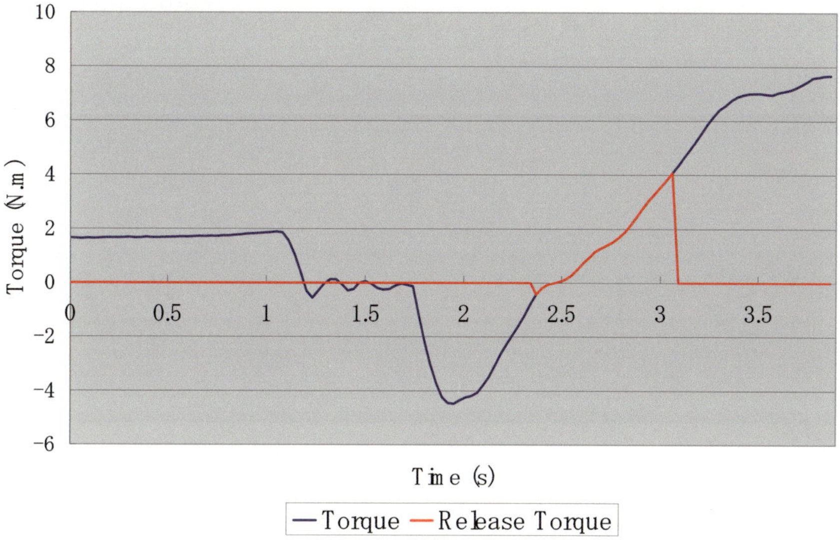

Fig. 14. Passive Joint Torque

Fig. 15. Simulation of A Semi-Passive Walking for Two Steps

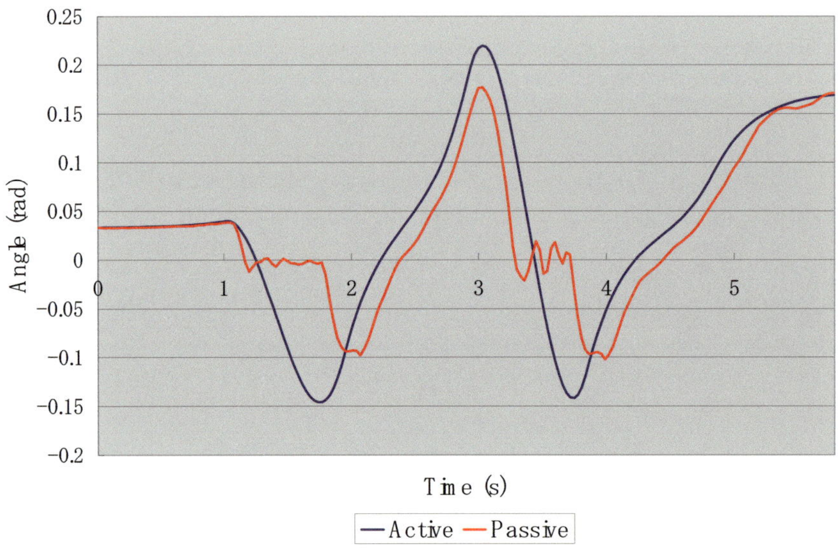

Fig. 16. Passive Joint Angle Measurement

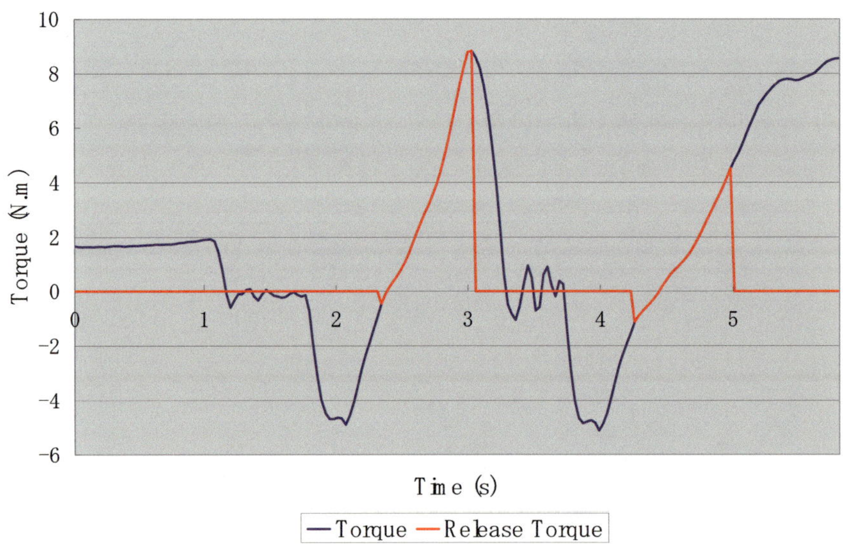

Fig. 17. Passive Joint Torque

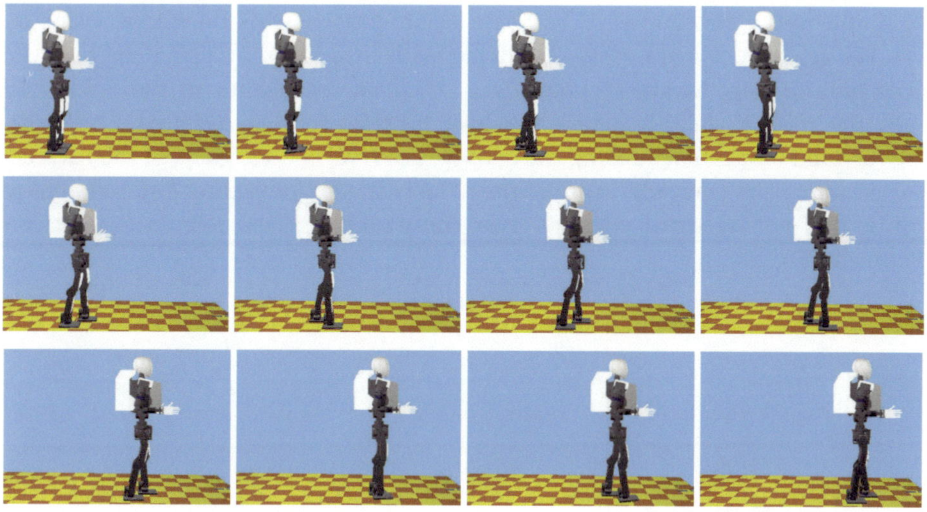

Fig. 18. Simulation for semi passive dynamic walking using RASAT mechanism

Fig. 19. Simulation for Semi-Passive Walking with a Back Actuator

5.3 With the back actuator

Several simulations were conducted to create a semi-passive walking with as many as steps possible. We achieved a semi-passive walking with four steps by using the back actuator controller only during the stance phase (Fig. 19). The compliancy of the ankle joint also improves the robot's stability when the heel touches the ground first, preventing the necessity of a flat foot contact during the collision. Several control gains of the back actuator were experimented by computer simulations. The best gain, providing a stable semi-passive walking, is 0.05 with a small velocity error compared with the reference velocity of the centre of mass (Fig. 20).

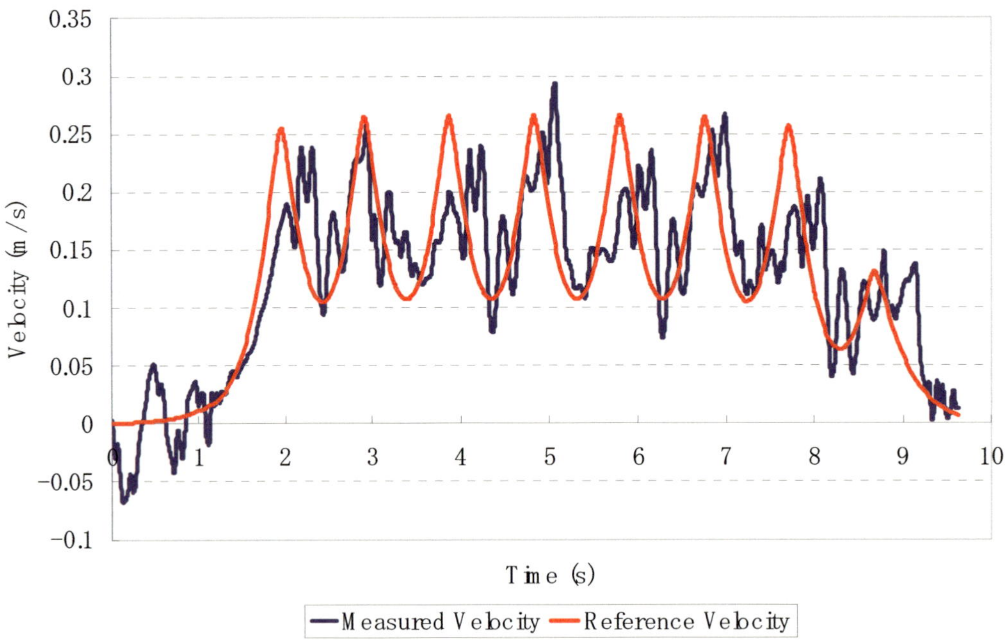

Fig. 20. Robot Velocity Measured Data

6. Conclusions

The semi-passive dynamic walking method is tested in WABIAN-2R using computer simulations. The design of the ankle joint of WABIAN-2R in the computer simulation is modified to include an elastic element in series with the pitch actuator. This allows the robot to perform a semi-passive dynamic walking. The results demonstrate that the semi-passive walking can be realized by using a 100 N.m/rad of torsion spring at the ankle joint for eight walking steps. However actuation and torque control of the ankle joint is necessary for lower stiffness values, 50 N.m/rad. In that case, different control gains are tested to obtain the best value. In addition, using a torque control at the hip joint is required to push the robot forward. Adjusting the stiffness of the ankle joint can be helpful in sustaining the semi-passive motion.

7. References

T. McGeer, "Passive dynamic walking", The International Journal of Robotics Research 9 (1990) 62–82.

M. Coleman, A. Ruina, An uncontrolled toy that can walk but cannot stand still, Physical Review Letters 80 (16) (1998) 3658–3661.

M.S. Garcia, Stability, scaling, and chaos in passive dynamic gait models, Ph.D. thesis, Theoretical and Applied Mechanics, Cornell University, May 1999.

S.H. Collins, M. Wisse, A. Ruina, A three-dimensional passive-dynamic walking robot with two legs and knees, International Journal of Robotics Research 20 (7) (2001) 607–615.

M. Wisse, J.V. Frankenhuyzen, Design and construction of mike; a 2d autonomous biped based on passive dynamic walking, Adaptive Motion of Animals and Machines (2006) 143–154.

(2005) The TRN homepage. [Online]. Available: http://www.trnmag.com/Stories/2005/022305/Humanoid_robots_walk_naturally_022305.html

M. Wisse, Essentials of dynamic walking: analysis and design of two-legged robots, Ph.D. thesis, Man Machine Systems, T.U. Delft, 2004.

R. Ghorbani, On controllable stiffness bipedal walking, Ph.D. thesis, Mechanical and Manufacturing Engineering, University of Manitoba, Canada, 2008.

Webots. http://www.cyberbotics.com. Commercial Mobile Robot Simulation Software.

Yu Ogura, Hiroyuki Aikawa, Kazushi Shimomura, Hideki Kondo, Akitoshi Morishima, Hun-ok Lim, and Atsuo Takanishi. "Development of a New Humanoid Robot WABIAN-2," Proceedings of ICRA2006, 2006, paper 76-81.

Yu Ogura, Kazushi Shimomura, Hideki Kondo, Akitoshi Morishima, Tatsu Okubo, Shimpei Momoki, Hun-ok Lim, and Atsuo Takanishi. "Human-like Walking with Knee Stretched, Heel-contact and Toe-off Motion by a Humanoid Robot". Proceedings of the 2006 IEEE/RSJ, IROS2006, 2006, paper 3976-3981.

Yu Ogura, Teruo Kataoka, Hiroyuki Aikawa, Kazushi Shimomura, Hun-ok Lim, and Atsuo Takanishi. "Evaluation of Various Walking Pattern of Biped Humanoid Robot," Proceedings of ICRA2005, 2005, paper 603-608.

Ghorbani, R. and Wu, Q., Conceptual Design of the Adjustable Stiffness Artificial Tendons for Legged Robotics. Mechanism and Machine Theory, Vol 44, issued 1 pp 140-161, 2009.

Masaki Ogino, Koh Hosoda, and Minoru Asada, Learning Energy-Efficient Walking with Ballistic Walking, Adaptive Motion of Animals and Machines Tokyo, Japan: Springer-Tokyo, 2006, paper 155-164.

Qian Yang, B.G. Shiva Prasad, Peter A. Engel, and Derek Woolatt. "Dynamic Response of Compressor Valve Springs to Impact Loading," Proceedings of the 1996 International Compressor Engineering Conference, 353-358. West Lafayette, IN, Purdue University.

7

Passive Dynamic Autonomous Control for the Multi-locomotion Robot

Tadayoshi Aoyama[1], Kosuke Sekiyama[2], Yasuhisa Hasegawa[3], and Toshio Fukuda[4]
[1,2,4]Nagoya University,
[3]University of Tsukuba
Japan

1. Introduction

Recently, skillful motions performed by animals are realized by used of actual robots (Hirai et al., 1998; McGeer, 1990; Nakanishi et al., 2000; Raibert, 1986; Saito & Fukuda, 1997). Most of these works focused on a single type of locomotion. On the other hand, animals such as primates move in several types of locomotion form and select suitable locomotion form depending on their surroundings, situation, and purpose. For example, a gorilla has high behavior flexibility in a forest by adopting bipedal walking in a narrow space, quadrupedal walking on rough terrain and brachiation in the forest canopy. Inspired by these high mobility of an animal, we have developed an anthropoid-like "Multi-locomotion robot" that can perform several types of locomotion and choose the proper one depending on robot's needs (Fig. 1, (Fukuda et al., 2009)). A development of a multi-locomotion robot which has plural locomotion types for high mobility is one of challenging issues, because a problem is remaining in addition to research issues on humanoid robot study. That is a control architecture that synthesizes several locomotion controllers. When we consider a transition connecting one locomotion to another, two independent controllers corresponding to each locomotion type are not enough. A control algorithm that covering control properties of multiple locomotion controllers should be developed because the transition motion cannot be realized by fusing control outputs from multiple controllers. Based on this notion, we have proposed a novel control method named Passive Dynamic Autonomous Control (PDAC) (Doi et al., 2004) that realize not only a bipedal walk (Aoyama et al., 2009) but also a quadrupedal

Fig. 1. Concept of the Multi-Locomotion Robot.

walk (Asano et al., 2007) and a brachiation (Fukuda et al., 2007). In this chapter, we focus on a bipedal walking control. We have already proposed 3-D biped control method based on PDAC (Aoyama et al., 2009). This chapter introduces another stabilizing method of the robot dynamics; then the stabilizing method is validated by the numerical simulation.
This chapter continues as follows. In Section 2, we introduce the Gorilla Robot III that has been developed as a prototype of the multi-locomotion robot. In Section 3, we explain about PDAC concisely. Section 4 describes the walking model, Section 5 introduces stabilization method, and Section 6 shows the results of the numerical simulation. Finally, we summarize this chapter in Section 7.

2. Multi–locomotion robot

This section introduces the "Gorilla Robot III" briefly. Gorilla robot III have been developed as a prototype of the Multi-locomotion Robot; details are found in (Fukuda et al., 2009). Figure 2 shows the overview of Gorilla Robot III and its link structure. This robot is about 1.0 m tall, weighs about 24.0 kg, and consists of 25 links and 24 AC motors including two grippers. The real-time operating system VxWorks (Wind River Systems Inc) runs on a Pentium III PC for processing sensory data and generating its behaviors. Two kinds of sensors are attached to each hand. The rate gyroscope, CRS03-04 manufactured by Silicon Sensing Systems Japan Ltd., measures the angular velocity around the contact bar to calculate the pendulum angle during the motion. The force sensor, IFS-67M25A made by NITTA CORPORATION, measures reaction forces from contact bars in order to judge whether the robot successfully grasps the bar or not.
This robot has been designed to perform biped locomotion, quadruped locomotion and brachiation. We designed the controller for all locomotion using the same algorithm "PDAC". The approach of PDAC is to describe the robot dynamics as an autonomous system around a contact point, using an interlocking so that the robot could keep the robot inherent dynamics. The PDAC is explained in next section. 3-D dynamic walking is achieved as shown in Fig.3 (Aoyama et al., 2009). We also designed a controller for a quadrupedal walk (Asano et

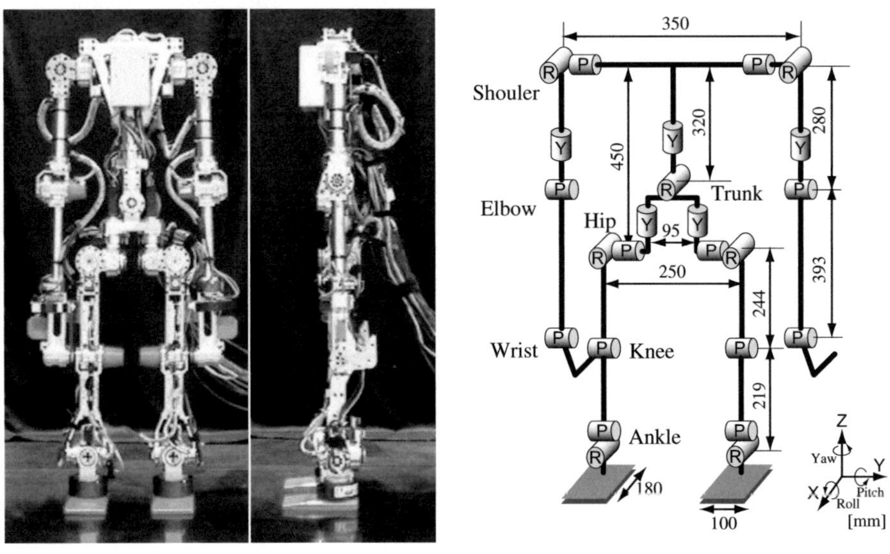

Fig. 2. Gorilla Robot III

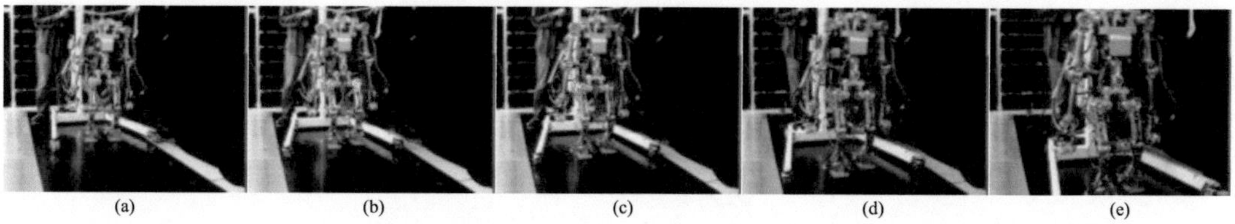

Fig. 3. Snapshots of the Bipedal Walking Experiment. Each figure shows the snapshots at (a)1st (b)7th (c)13th (d)19th (e)25th step. (Aoyama et al., 2009)

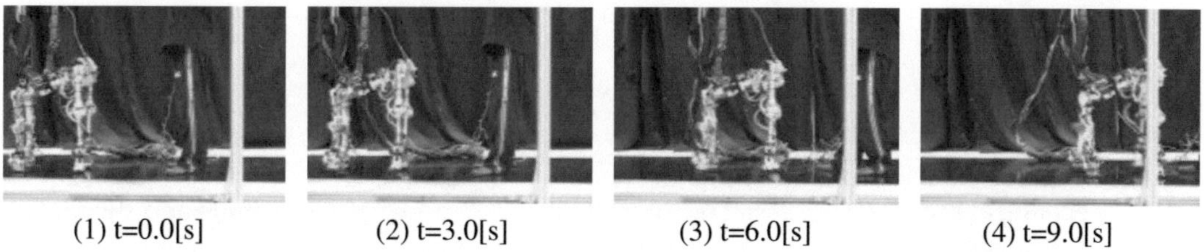

Fig. 4. Snapshots of the quadrupedal walking using PDAC. (Asano et al., 2007)

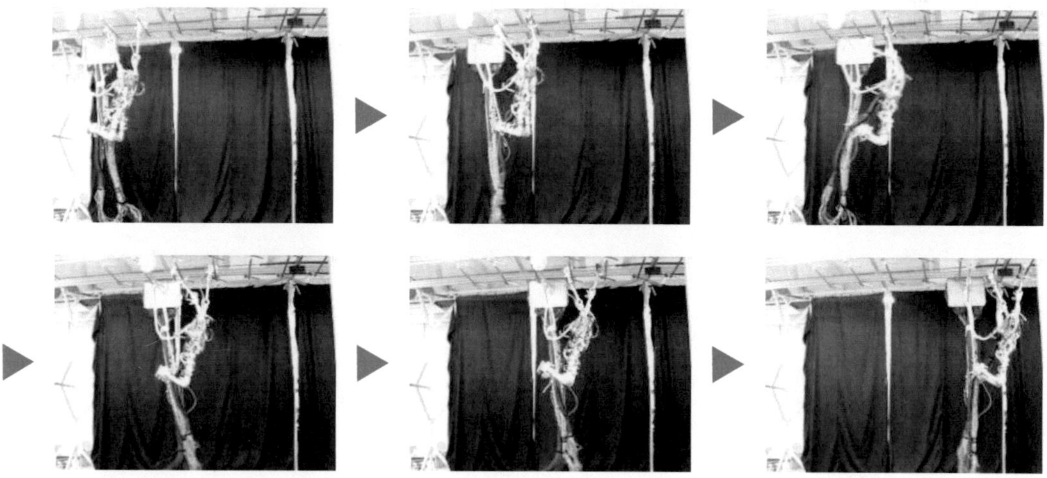

Fig. 5. Snapshots of the brachiation using PDAC. (Fukuda et al., 2007)

al., 2007), and brachiation (Fukuda et al., 2007) using the same PDAC. The snapshot of the quadrupedal walk is shown in Fig.4 and the brachiation is shown in Fig.5.

3. Passive dynamic autonomous control (PDAC)

3.1 Converged dynamics

This section gives explanation of PDAC that was proposed previously by Doi et.al. based on two concepts, i.e. the point-contact and the virtual constraint (Doi et al., 2004). The point-contact means that a robot contacts a ground at a point, that is, the first joint is passive. The virtual constraint has been proposed by Grizzle and Westervelt et al. (Grizzle et al., 2001; Westervelt et al., 2004) as a set of holonomic constraints on the robot's actuated DOF

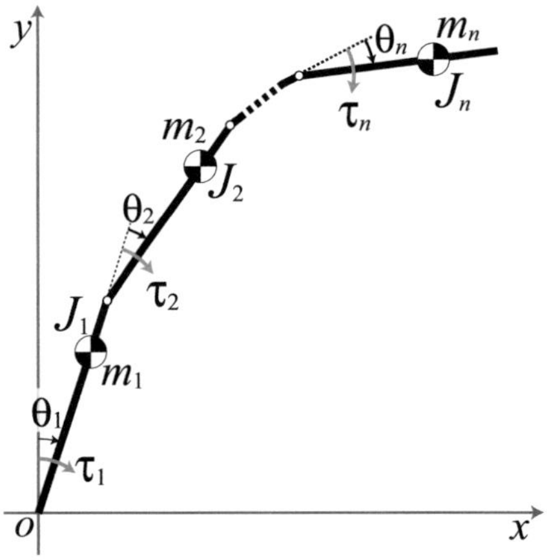

Fig. 6. Mechanical model of the serial n-link rigid robot. θ_i and τ_i are the angle and the torque of ith joint respectively. m_i and J_i are the mass and the moment of inertia of ith link respectively.

parameterized by the robot's unactuated DOF. Assuming that PDAC is applied to a serial n-link rigid robot as shown in Fig. 6, these two concepts are expressed as follows:

$$\tau_1 = 0, \tag{1}$$

$$\begin{aligned}\boldsymbol{\Theta} &= [\theta_1, \theta_2, \cdots, \theta_n]^T = [f_1(\theta), f_2(\theta), \cdots, f_n(\theta)]^T \\ &:= \boldsymbol{f}(\theta),\end{aligned} \tag{2}$$

where θ is the angle around the contact point with respect to the absolute coordinate system, that is, $\theta_1 = f_1(\theta) = \theta$.

The dynamic equations of this model are given by

$$\frac{d}{dt}\left(\boldsymbol{M}(\boldsymbol{\Theta})\dot{\boldsymbol{\Theta}}\right) - \frac{1}{2}\frac{\partial}{\partial \boldsymbol{\Theta}}\left(\dot{\boldsymbol{\Theta}}^T \boldsymbol{M}(\boldsymbol{\Theta})\dot{\boldsymbol{\Theta}}\right) - G(\boldsymbol{\Theta}) = \boldsymbol{\tau}, \tag{3}$$

where

$\boldsymbol{M}(\boldsymbol{\Theta}) := [\boldsymbol{m}_1(\boldsymbol{\Theta})^T, \boldsymbol{m}_2(\boldsymbol{\Theta})^T, \cdots, \boldsymbol{m}_n(\boldsymbol{\Theta})^T]^T$, $\boldsymbol{G}(\boldsymbol{\Theta}) := [G_1(\boldsymbol{\Theta}), G_2(\boldsymbol{\Theta}), \cdots, G_n(\boldsymbol{\Theta})]^T$, $\boldsymbol{\Theta} := [\theta_1, \theta_2, \cdots, \theta_n]^T$, $\boldsymbol{\tau} := [\tau_1, \tau_2, \cdots, \tau_n]^T$, and $\frac{\partial}{\partial \boldsymbol{\Theta}} = [\frac{\partial}{\partial \theta_1}, \frac{\partial}{\partial \theta_2}, \cdots, \frac{\partial}{\partial \theta_n}]^T$.

Since the dynamic equation around the contact point has no term of the Coriolis force in this model, it is given as

$$\frac{d}{dt}\left(\boldsymbol{m}_1(\boldsymbol{\Theta})^T \dot{\boldsymbol{\Theta}}\right) - G_1(\boldsymbol{\Theta}) = \tau_1. \tag{4}$$

By differentiating Eq. (2) with respect to time, the following equation is acquired,

$$\dot{\boldsymbol{\Theta}} = \frac{\partial \boldsymbol{f}(\theta)}{\partial \theta}\dot{\theta} = \left[\frac{\partial f_1(\theta)}{\partial \theta}, \frac{\partial f_2(\theta)}{\partial \theta}, \cdots, \frac{\partial f_n(\theta)}{\partial \theta}\right]^T \dot{\theta}. \tag{5}$$

Substituting Eqs. (1), (2) and (5) into Eq. (3) yields the following dynamic equation,

$$\frac{d}{dt}\left(M(\theta)\dot{\theta}\right) = G(\theta), \tag{6}$$

where

$$M(\theta) := \boldsymbol{m}_1 \Big(\boldsymbol{f}(\theta)\Big)^T \frac{d\boldsymbol{f}(\theta)}{d\theta}, \tag{7}$$

$$G(\theta) := G_1 \Big(\boldsymbol{f}(\theta)\Big). \tag{8}$$

By multiplying both sides of Eq. (6) by $M(\theta)\dot{\theta}$ and by integrating with respect to time, the dynamics around the contact point is obtained as follows:

$$\int \Big(M(\theta)\dot{\theta}\Big) \frac{d}{dt}\left(M(\theta)\dot{\theta}\right) dt = \int M(\theta)G(\theta)\dot{\theta}\, dt \tag{9}$$

$$\Longleftrightarrow \frac{1}{2}\Big(M(\theta)\dot{\theta}\Big)^2 = \int M(\theta)G(\theta)\, d\theta. \tag{10}$$

Therefore, the whole robot dynamics is expressed as the following one-dimensional autonomous system,

$$\dot{\theta} = \frac{1}{M(\theta)}\sqrt{2\int M(\theta)G(\theta)\, d\theta} \tag{11}$$

$$:= \frac{1}{M(\theta)}\sqrt{2(D(\theta)+C)} \tag{12}$$

$$:= F(\theta). \tag{13}$$

In this chapter, we term Eqs. (12) and (13) as the Converged dynamics.

4. Dynamics and walking model

In this section, we describe biped walking model by means of 3-D inverted pendulum that is same as our previous work (Aoyama et al., 2009). At first, the dynamics of the model representing the robot dynamics in single-support phase is obtained. Next, by describing the pendulum length as a function of its angle, we express the whole robot dynamics as the 2-D autonomous system under the constraint that the trunk inclination is kept in the gravitational direction during the walk.

4.1 3-D inverted pendulum model

In this chapter, a robot is modeled as a 3-D inverted pendulum shown in Fig. 7(a). Since walking motion is symmetrical, the left-handed system is used in the left-leg supporting phase and vice versa as shown in Fig. 7(b) so that it is possible to describe the robot dynamics in both supporting phases as single dynamics. We apply the assumption of point-contact to this pendulum later in accordance with PDAC, hence it is possible to choose the axes of pendulum angle around contact point to express its dynamics. In this chapter, we utilize the polar coordinate system. The state variables and parameters are shown in Fig. 8(b). By use of the six variables q_1 to q_5 and l, it is possible to express any state of the robot.
Let mass of the robot be m and let tensor of inertia be

$$\boldsymbol{I} := \begin{pmatrix} I_{xx} & I_{xy} & I_{xz} \\ I_{yx} & I_{yy} & I_{yz} \\ I_{zx} & I_{zy} & I_{zz} \end{pmatrix}. \tag{14}$$

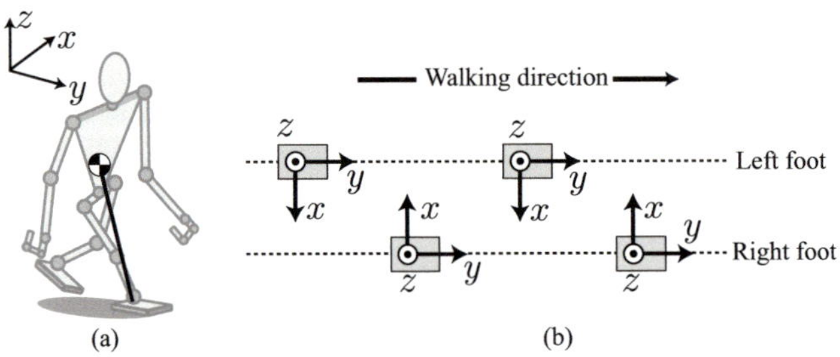

Fig. 7. (a) 3-D inverted pendulum model. (b) Definition of coordinate system. The left-handed system is used in the left-leg supporting phase and vice versa in order to facilitate the dynamics description of walking motion. Note that this figure shows just a coordinate system definition and doesn't mean that foot placement is in alignment.

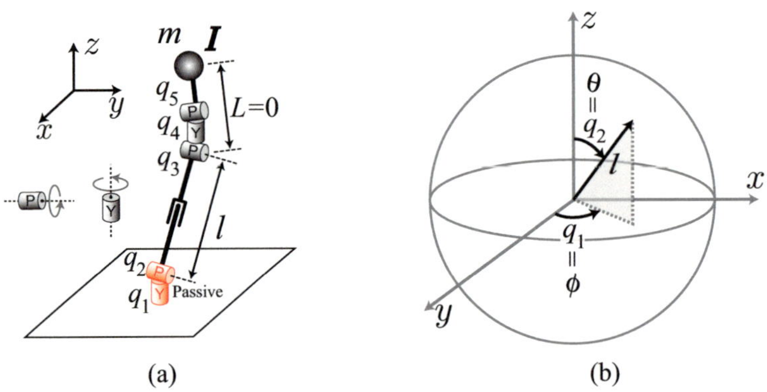

Fig. 8. (a) Parameters and variables of the 3-D inverted pendulum model. q_1 and q_2 are the variables of the pendulum angle around the contact point. q_3, q_4 and q_5 decide the upper body posture. l is the variable of the pendulum length. L is the virtual value for convenience of description and equals zero. (b) Polar coordinate system around contact point

The configuration of humanoid robots is basically symmetrical, hence generality is not lost by assuming that all products of inertia are zero, i.e. $I_{xy} = I_{yx} = I_{yz} = I_{zy} = I_{zx} = I_{xz} = 0$,

$$\begin{aligned} \boldsymbol{I} &= diag(I_{xx}, I_{yy}, I_{zz}) \\ &:= diag(I_x, I_y, I_z). \end{aligned} \tag{15}$$

Note that this tensor of inertia is parameter in the local coordinate system that is attached to the robot body, not in the global one.
In this chapter, the trunk inclination is kept in the gravitational direction and the upper body does not rotate around yaw-axis, that is,

$$q_3 = -q_2 \tag{16}$$

$$q_4 = -q_1 \tag{17}$$

$$q_5 = 0. \tag{18}$$

Then, the dynamic equations with respect to q_1, q_2, and l are obtained respectively as follows:

$$\frac{d}{dt}\left(ml^2 \sin^2 q_2 \dot{q_1}\right) = \tau_1, \tag{19}$$

$$\frac{d}{dt}\left(ml^2 \dot{q_2}\right) - ml^2 \dot{q_1}^2 \sin q_2 \cos q_2 - mgl \sin q_2 = \tau_2, \tag{20}$$

$$\frac{d}{dt}\left(m\dot{l}\right) - ml(\dot{q_1}^2 \sin^2 q_2 + \dot{q_2}^2) - mg \cos q_2 = f, \tag{21}$$

where (τ_1, τ_2, f) are the torques and force corresponding to the variables (q_1, q_2, l).

4.2 Converged dynamics

In order to control the 3-D inverted pendulum by means of PDAC, the assumption of point-contact is applied, i.e.

$$\tau_1 = \tau_2 = 0. \tag{22}$$

For simplicity of description, we describe q_1 and q_2 as ϕ and θ respectively in the below. From Eq. (16)-(22), Eq. (19) and (20) are expressed as follows:

$$\frac{d}{dt}\left(ml^2 \sin^2 \theta \dot{\phi}\right) = 0, \tag{23}$$

$$\frac{d}{dt}\left(ml^2 \dot{\theta}\right) = ml^2 \dot{\phi}^2 \sin\theta \cos\theta + mgl \sin\theta. \tag{24}$$

By multiplying both sides of Eq. (23) by $ml^2 \sin^2 \theta \dot{\phi}$, and integrating with respect to time, the following constraint equation is obtained,

$$\dot{\phi} = \frac{\sqrt{2C_1}}{ml^2 \sin^2 \theta} \tag{25}$$

$$:= F_1(\theta), \tag{26}$$

where C_1 is the integral constant which is decided by initial status just after foot-contact. Substituting Eq. (25) into Eq. (24) results in

$$\dot{\theta} = \frac{1}{ml^2}\sqrt{2\int\left(\frac{2C_1 \cos\theta}{\sin^3\theta} + m^2 g l^3 \sin\theta d\theta\right)} \tag{27}$$

$$:= \frac{1}{M(\theta)}\sqrt{2(D(\theta) + C_2)} \tag{28}$$

$$:= F_2(\theta). \tag{29}$$

Next, in accordance with PDAC, the pendulum length is described as the function of θ,

$$l := \lambda(\theta). \tag{30}$$

In this chapter, λ is defined as the following function of θ,

$$\lambda(\theta) =: \sqrt[3]{p_1\theta^3 + p_2\theta^2 + p_3\theta + p_4} \tag{31}$$

$$=: \sqrt[3]{f(\theta)}. \tag{32}$$

By substituting this equation into Eq. (28), converged dynamics is derived,

$$M(\theta) = mf(\theta)^{2/3}, \tag{33}$$

$$D(\theta) = -\frac{C_1}{\sin^2\theta} - m^2 g\Big(\big(f(\theta) - f''(\theta)\big)\cos\theta - \big(f'(\theta) - f'''(\theta)\big)\sin\theta\Big). \tag{34}$$

4.3 Design of walking cycle

In this subsection, the actual motion of the robot is designed. Figure 9 shows the schematics of the pendulum motion and the COG trajectory. The continuous line shows a trajectory of the COG in the right-leg support phase and the dotted line shows in the left-leg support phase. The dot on the edge of both the continuous line and the dotted one means a foot-contact. Figure 10 shows the parameters and variables of the pendulum motion. S_0 and S_2 denote moments just before and after a foot-contact, and S_1 is a moment at $\dot{\theta} = 0$. θ_i, ϕ_i, and l_i denote the roll angle, yaw angle, and pendulum length at S_i ($i = 0, 1, 2$) respectively. During a cycle of walking motion, ϕ is monotonically increasing. Meanwhile, θ decreases at first, and then increases, after posing for a moment at θ_1. Thus, we compartmentalize a walking cycle from a foot-contact to the next foot-contact into two phases—Phase (A): from S_0 to S_1 ($\dot{\theta} < 0$), Phase (B): from S_1 to S_2 ($\dot{\theta} > 0$). In the phase (A), the pendulum length is constant, thus the

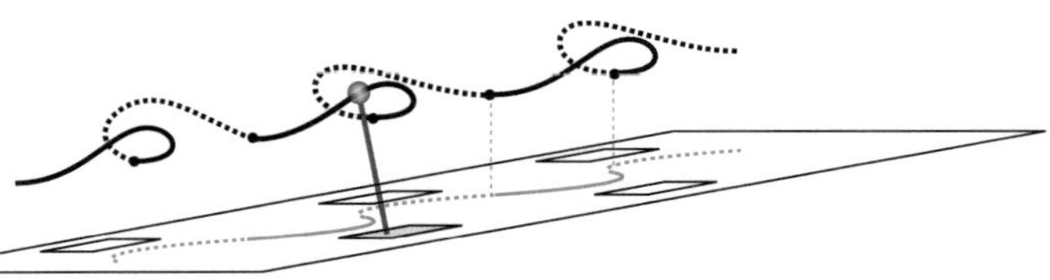

Fig. 9. Motion of the inverted pendulum. The continuous line shows the COG trajectory in the right-leg support phase and the dotted line shows in the left-leg support phase. The dot on the edge of both the continuous line and the dotted one depicts means foot-contact.

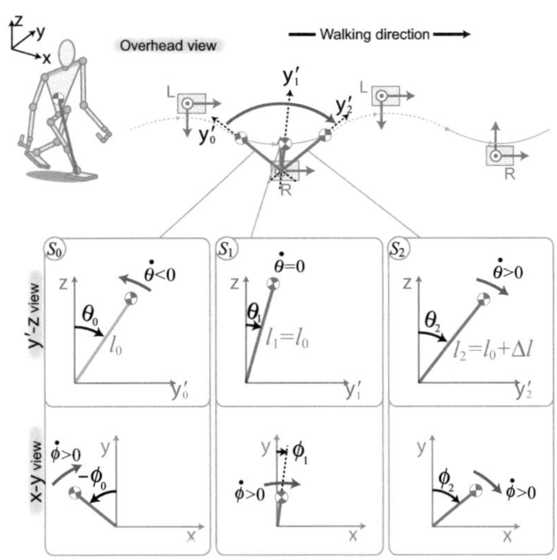

Fig. 10. Parameters and variables of dynamic walking based on 3D inverted pendulum model.

coefficients p_1-p_4 in Eq. (31) are

$$p_1 = p_2 = p_3 = 0 \tag{35}$$

$$p_d = l_0^3. \tag{36}$$

In the phase (B), the coefficients p_1-p_4 are decided so that the following four conditions are satisfied,

$$f(\theta_1) = l_1^3, \tag{37}$$

$$f(\theta_2) = l_2^3, \tag{38}$$

$$f'(\theta_2) = 0, \text{ and} \tag{39}$$

$$-f''(\theta_1)\cos\theta_1 + \big(-f'(\theta_1) + f'''(\theta_1)\big)\sin\theta_1 = 0. \tag{40}$$

Eqs. (37) and (38) signify the condition of pendulum length continuity, and Eq. (39) is the condition that the velocity of pendulum along l is 0 at a foot-contact. The objective of Eq. (40) is to match PDAC constants of the phase (A) and (B).
From Eqs. (37)-(40), the coefficients p_1-p_4 are derived as follows:

$$p_1 = -\frac{l_2^3 - l_0^3}{(\theta_2-\theta_1)^2}\frac{u_3}{u_1u_3-u_2}, \tag{41}$$

$$p_2 = -\frac{l_2^3 - l_0^3}{(\theta_2-\theta_1)^2}\frac{u_2}{u_1u_3-u_2}, \tag{42}$$

$$p_3 = -3p_1\theta_2^2 - 2p_2\theta_2, \text{ and} \tag{43}$$

$$p_4 = l_2^3 - p_1\theta_2^3 - p_2\theta_2^2 - p_3\theta_2, \tag{44}$$

where

$$u_1 = 2\theta_2 + \theta_1, \tag{45}$$

$$u_2 = -6\theta_1\cos\theta_1 - 3\theta_1^2\sin\theta_1 + 6\sin\theta_1 + 3\theta_2^2\sin\theta_1, \text{and} \tag{46}$$

$$u_3 = -2\cos\theta_1 - 2\theta_1\sin\theta_1 + 2\theta_2\sin\theta_1. \tag{47}$$

4.4 Foot-contact model

In this chapter, the impact between the foot and a ground is assumed to be perfectly inelastic. Thus the angular momentum around a new contact point is conserved. Assuming that ϕ_0 is the angle of ϕ just before a foot-contact, the position vector of the pendulum after impact, $\boldsymbol{L}$ is

$$\boldsymbol{L} = [l_0\sin\phi_0\sin\theta_0, l_0\cos\phi_0\sin\theta_0, l_0\cos\theta_0]^T, \tag{48}$$

where ϕ_0 and θ_0 are angles in the coordinate system of next step.
The vector of velocity immediately prior to a foot-contact, $\boldsymbol{V_1}$, is calculated as follows:

$$\boldsymbol{V_1} = [v_x, v_y, v_z]^T, \tag{49}$$

$$\text{where } v_x = l_2(\dot\phi_2\cos\phi_2\sin\theta_2 + \dot\theta_2\sin\phi_2\cos\theta_2) + \dot l_2(\sin\phi_2\sin\theta_2)$$

$$v_y = l_2(-\dot\phi_2\sin\phi_2\sin\theta_2 + \dot\theta_2\cos\phi_2\cos\theta_2) + \dot l_2(\cos\phi_2\sin\theta_2)$$

$$v_z = -l_2\dot\theta_2\sin\theta_2 + \dot l_2(\cos\theta_2),$$

where ϕ_2 is the angle of ϕ just before an impact.
The velocity vector after the impact, $\boldsymbol{V_0}$, is found by the following calculation,

$$\boldsymbol{V_0} = \frac{\boldsymbol{V_1} \cdot (\boldsymbol{L} \times (\boldsymbol{V_1} \times \boldsymbol{L}))}{|\boldsymbol{L} \times (\boldsymbol{V_1} \times \boldsymbol{L})|} (\boldsymbol{L} \times (\boldsymbol{V_1} \times \boldsymbol{L})) \tag{50}$$

$$= \frac{\boldsymbol{L} \times (\boldsymbol{V_1} \times \boldsymbol{L})}{l^2} \tag{51}$$

$$:= [v'_x, v'_y, v'_z]^T. \tag{52}$$

Note that in the above calculation of $\boldsymbol{V_0}$, $\boldsymbol{V_1}$ must be treated as $\boldsymbol{V_1} = [-v_x, v_y, v_z]$ since left- and right-handed systems are switched at the foot-contact.
From Eq. (52), $\dot{\theta}_0$ and $\dot{\phi}_0$ are decided,

$$\dot{\theta}_0 = \frac{l_2}{l_0}\Big(\dot{\phi}_2 \cos\theta_0 \sin\theta_2 \sin(\phi_0 + \phi_2) + \dot{\theta}_2\big(\sin\theta_2 \sin\theta_0 - \cos\theta_0 \cos\theta_2 \cos(\phi_0 + \phi_2)\big)\Big), \tag{53}$$

$$\dot{\phi}_0 = \frac{l_2}{l_0 \sin\theta_0}\Big(\dot{\theta}_2 \cos\theta_2 \sin(\phi_0 + \phi_2) + \dot{\phi}_2 \sin\theta_2 \cos(\phi_0 + \phi_2)\Big). \tag{54}$$

5. Stabilization

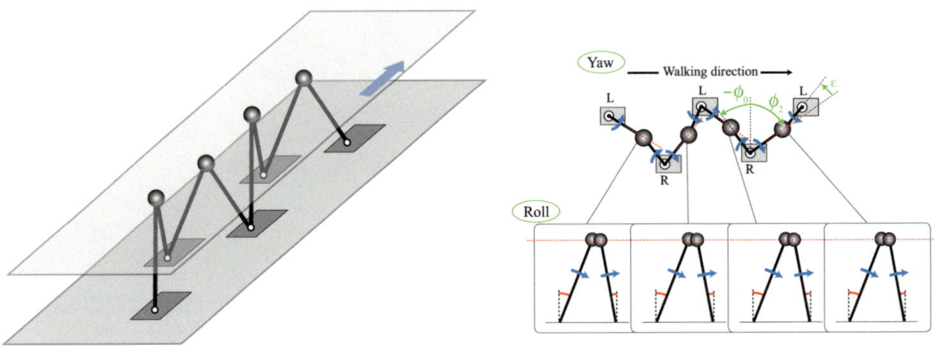

Fig. 11. Geometrical constraints at foot-contact

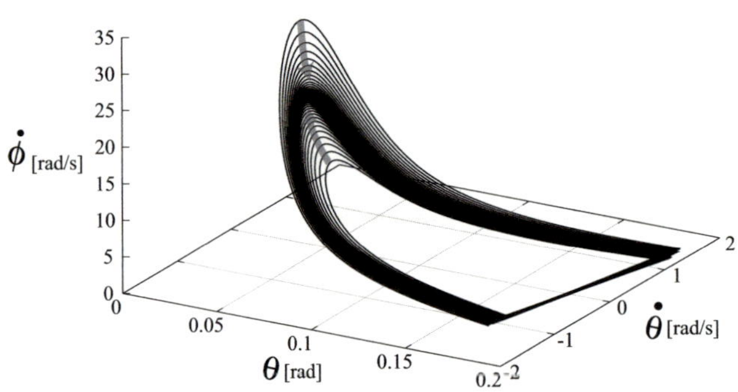

Fig. 12. Simulation results under the condition of ϵ=0.018[rad]=const, l_0 =0.51[m], Δl =0.007[m]

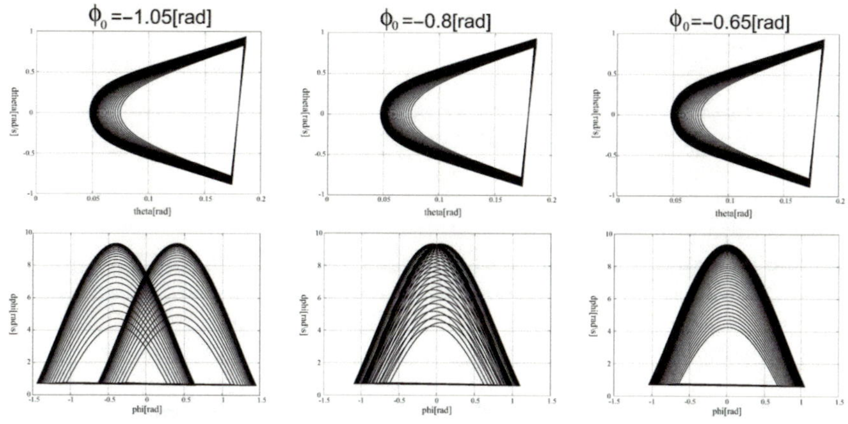

Fig. 13. Phase portrait around yaw- and roll axis under the condition of ϵ = const. The three instance of various ϕ_0 is shown.

5.1 Geometrical constraints

In order to stabilize walking, some constraints are given. At first, the lengthening value of pendulum is fixed at constant value. In this constraint, supplied energy is nearly constant. In addition, the following two constraints at foot-contact, are designed as shown in Fig. 11,

- COG height h at a foot-contact is constant, i.e. roll angles of stance- and swing-leg are constant at foot-contact.
- Yaw angle of swing-leg is shifted by ϵ from the symmetrical position with stance-leg at foot-contact, i.e. it is $\phi_0[k+1] = -\phi_2[k] + \epsilon$ where $\phi_0[k+1]$ and $\phi_2[k]$ denote ϕ_0 and ϕ_2 at $k+1$th and kth step respectively.

5.2 Convergence analysis of the dynamics

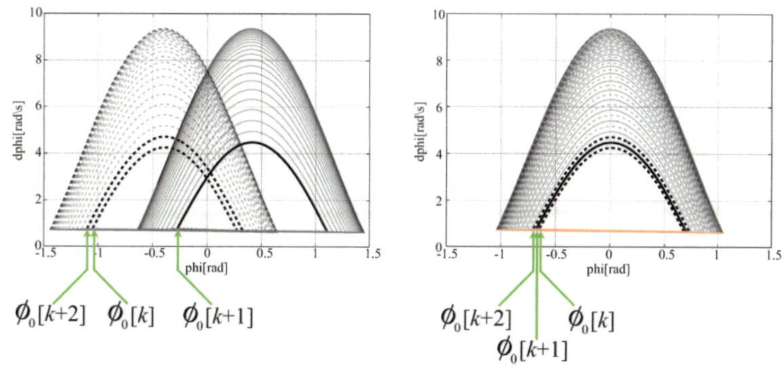

Fig. 14. Nested structure of the trajectory. $\phi_0[k]$ denotes the yaw angle at the beginning of stance-leg phase.

In this subsection, the convergence analysis is conducted. Fig. 12 depicts the simulation result under the condition of ϵ=0.018[rad]=const, l_0 =0.51[m], and Δl =0.007[m]. As can be seen in this figure, θ and $\dot{\theta}$, $\dot{\phi}$ are converged on a certain fixed point. However, it can be seen in Fig. 13, which shows the phase portrait around yaw- and roll axis, that the trajectory is not converged

on a unique one and two-cycle trajectory appears. If two-cycle occurs, a robot walks leftwards and cannot walk straight.
In this simulation, the initial angle around yaw-axis, ϕ_0, is varied without changing any other state and condition. From this figure, it can be seen that the converged trajectory seems to be depending on the initial state and that there is the certain set of initial state converging a unique trajectory.
Here, the phase portrait around yaw-axis is focused on. Fig. 14 shows the ones under the left and right condition in Fig. 13 in which the left- and right-leg supporting phases are depicted by the dotted and continuous line respectively. As shown in this figure, in the condition of unique-trajectory-convergence, the phase portrait around yaw axis possesses the nested structure, i.e. the trajectories in left- and right-leg supporting phase are nested each other—$\phi_0[k+1]$ is in between $\phi_0[k]$ and $\phi_0[k+2]$—and gradually attracted. Note that $\phi_0[k]$ is ϕ_0 value at kth step. Hence, it is contemplated that by the controller adjusting ϵ so as to achieve the nested structure of phase portrait around yaw-axis, it is possible to converge the dynamics on a unique trajectory.

5.3 Landing position control

In this subsection, we design the stabilizing controller that adjusts the landing position of stance-leg foot base on the nested structure.
If the present state is inside the converged trajectory, the condition to achieve the nested structure is described as

$$(\phi_0[k] > \phi_0[k+1]) \wedge (\phi_0[k+1] > \phi_0[k+2]). \tag{55}$$

Meanwhile, it the present state is outside, it is

$$(\phi_0[k] < \phi_0[k+1]) \wedge (\phi_0[k+1] < \phi_0[k+2]). \tag{56}$$

In order to build the stabilizing controller making the yaw dynamics attract to the nested structure, we define the distance between present state and the nested structure and design the stabilizing method minimizing this distance as follows:

$$\min(\sqrt{(\phi_0[k+2] - \phi_0[k+1])^2 + (\phi_0[k+1] - \phi_0[k])^2}). \tag{57}$$

By adjusting the landing position according to this equation, it is conceivable that robot dynamics is attracted to the nested structure and consequently converged on a unique trajectory.

6. Walking simulation

The validity of proposed control was tested by numerical simulation. Fig. 15 and Fig. 16 show the simulation results of the proposed control. Fig. 15(b) depicts the result of the controller embedded the above-mentioned stabilization. In Fig. 15(a) and (b), same initial condition is employed. Note that, as shown in Fig. 15(a), the two-cycle trajectory appears without stabilization, however in Fig. 15(b), the dynamics is converged on unique trajectory with stabilization.
The proposed stabilizing method succeeded in convergence the two-cycle trajectory on a unique trajectory. Fig. 17 shows the simulation snapshots of dynamic walking control. Without stabilization based on landing position control, 3-D walking was converged on not unique trajectory but two-cycle one, hence the robot could not walk straight and walked leftwards as can be seen in Fig. 17. Meanwhile, with stabilization, it was confirmed that the

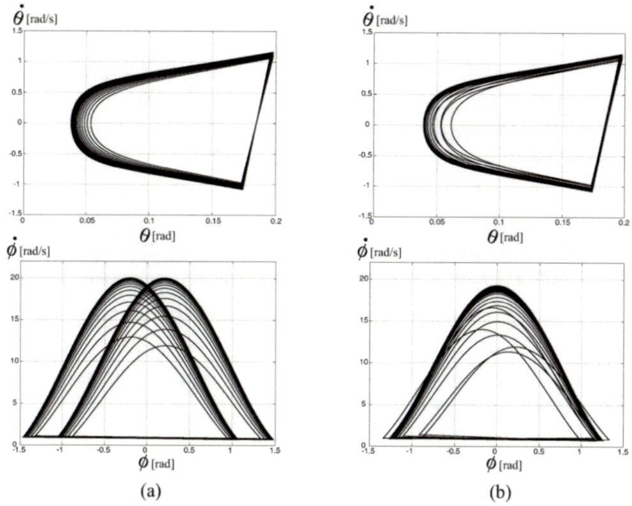

Fig. 15. Simulation results of the proposed stabilizing control.

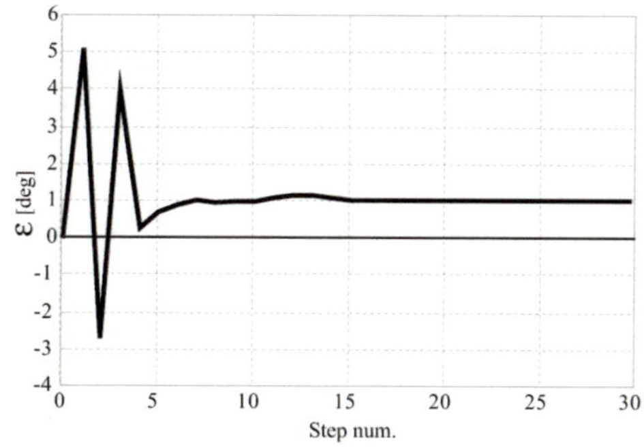

Fig. 16. Alteration of ϵ in landing position control

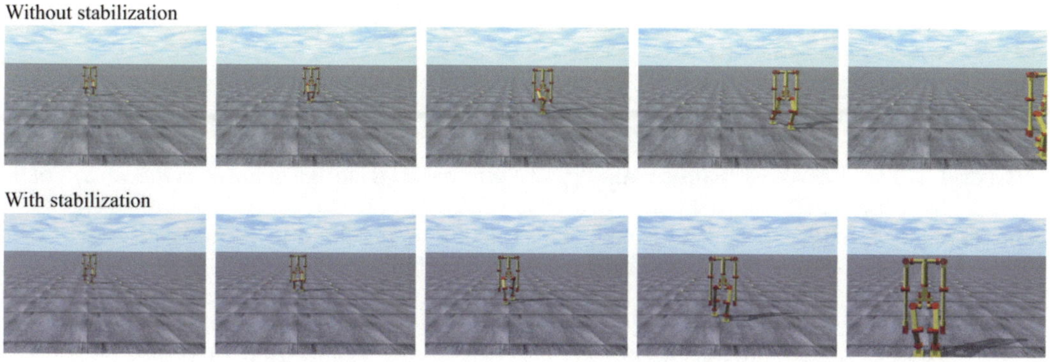

Fig. 17. Simulation snapshots of biped walking based on 3D dynamics

robot can walk straight. After convergence, step-length, walking period, and walking velocity were 0.18[m], 0.65[s], and 0.28[m/s] respectively.

7. Conclusion

In this chapter, the concept of the multi-locomotion robot that has the high mobilization capacity by achieving several kinds of locomotion independently was introduced at first. In addition, the Gorilla Robot III was introduced as hardware of the multi-locomotion robot. Second, the Passive Dynamic Autonomous Control (PDAC) which has proposed previously was explained. Not only biped walk but also quadruped walk and brachiation have been realized in our previous work. Third, we proposed the stabilizing control method that realizes 3-D biped walking based on the assumption of point-contact. The proposed method described the robot dynamics by use of a 3-D inverted pendulum model in the polar coordinate system. We applied the PDAC concept to the robot dynamics and expressed the 3-D dynamics as the 2-D autonomous system. In addition, the stabilizing controller adjusting landing position to make yaw dynamics attract to the nested structure was designed. Finally, the validity of proposed controller is tested by numerical simulation.

8. References

Aoyama, T.; Hasegawa, Y.; Sekiyama K. & FukudaT. (2009). Stabilizing and Direction Control of Efficient 3-D Biped Walking Based on PDAC. *IEEE/ASME Transactions on Mechatronics*, Vol.14, No.6, pp.712-718.

Asano, Y.; Doi, M.; Hasegawa, Y.; Matsuno, T. & Fukuda, T. (2007). Quadruped Walking by Joint-interlocking Control Based on the Assumption of Point-contact. *Transactions of the Japan Society of Mechanical Engineers, Series C*, Vol.73, No.727, pp.230-236 (in Japanese).

Doi, M.; Hasegawa, Y. & Fukuda, T. (2004b). Passive Dynamic Autonomous Control of Bipedal Walking. *Proceedings of IEEE-RAS/RSJ International Conference on Humanoid Robots*, Paper no.72, Los Angeles.

Fukuda, T.; Kojima, S.; Sekiyama, K. & Hasegawa, Y. (2007). Design Method of Brachiation Controller based on Virtual Holonomic Constraint. *Proceedings of the IEEE/RSJ International Conference on Intelligent Robots and Systems*, pp.450-455, San Diego.

Fukuda, T.; Aoyama, T.; Hasegawa, Y. & Sekiyama, K. (2009). Multilocomotion Robot:Novel Concept, Mechanism, and Control of Bio-inspired Robot. *Arti□cial Life Models in Hardware*, pp. 65-86, Springer-Verlag.

Grizzle, J. W.; Abba, G. & Plestan, F. (2001). Asymptotically Stable Walking for Biped Robots: Analysis via Systems with Impulse Effects. *IEEE Transactions on Automatic Control*, Vol.46, No.1, pp.56-64.

Hirai, K.; Hirose, M.; Haikawa, Y. & Takenaka, T. (1998). The Development of Honda Humanoid Robot. *Proceedings of the IEEE International Conference on Robotics and Automation*, pp.1321-1326, Leuven.

McGeer, T. (1990). Passive dynamic walking. *The International Journal of Robotics Research*, Vol.9, No.2, pp.62-82.

Nakanishi, J.; Fukuda, T. & Koditschek, D.E. (2000). A brachiating robot controller. *IEEE Transactions on Robotics and Automation*, Vol.16, No.2, pp.109-123.

Raibert, M.H.; (1986). Legged Robots that Balance. Cambridge, MA:MIT Press.

Saito, F.; & Fukuda, T. (1997). A First Result of The Brachiator III -A New Brachiation Robot Modeled on a Siamang. *Artificial Life V*, pp.354-361, Cambridge, MA:MIT Press.

Westervelt, E. R.; Buche, G. & Grizzle, J. W. (2004). Experimental Validation of a Framework for the Design of Controllers that Induce Stable Walking in Planar Bipeds. *The International Journal of Robotics Research*, Vol.24, No.6, pp.559-582.

8

Section-Map Stability Criterion for Biped Robots

Chenglong Fu, Zhao Liu and Ken Chen
Tsinghua University
China

1. Introduction

The importance of stability for dynamical systems is well-known. Any real system, including biped robots, need to be working under all kinds of disturbances. Whether the biped robot can effectively keep the planned motion under these disturbances is a fundamental property, and that is the explanation of stability intuitively. Stability of biped walking is the key problem in the theoretical framework of biped robots. Roughly speaking, the research of biped robots can be classified as the following three aspects: stability criterion, walking pattern planning, and walking pattern control. The purpose of stability criterion is to give the condition that the robot can realize stable walking under some control strategy. The purpose of walking pattern planning is to generate a desired gait offline or online, and it plays the role of feed-forward (Huang et al., 2001). The purpose of walking pattern control is to modify the planning walking pattern based on sensory information, and it plays the role of feedback (Huang & Nakamura, 2005). Among the above three aspects, stability criterion is the most fundamental and important, and it is the basis of walking pattern planning and real-time control. Although some researchers have proposed several walking control methods which are not based on stability criterion (Raibert, 1986; Geng et al., 2006); however if these methods can not ensure walking stability from the aspect of theory, then it will need many trials on hardware before success, and it is difficult to generate them to other platforms. Presently, there are the following three stability criteria for biped walking.

The first criterion is zero moment point (ZMP) criterion. The ZMP was originally defined as the point in the ground plane about which the net moments due to ground contacts become zero in the plane of ground (Vukobratovic & Juricic, 1969). As long as the ZMP lies strictly inside the support polygon of the foot, then the support foot will not rotate about its extremities, and the desired trajectories of the robot's joints are dynamically feasible, just like a stationary manipulator. Takanishi et al. (1985) and Hirai et al. (1998) have proposed the methods of pattern synthesis based on ZMP offline. Recently, Kajita et al. (2001), Lim et al. (2002), and Nishiwaki et al. (2002) discussed the methods of online pattern generation. The ZMP criterion is not a necessary condition for stable walking. The ZMP criterion results in a flat-footed and short-step walking style which is less dynamic than human beings. During normal walking, human do not always obeys the ZMP requirement and the foot does not always remain flat on the ground. Humans, even with prosthetic legs, use foot rotation to decrease energy loss at impact (Kuo, 2002). Based on the ZMP criterion, the robot can only realize static walking or quasi-dynamic walking, as shown in Fig. 1(a) and (b). During the dynamic walking of human beings, the under-actuated degree-of-freedom (DOF) emerges between the support foot and the ground, as shown in Fig.1 (c).

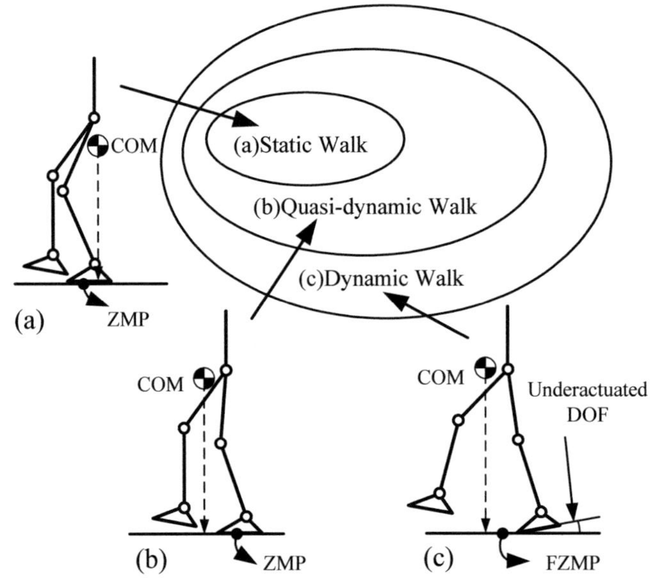

Fig. 1. Classification of biped locomotion. In case (a), the robot's nominal trajectory has been planned so that the center of mass (COM) and ZMP are both within the interior of the footprint. In case (b), COM has moved out of the footprint while ZMP still keeps within the interior of the footprint. In both case (a) and (b), the foot will not rotate, and thus the foot is acting as a base, just like a normal robotic manipulator. In case (c), however, both COM and ZMP has moved out the interior of the footprint, allowing the foot to rotate

The second stability criterion is Poincare return maps (Guckenheimer & Holmes, 1985), which is a technique for determining the existence of periodic orbits and their stability properties. With this method, the system is assumed to have a periodic limit cycle. Small deviations from the cycle follow the linear relation

$$X_{n+1} = KX_n \tag{1}$$

where X_n is the vector of deviations from the fixed point, K is a linear return map, and X_{n+1} is the vector of deviations in the following cycle. If the eigenvalues of K have moduli less than one, then the limit cycle is stable. Hurmuzlu and Moskowitz (1993) first applied the Poincare map to the locomotion systems, McGeer (1990) and Goswami et al. (1996) used this technique to analyze stability issues of passive walking robots. Grizzle et al. (2001) developed an extension of Poincare method that reduces the stability calculation to a one-dimensional map, and Westervelt et al. (2003a) used this method to design automated control for an under-actuated planar biped robot (Chevallereau et al., 2003). However, Using Poincare return maps as a stability criterion of biped walking has two serious limitations. Firstly, they are only applicable for periodic bipedal walking. There is nothing periodic about walking across unevenly spaces obstacles, or changing walking speed. Secondly, using eigenvalues of Poincare return maps is valid only for small deviations from a limit cycle. Large disruptions from a limit cycle, such as when being pushed, cannot be analyzed using this technique. Therefore, Poincare return maps are not necessary for analysing bipedal walking in general.

The third stability criterion is motivated by observation that human beings keep the relative small size of angular momentum about the center of mass (CoM) during walking. In the book Legged Robots that Balance, Raibert (1986) speculated that a control system that keeps angular momentum during stance could achieve higher efficiency and better performance. Popovic and Englehart (2004) have suggested that humanoid control systems should explicitly minimize global spin angular momentum during steady state forward walking. However, minimizing angular momentum is not a necessary condition for stable walking. Human can walk can walk while swinging his or her upper body which makes the global spin angular momentum larger than zero. Minimizing angular momentum is not a sufficient condition for stable walking, as a biped robot can fall down the ground while maintaining an angular zero momentum (Pratt & Tedrake, 2005). Therefore, angular momentum about the Center of Mass is not a good stability criterion for biped walking.

In fact, the desirable characteristics of an ideal stability criterion for biped walking may include:

1. Universal. The ideal stability criterion should be applicable not only to static walking, but also to dynamic walking. The ideal stability criterion should be applicable not only to periodic walking, but also to non-periodic walking.
2. Sufficient and Necessary. If the stability margin is outside an acceptable threshold of values, the robot will fall down. If the stability margin is inside an acceptable threshold of values, the robot will walk stably.
3. Comparable and Measurable. Two walking patterns should be comparable for stability based on their relative stability margins. One should be able to measure the relevant state variables and estimate the stability margin on-line in order to use it for control purposes (Pratt & Tedrake, 2005).
4. Simple and Convenient. The ideal stability criterion should be easy to compute, and convenient to be used in analyzing and controlling robots.

This chapter explores such a coherent stability criterion based on the description of biped walking from a global point of view. The organization of this chapter is organized as follows. Section II proposed an overall mathematical modeling method for biped walking is based on dimensional-variant hybrid automata. Section III presented a rigorous definition of biped walking stability by combining the character of biped locomotion with the notion of classical stability, and pointed out that the model in the task space is a length-varying and inertia-varying inverted pendulum. Section IV presented a stability criterion in task space of biped walking. Section V introduced application methods of the proposed criterion. Section VI provided the experimental results of a planar biped robot. Section VII concluded the chapter.

2. Overall mathematical model for dynamic biped walking

2.1 Assumption of dynamic walking

During biped locomotion, two legs alternately contact the ground. When only one leg contacts the ground, the robot is called in single support phase. When both legs contact the ground, the robot is called in double support phase. The overall biped walking consists of single support phase and double support phase.

In the field of biped locomotion, there's still no accurate and rigorous definition for dynamic walking (Goswami & Kallem, 2004). For the purposes of this chapter, to eliminate complications, we assume that dynamic walking should satisfy the following two requirements:

1. The under-actuated DOF emerges between the support foot and the ground during dynamic walking.
2. The double support phase is instantaneous and can be modeled as a rigid contact.

The robot is said to be in under-actuated phase when the robot is in the mode of toe or heel contact, and in fully-actuated phase when the robot is in the mode of full sole contact. A typical dynamic walking for biped robots with feet is shown in Fig.2.

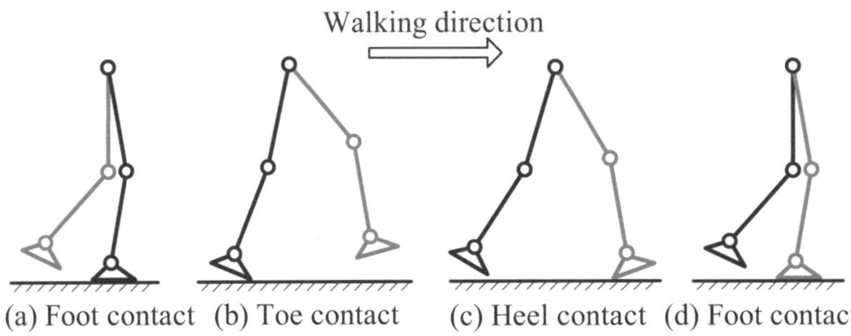

Fig. 2. A typical dynamic walking for biped robots with feet

Since the assumption of instantaneous double support phase, each discrete phase can be modeled as an N-link rigid body open-chain robot with one-DOF revolute joints. The equations of motion are given by the following general form:

$$D(q)\ddot{q} + C(q,\dot{q})\dot{q} + G(q) = Bu \tag{2}$$

where $q := (q_1;\cdots;q_N) \in Q$ are the joint angles, Q is a simply-connected, open subset of $[0,2\pi)^N$ corresponding to physically reasonable configuration of the robot. The matrix $D(q)$ is the inertia matrix, the matrix $C(q)$ contains Coriolis and centrifugal terms, $G(q)$ is the gravity vector, and B is an input matrix.

Defining $x := (q;\dot{q})$, the model in each phase can be written in state space form

$$\dot{x} = \begin{bmatrix} \dot{q} \\ D^{-1}[-C\dot{q} - G] \end{bmatrix} + \begin{bmatrix} 0 \\ D^{-1}B \end{bmatrix} u =: f(x) + g(x)u \tag{3}$$

with state space $TQ := \{(q;\dot{q}) \mid q \in Q, \dot{q} \in R^N\}$.

2.2 Overall biped model based on dimension-variant hybrid automata

The overall biped model is hybrid and dimension-variant in nature, consisting of some continuous dynamics and re-initialization rules at the contact event. We propose an overall mathematical modeling method for biped walking based on dimension-variant hybrid automata. This method expresses the overall biped walking model as an 8-tuple

$$\mathbf{H} = (\boldsymbol{V}, \boldsymbol{X}, \boldsymbol{N}, \boldsymbol{F}, \boldsymbol{D}, \boldsymbol{E}, \boldsymbol{S}, \Delta) \tag{4}$$

where

$\boldsymbol{V} = \{\boldsymbol{foot}, \boldsymbol{toe}, \boldsymbol{heel}, \cdots\}$ is the collection of discrete states;

$\boldsymbol{X} = \{\boldsymbol{x}_i : i \in \boldsymbol{V}\}$ is the collection of continuous states;

$N = \{\dim(\boldsymbol{x}_i) : i \in \boldsymbol{V}\}$ is the dimension of X;

$\boldsymbol{F} = \{\boldsymbol{F}(i,\boldsymbol{x}) : \boldsymbol{V} \times \boldsymbol{X} \to T\boldsymbol{X}\}$ is the vector fields;

$\boldsymbol{D} = \{\boldsymbol{TQ}_i : i \in \boldsymbol{V}\}$ is the collection of domains;

$\boldsymbol{E} \subseteq \boldsymbol{V} \times \boldsymbol{V}$ is the collection of edges;

$\boldsymbol{S} = \{\boldsymbol{S}_e : e \in \boldsymbol{E}\}$ is the collection of transition sections;

$\Delta = \{\Delta_e : e \in \boldsymbol{E}\}$ is the collection of transition rules;

Let $(\cdot)^-$ and $(\cdot)^+$ donate quantities immediately just before and after transition. Given $\mathbf{H}$, the basic idea is that starting from a point in some domain $\boldsymbol{TQ}_i$, we flow according to $\boldsymbol{F}_i$ until (and if) we reach some transition section $\boldsymbol{S}_i^j$, then switch via the transition rule Δ_i^j, continue flowing in $\boldsymbol{TQ}_j$, according to $\boldsymbol{F}_j$ and so on, as shown in Fig. 3.

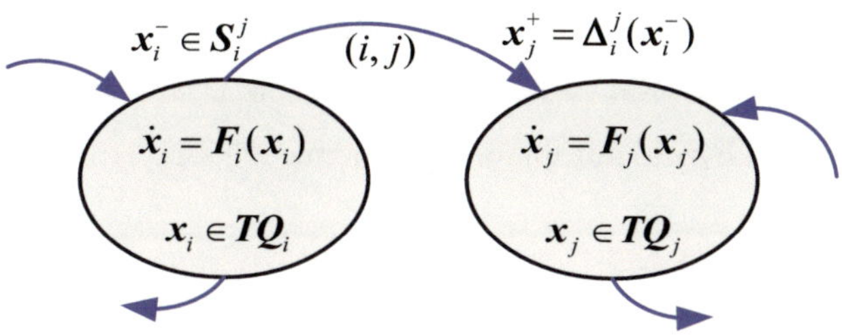

Fig. 3. Diagram of dimension-variant hybrid automata

A typical dynamic walking with feet shown in Fig.2 can be modeled as a dimension-variant hybrid automaton, as shown in Fig. 4. This modelling method can reflect all kinds of continuous and discrete properties of biped walking, which makes it possible to study stability and design control strategy for biped locomotion from a global point of view.

It should be noted that the solution $\boldsymbol{\varphi}^t(\boldsymbol{x}_0)$ of dimension-variant hybrid automata is piecewise continuous and hybrid, as shown in Fig.5.

3. Stability definition and task space

3.1 Stability definition of biped walking

The manuscript has to be submitted in MS Word (*.doc) and PDF format. If you use other word editors and can not transfer it in Word and PDF please contact us. The most intuitive definition of biped stability is likely that "the biped does not fall". This section will give a sequence of preliminary definitions leading to a rigorous mathematical definition of biped walking stability by combining the character of biped locomotion with the notion of classical stability from the view of hybrid automata.

Since the main destination of biped walking is to avoid fall, following (Pratt & Tedrake, 2005), we define a fall in this chapter as follows.

Definition 4.1 (Fall) Let $\boldsymbol{Q}_{\text{Fall}}$ be a set of the robot's configuration in which a point on the biped, other than a point on the feet, touches the ground.

There are three modes of fall for biped robots considering in this chapter as shown in Fig. 6. Let $\boldsymbol{q}$ and $\dot{\boldsymbol{q}}$ denote the vector of generalized position and velocity respectively. $\boldsymbol{Q}_{\text{Fall}}$ can be expressed as

$$\boldsymbol{Q}_{\text{Fall}} = \{\boldsymbol{q} \mid y_{\text{torso}}(\boldsymbol{q}) = 0\} \cup \{\boldsymbol{q} \mid y_{\text{hip}}(\boldsymbol{q}) = 0\} \cup \{y_{\text{knee}}(\boldsymbol{q}) = 0\} \quad (5)$$

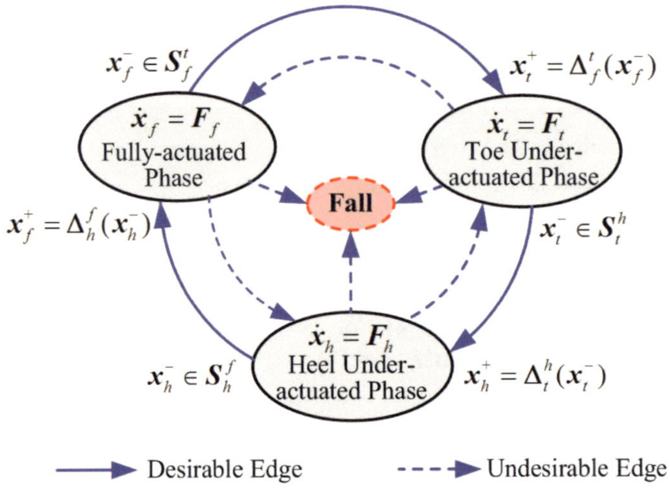

Fig. 4. Dimension-variant hybrid automata for dynamic walking wirh feet

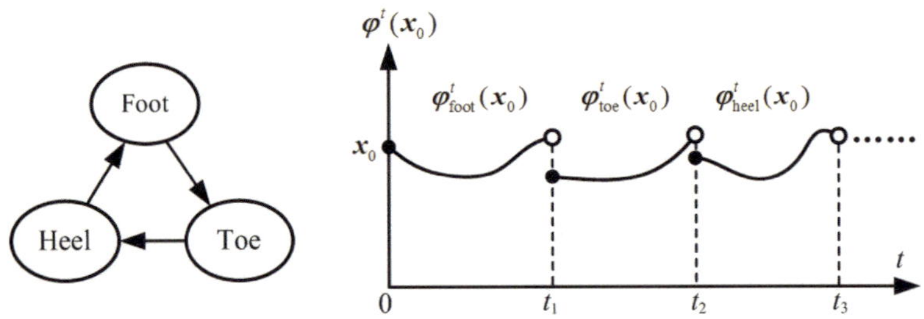

Fig. 5. Solution of dimension-variant hybrid automata

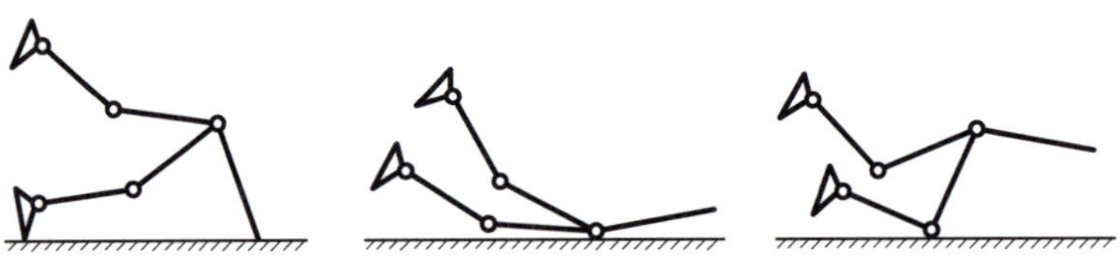

Fig. 6. Three modes of fall configuration

Introducing the state vector $\boldsymbol{x} = (\boldsymbol{q};\dot{\boldsymbol{q}})$, the solution starting from $\boldsymbol{x}_0$ can be donated as $\boldsymbol{\varphi}^t(\boldsymbol{x}_0)$, and it is hybrid and piecewise continuous. Let $\mathbf{Orb}(\boldsymbol{x}_0) = \{\boldsymbol{\varphi}^t(\boldsymbol{x}_0) \mid 0 \le t < \infty\}$ donate the hybrid orbit of biped walking. It should be noted that $\mathbf{Orb}(\boldsymbol{x}_0)$ can be not only periodic walking (Gizzle et al., 2001), but also non-periodic walking.

Definition 4.2 (Feasible Orbit) Let $\mathbf{Orb}(\boldsymbol{x}_0)$ be a hybrid orbit starting from $\boldsymbol{x}_0$. If $\forall(\boldsymbol{q};\dot{\boldsymbol{q}}) \in \mathbf{Orb}(\boldsymbol{x}_0)$, satisfying

$$\boldsymbol{q} \notin \boldsymbol{Q}_{\text{Fall}} \tag{6}$$

then $\mathbf{Orb}(\boldsymbol{x}_0)$ is a feasible orbit.

Definition 4.3 (Distance between a Point and an Orbit) Given a norm $||\cdot||$, the distance between a point $\boldsymbol{x}_i \in \boldsymbol{TQ}_i$ and an orbit $\mathbf{Orb}(\boldsymbol{x}_0)$ can be defined as

$$\mathrm{dist}(\boldsymbol{x}_i, \mathbf{Orb}(\boldsymbol{x}_0)) := \inf_{\boldsymbol{y} \in \mathbf{Orb}(\boldsymbol{x}_0) \cap \boldsymbol{TQ}_i} ||\boldsymbol{x}_i - \boldsymbol{y}|| \tag{7}$$

Definition 4.4 (Open Neighborhood of an Orbit) Let $\mathbf{Orb}(\boldsymbol{x}_0)$ be an orbit starting from $\boldsymbol{x}_0$. Given a norm $||\cdot||$, the open neighborhood of an orbit $\mathbf{Orb}(\boldsymbol{x}_0)$ can be defined as

$$\boldsymbol{\Omega}_\delta(\mathbf{Orb}(\boldsymbol{x}_0)) = \{\boldsymbol{x} \,|\, \mathrm{dist}(\boldsymbol{x}, \mathbf{Orb}(\boldsymbol{x}_0)) < \delta\} \tag{8}$$

Definition 4.5 (Stable Walking) Let $\mathbf{Orb}(\boldsymbol{x}_0)$ be a feasible orbit of biped walking. If $\forall \varepsilon > 0$, $\exists \delta(\varepsilon) > 0$ which determines an open neighbourhood $\boldsymbol{\Omega}_\delta(\mathbf{Orb}(\boldsymbol{x}_0))$ such that for every $\boldsymbol{x} \in \boldsymbol{\Omega}_\delta(\mathbf{Orb}(\boldsymbol{x}_0))$, satisfying $\boldsymbol{\varphi}^t(\boldsymbol{x}) \in \boldsymbol{\Omega}_\varepsilon(\mathbf{Orb}(\boldsymbol{x}_0))$ for all $t \geq 0$, then the biped walking is stable.

Definition 4.6 (Attractive Walking) Let $\mathbf{Orb}(\boldsymbol{x}_0)$ be a feasible orbit of biped walking. If $\exists \delta > 0$ which determines an open neighbourhood $\boldsymbol{\Omega}_\delta(\mathbf{Orb}(\boldsymbol{x}_0))$ such that for every $\boldsymbol{x} \in \boldsymbol{\Omega}_\delta(\mathbf{Orb}(\boldsymbol{x}_0))$, satisfying $\lim_{t\to\infty} \boldsymbol{\varphi}^t(\boldsymbol{x}) \in \mathbf{Orb}(\boldsymbol{x}_0)$, then the biped walking is attractive.

Definition 4.7 (Asymptotically Stable Walking) If a biped walking is both stable and attractive, then it is asymptotically stable.

Definition 4.8 (Exponentially Stable Walking) If there exists $\delta > 0$, $\gamma > 0$, and $\beta > 0$ such that, $\forall t > 0$,

$$\mathrm{dist}(\boldsymbol{\varphi}^t(\boldsymbol{x}), \mathbf{Orb}(\boldsymbol{x}_0)) \leq \gamma e^{-\beta t} \mathrm{dist}(\boldsymbol{x}, \mathbf{Orb}(\boldsymbol{x}_0)) \tag{9}$$

whenever $\boldsymbol{x} \in \boldsymbol{\Omega}_\delta(\mathbf{Orb}(\boldsymbol{x}_0))$.

3.2 Biped model in the task space

The definition of biped stability is established in high-DOF space; however it is difficult to study the stability in this high-DOF space directly. Although biped robots are typically high DOF mechanisms, the task of biped walking is inherent a low DOF task. Considering planar biped walking, the task space is only 1-DOF problem, as shown in Fig. 7.

In fact, stability of biped walking can be studied in the low-DOF task space under virtual constraint control strategy (Gizzle et al., 2001; Canudas-de-Wit, 2004), as shown in Fig. 8.

Inspired by the work of Westervelt et al. (2003a), let θ and σ donate the position and angular momentum around the pivot point of the stance leg respectively. Let $z_i = (\theta_i; \sigma_i) \in \boldsymbol{Z}_i \subset \mathbb{R}^2$ donate the state variable of the task space. According to the angular momentum balance theorem, the model of the task space has the following special form

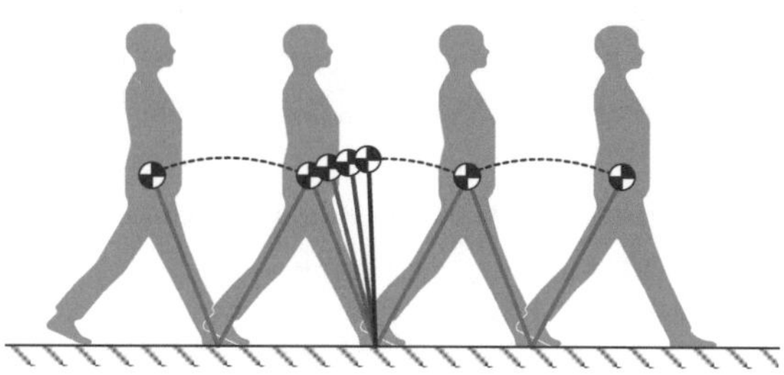

Fig. 7. Task Space of biped locomotion in sagittal plane

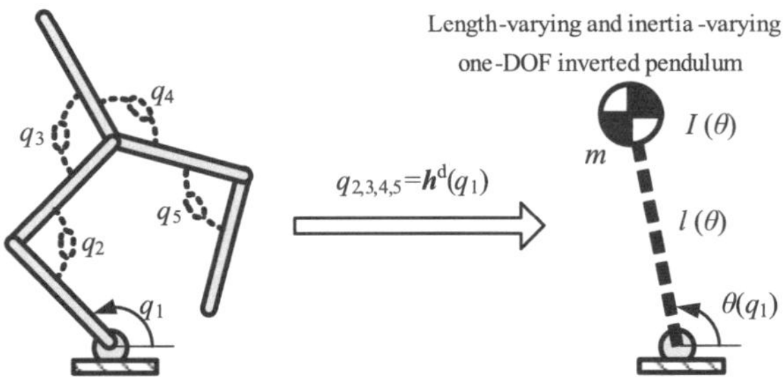

Fig. 8. Example of virtually constrainted system

$$\begin{pmatrix} \dot{\theta}_i \\ \dot{\sigma}_i \end{pmatrix} = \begin{pmatrix} \dfrac{\sigma_i}{I_i(\theta_i)} \\ J_i(\theta_i) \end{pmatrix} \tag{10}$$

where I_i plays the role of an inertia, and J_i plays the role of a net moment around the pivot.

3.3 Basic definitions for the biped model in the task space

Let $\boldsymbol{F}_i(\theta_i,\sigma_i) := \begin{pmatrix} \dfrac{\sigma_i}{I_i(\theta_i)} \\ J_i(\theta_i) \end{pmatrix}$ donate the vector field in task space. We assume the following conditions are satisfied:

H1) $\boldsymbol{Z}_i \subset \mathbb{R}^2$ is open and connected;

H2) $\boldsymbol{F}_i : \boldsymbol{Z}_i \rightarrow \mathbb{R}^2$ is C^1;

H3) A solution $\boldsymbol{\varphi}^t(z_i)$ is right continuous on t, and depends continuously on the initial condition z_i;

H4) Transition section is designed as $\boldsymbol{S}_i^{i+1} = \{(\theta_i,\sigma_i) \mid \theta_i = \theta_i^-\}$

H5) v is C^1, and $\Delta_i^{i+1}(\boldsymbol{S}_i^{i+1}) \cap \boldsymbol{S}_i^{i+1} = \varnothing$;

H6) $\sigma_i > 0$ during normal forward walking.

Definition 4.9 (Time Function) $\mathrm{T} : \boldsymbol{Z}_i \rightarrow \mathbb{R}$ is defined as

$$\mathrm{T}(z_i) := \{t \geq 0 \mid \boldsymbol{\varphi}^t(z_i) \in \boldsymbol{S}_i^{i+1}\} \tag{11}$$

The meaning of time function is the time to the transition section at the first time.

Definition 4.10 (Distance Function) $\mathrm{d} : \boldsymbol{Z}_i \rightarrow \mathbb{R}$ is defined as

$$\mathrm{d}(z_i) := \sup_{0 \leq t < \mathrm{T}(z_i)} \mathrm{dist}(\boldsymbol{\varphi}^t(z_i), \mathbf{Orb}(z_0^*)) \tag{12}$$

The meaning of distance function is the maximum distance between $\mathbf{Orb}(z_0^*)$ and solution $\boldsymbol{\varphi}^t(z_i)$ before the first time to impact section.

Definition 4.11 (Total Distance Function) $\mathrm{D} : \boldsymbol{Z}_i \rightarrow \mathbb{R}$ is defined as

$$\mathrm{D}(z_i) := \sup_{0 \le t < \infty} \mathrm{dist}(\boldsymbol{\varphi}^t(z_i), \mathbf{Orb}(z_0^*)) \qquad ()$$

The meaning of distance function is the maximum distance between $\mathbf{Orb}(z_0^*)$ and solution $\boldsymbol{\varphi}^t(z_i)$ while $t \in (0, \infty]$.

Lemma 1: [(Grizzle, 2001), Lemma3 and 4] Suppose that H1-H4 hold, then $\mathrm{T}(z_i)$ and $\mathrm{d}(z_i)$ is continuous.

Lemma 2: Suppose that H1-H5 hold, then $\mathrm{D}(z_i)$ is continuous.

Proof: According to the definition of total distance function, $\mathrm{D}(z_i)$ can be written as

$$\mathrm{D}(z_i) = \sup\{\mathrm{d}(z_i),\ \mathrm{d}(z_{i+1}^+),\ \mathrm{d}(z_{i+2}^+),\ \cdots\} \qquad (14)$$

where $\mathrm{d}(z_{i+1}^+) = \mathrm{d} \circ \Delta_i^{i+1} \circ \lim_{t \to \mathrm{T}(z_i)} \boldsymbol{\varphi}^t(z_i)$. By H3), H5), and Lemma 2, $\mathrm{d}(z_{i+1}^+)$ is continuous. In the same way, any item in the right side of the equation (14) is continuous. Therefore, $\mathrm{D}(z_i)$ is continuous. ♠

4. Stability criterion in the task space

4.1 Section Sequence and its Stability Equivalence to Orbit

Definition 5.1 (Section Sequence) $\{\overline{z}_i^*\}_{i=0}^{\infty}$ is defined as a set of intersection point between $\mathbf{Orb}(z_0^*)$ and transition sections as shown in Fig. 9, and $\overline{z}_i^*$ can be written as

$$\overline{z}_i^* := \overline{\mathbf{Orb}}(z_0^*) \cap \boldsymbol{S}_i^{i+1} \qquad (15)$$

where $\overline{\mathbf{Orb}}(z_0^*)$ is the set closure of $\mathbf{Orb}(z_0^*)$.

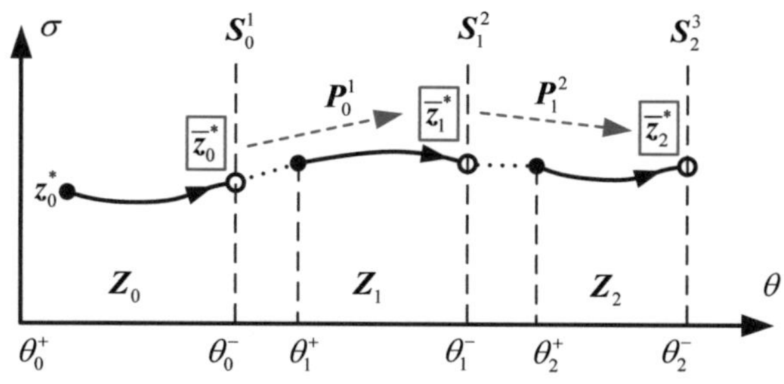

Fig. 9. Section sequence and section map

Theorem 1 Under H1)-H5), if section sequence $\{\overline{z}_i^*\}_{i=0}^{\infty}$ is stable (resp., asymptotically stable, or exponentially stable), orbit $\mathbf{Orb}(z_0^*)$ is stable (resp., asymptotically stable, or exponentially stable).

Proof: The process can be summed up into the following three parts:

1. **Proving stable**

Since section sequence $\{\overline{z}_i^*\}_{i=0}^{\infty}$ is stable, then $\forall \overline{\varepsilon} > 0$, $\exists \overline{\delta}(\overline{\varepsilon}) > 0$ such that for every $\overline{z}_0 \in \boldsymbol{B}_{\overline{\delta}(\overline{\varepsilon})}(\overline{z}_0^*)$, satisfying $\overline{z}_i \in \boldsymbol{B}_{\overline{\varepsilon}}(\overline{z}_i^*)$ for all $i \ge 0$. This implies that $\forall \overline{z}_0 \in \boldsymbol{B}_{\overline{\delta}(\overline{\varepsilon})}(\overline{z}_0^*)$, there exists a solution $\boldsymbol{\varphi}^t(z_0)$ defined on $[0, \infty)$ with the initial value z_0. Moreover, following [(Grizzle, 2001), Equation (55)], an upper bound on how far the solution $\boldsymbol{\varphi}^t(z_0)$ wanders from the orbit is given by

$$\sup_{t\geq 0} \mathrm{dist}(\boldsymbol{\varphi}^t(z_0), \mathbf{Orb}(z_0^*)) \leq \sup_{z\in\boldsymbol{\Omega}_{\bar{\varepsilon}}(\overline{\mathbf{Orb}(z_0^*)})\cap \boldsymbol{S}} \mathrm{D}(z) \tag{16}$$

According to Lemma 2, $\mathrm{D}(z_i)$ is continuous; moreover, for $z \in \overline{\mathbf{Orb}}(z_0^*) \cap \boldsymbol{S}$, $\mathrm{D}(z) = 0$. This implies that $\forall \varepsilon > 0$, $\exists \bar{\varepsilon} > 0$, satisfying

$$\sup_{z\in\boldsymbol{\Omega}_{\bar{\varepsilon}}(\overline{\mathbf{Orb}(z_0^*)})\cap \boldsymbol{S}} \mathrm{D}(z) < \varepsilon \tag{17}$$

Take (17) into (16), and yield

$$\sup_{t\geq 0} \mathrm{dist}(\boldsymbol{\varphi}^t(z_0), \mathbf{Orb}(z_0^*)) < \varepsilon \tag{18}$$

According to definition 4.4, (18) can be written as

$$\boldsymbol{\varphi}^t(z_0) \in \Omega_{\varepsilon}(\mathbf{Orb}(z_0^*)) \tag{19}$$

By H3) and H4), It is easy to construct a small enough open neighborhood $\boldsymbol{\Omega}_{\delta}(\mathbf{Orb}(z_0^*))$ satisfying that when $z \in \boldsymbol{\Omega}_{\delta}(\mathbf{Orb}(z_0^*)) \cap \boldsymbol{Z}_0$, $\bar{z}_0 \in \boldsymbol{B}_{\bar{\delta}}(\bar{z}_0^*)$, which proves that the orbit is stable

2. Proving asymptotic stable

Since section sequence $\{\bar{z}_i^*\}_{i=0}^{\infty}$ is asymptotically stable, then there exists $\delta > 0$ such that for every $\bar{z}_0 \in \boldsymbol{B}_{\bar{\delta}(\bar{\varepsilon})}(\bar{z}_0^*)$, satisfying $\lim_{i\to\infty} \bar{z}_i \in \overline{\mathbf{Orb}}(z_0^*)$. According to definition 4.3, we have $\lim_{i\to\infty} \mathrm{D}(\bar{z}_i) = 0$. This implies that there exists a solution $\boldsymbol{\varphi}^t(z_0)$ satisfying $\lim_{t\to\infty} \mathrm{dist}(\boldsymbol{\varphi}^t(z_0), \mathbf{Orb}(z_0^*)) = 0$, which proves that the orbit is asymptotically stable.

3. Proving exponentially stable

Since section sequence $\{\bar{z}_i^*\}_{i=0}^{\infty}$ is exponentially stable, then there exists $\delta > 0$, $\gamma > 0$ and $\beta > 0$ such that, for all $i \geq 0$,

$$\| \bar{z}_i - \bar{z}_i^* \| \leq \gamma e^{-\beta i} \| \bar{z}_0 - \bar{z}_0^* \| \tag{20}$$

whenever $\bar{z}_0 \in \boldsymbol{\Omega}_{\delta}(\overline{\mathbf{Orb}}(z_0^*)) \cap \boldsymbol{S}_0^1$

According to Lemma 1 and H5), for any $i \geq 0$, $\mathrm{T}(z_i)$ $\mathrm{T}(z_i)$ and $\Delta_i^{i+1}(\bar{z}_i)$ are all continuous; therefore, there exists an open ball $\boldsymbol{B}_r(\bar{z}_i^*)$, $\mathrm{T}_{\min} > 0$, $\mathrm{T}_{\max} > 0$, such that for $\bar{z}_i \in \boldsymbol{B}_r(\bar{z}_i^*) \cap \boldsymbol{S}_i^{i+1}$,

$$0 < \mathrm{T}_{\min} \leq \mathrm{T} \circ \Delta_i^{i+1}(\bar{z}_i) \leq \mathrm{T}_{\max} < \infty \tag{21}$$

Since exponential stability of $\{\bar{z}_i^*\}_{i=0}^{\infty}$ implies stability of $\{\bar{z}_i^*\}_{i=0}^{\infty}$, $\mathbf{Orb}(z_0^*)$ is also stable. Thereby, there exists $\delta > 0$, such that for $z_0 \in \boldsymbol{\Omega}_{\delta}(\mathbf{Orb}(z_0^*))$, satisfying $\boldsymbol{\varphi}^t(z_0) \in \boldsymbol{\Omega}_r(\mathbf{Orb}(z_0^*))$ for all $t \geq 0$. According to H3), H4) and standard bounds for the Lipschitz dependence of the solution w.r.t. its initial condition, it follows that for $\bar{z}_i \in \boldsymbol{B}_{\delta}(\bar{z}_i^*) \cap \boldsymbol{S}_i^{i+1}$,

$$\sup_{0\leq t\leq \mathrm{T}\circ\Delta_i^{i+1}(\bar{z}_i)} \mathrm{dist}(\boldsymbol{\varphi}^t \circ \Delta_i^{i+1}(\bar{z}_i), \mathbf{Orb}(z_0^*)) \leq \sup_{0\leq t\leq \mathrm{T}_{\max}} \| \boldsymbol{\varphi}^t \circ \Delta_i^{i+1}(\bar{z}_i) - \boldsymbol{\varphi}^t \circ \Delta_i^{i+1}(\bar{z}_i^*) \| \leq L_i \| \bar{z}_i - \bar{z}_i^* \| \tag{22}$$

where L_i is the Lipschitz constant of continuous function $\boldsymbol{\varphi}^t \circ \Lambda_i^{i+1}(\cdot)$.

According to (20) and (22), for $i \geq 0$,

$$\sup_{0\leq t\leq \mathrm{T}\circ\Delta_i^{i+1}(\bar{z}_i)} \mathrm{dist}(\boldsymbol{\varphi}^t \circ \Delta_i^{i+1}(\bar{z}_i), \mathbf{Orb}(z_0^*)) \leq L_i \gamma e^{-\beta i} \| \bar{z}_0 - \bar{z}_0^* \| \tag{23}$$

Define $L = \sup_i L_i$, and considering (21) and (23), for all $t \geq 0$,

$$\mathrm{dist}(\boldsymbol{\varphi}^t(z_0), \mathbf{Orb}(z_0^*)) \leq L\gamma e^{\beta} \cdot \mathrm{e}^{-\frac{\beta}{\mathrm{T}_{\max}}t} \,||\, \overline{z}_0 - \overline{z}_0^* \,||| \tag{24}$$

By H3) and H4), there exists $c > 0$ such that

$$||\, \overline{z}_0 - \overline{z}_0^* \,|| \leq c \cdot \mathrm{dist}(z_0, \mathbf{Orb}(z_0^*)) \tag{25}$$

The proof of asymptotical stability can be finished by taking (25) into (24).

4.2 Section map and its analytical form

Definition 5.2 (Section Map) $\boldsymbol{P}_i^{i+1} : \boldsymbol{S}_i^{i+1} \cap \boldsymbol{Z}_i \to \boldsymbol{S}_{i+1}^{i+2} \cap \boldsymbol{Z}_{i+1}$ is defined as

$$\boldsymbol{P}_i^{i+1}(\overline{z}_i^*) := \varphi^{\mathrm{T}(\Delta(\overline{z}_i^*))}(\Delta(\overline{z}_i^*)) \tag{26}$$

The meaning of section map is the map between two contiguous section sequences, as shown in Fig. 9.

Remark 1: It should be noted that section map does not need the system is periodic.

Since $\boldsymbol{Z}_i$ is two-dimensional, $\boldsymbol{S}_i^{i+1} = \{(\theta_i;\sigma_i) \mid \theta_i = \theta_i^-\}$ is a one-dimensional restriction; therefore, section map and section sequence are both one-dimensional essentially. By H6), $\rho_i^{i+1} : (\sigma_i^-)^2 \to (\sigma_{i+1}^-)^2$ is homeomorphous with $\boldsymbol{P}_i^{i+1}$, and section map $\{\sigma_i^-\}_{i=0}^{\infty}$ is homeomorphous to $\{(\sigma_i^-)^2\}_{i=0}^{\infty}$, which can be written as

$$\{(\sigma_i^-)^2\}_{i=0}^{\infty} = \{\rho_{i-1}^i \circ \cdots \circ \rho_0^1((\sigma_0^-)^2)\}_{i=0}^{\infty} \tag{27}$$

Thereby, the stability of section sequence is determined by the form of ρ_i^{i+1}.

By (10), section map $\rho_{i-1}^i : (\sigma_{i-1}^-)^2 \to (\sigma_i^-)^2$ can be written as

$$(\sigma_i^-)^2 = (\delta_{i-1}^i \cdot \sigma_{i-1}^-)^2 + 2\int_{\theta_i^+}^{\theta_i^-} I_i(\theta_i) J_i(\theta_i) d\theta_i \tag{28}$$

where $\delta_{i-1}^i := \sigma_i^+ / \sigma_{i-1}^-$ is called section-map factor.

It should be noted that (28) is a one-dimensional linear time-invariant map.

4.3 Section-map stability criterion

Theorem 2 (Section-map Stability Criterion) Under H1)–H6), if $0 < \sup_{i>0} \delta_{i-1}^i < 1$, $\mathbf{Orb}(z_0^*)$ is exponentially stable; moreover, the smaller $\sup_{i>0} \delta_{i-1}^i$ is, the faster $\mathbf{Orb}(z_0^*)$ converges.

Proof: According to Theorem 1, the exponential stability of $\mathbf{Orb}(z_0^*)$ lies on the exponential stability of $\{\overline{z}_k^*\}_{k=0}^{\infty}$. Since $\{\overline{z}_k^*\}_{k=0}^{\infty}$ and $\{(\sigma_i^-)^2\}_{i=0}^{\infty}$ is homeomorphous, the following will prove the exponential stability of $\{(\sigma_i^-)^2\}_{i=0}^{\infty}$.

Define $\xi_i = (\sigma_i^-)^2$, and $\{\xi_i^*\}_{i=0}^{\infty}$ can be written as

$$\{\xi_i^*\}_{i=0}^{\infty} = \{\xi_0^*,\ \rho_0^1(\xi_0^*),\ \rho_1^2 \circ \rho_0^1(\xi_0^*),\ \cdots,\ \rho_{i-1}^i \circ \cdots \circ \rho_0^1(\xi_0^*),\ \cdots\} \tag{29}$$

When there exists an initial perturbation $||\, \xi_0 - \xi_0^* \,|| < \delta$, according to (28), for all $i \geq 0$,

$$||\xi_i - \xi_i^*|| = ||\rho_{i-1}^i(\xi_{i-1}) - \rho_{i-1}^i(\xi_{i-1}^*)|| = (\delta_{i-1}^i)^2 ||\xi_{i-1} - \xi_{i-1}^*|| = (\delta_{i-1}^i)^2 \cdots (\delta_0^1)^2 \cdot ||\xi_0 - \xi_0^*|| \quad (30)$$

Moreover,

$$(\delta_{i-1}^i)^2 \cdots (\delta_0^1)^2 \le (\delta_{\max})^{2i} = (e^{\ln \delta_{\max}})^{2i} = e^{2(\ln \delta_{\max})\cdot i} \quad (31)$$

Take (31) into (30), and yield

$$||\xi_i - \xi_i^*|| \le e^{2(\ln \delta_{\max})\cdot i} ||\xi_0 - \xi_0^*|| \quad (32)$$

Since $0 < \sup_{i>0} \delta_{i-1}^i < 1$, then $\ln(\sup_{i>0} \delta_{i-1}^i) < 0$, which proves that $\{\xi_i^*\}_{i=0}^{\infty}$ is exponential stable. ♠

Remark 2: Theorem 2 can be intuitively explained as: If the error arising from an initial perturbation can be shrinked at each impact, then the stability of the orbit can be achieved, vice versa, as shown in Fig. 10.

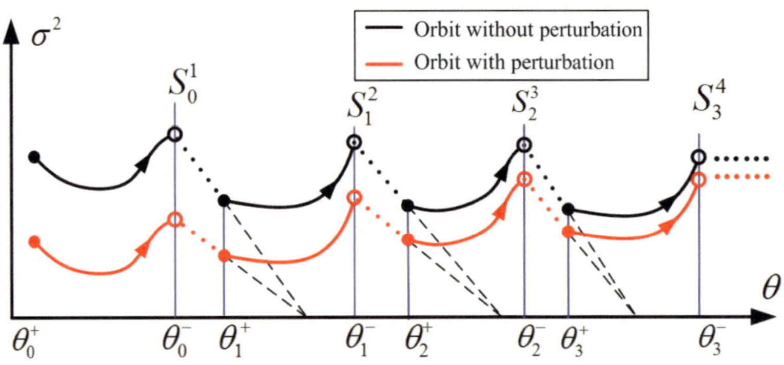

(a) Stable $(\delta_i^{i+1} < 1, i \ge 0)$

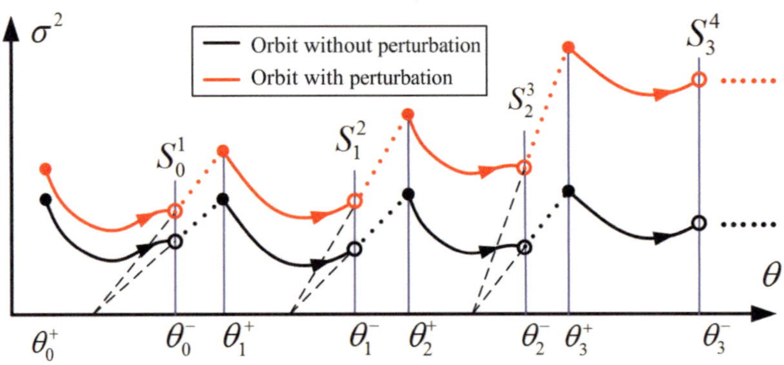

(b) Unstable $(\delta_i^{i+1} > 1, i \ge 0)$

Fig. 10. Intuitive explaination of section-map stability criterion

Remark 3: Section-map Stability Criterion is an extension of Poincare return maps, and it is applicable to non-periodic walking which Poincare return map criterion can not solve.

In Section I, we have asserted that a desirable stability criterion for biped walking should satisfy four characteristics. In followings, we will give an explanation for section-map stability criterion about the above four characteristics.

1. Universal. Section-map stability criterion is established on a rigorous definition of biped walking; therefore it can not only be applicable to static walking, but also to dynamic walking, and can not only be used to study periodic walking, but also to non-periodic walking.
2. Necessary and Sufficient. Section-map stability criterion is a sufficient condition for biped walking, is a necessary condition for periodic walking, and is a quasi-necessary condition for non-periodic walking. The quasi-necessary condition means that the condition that all section-map factors are less than one is not necessary for non-periodic walking, but the number of section-map factors which is less than one should be larger than the number of section-map factors which is more than one.
3. Comparable and Measurable. By comparing the section-map factors of two walking patterns, one can determine which pattern is more stable. The lower δ, the faster the convergence toward the reference trajectory after perturbation. One can measure the relevant state variables and calculate or estimate the stability margin on-line in order to use it for control purposes.
4. Simple and Convenient. Comparing with ZMP criterion, it is not necessary to calculate all points of trajectories, and only transition points need to be calculated. Comparing with Poincare methods, the proposed criterion study biped walking in low-dimension task space and has a concise form; therefore section-map stability criterion is easy to compute, and convenient to be used in analyzing and controlling biped walking.

5. Applications of section-map stability criterion to planar biped walking

5.1 Planar biped robot THR-I

To test the validation of the proposed criterion, a planar biped robot called THR-I has been developed, and this robot has five links which are connected by revolute joints. To constrain motions in the frontal plane, THR-I was constructed with a boom attached at the hip joint, as shown in Fig. 11. The boom constrains the sagittal plane to be tangent to a sphere centered at the universal joint, and still allows the robot to freely trip or fall forward or backward. The material of the boom is made of carbon fiber which is rather light, and the length of the boom is more than 5 times leg length of THR-I; therefore, the influence of the boom on THR-I′ dynamics in the sagittal plane is very small.

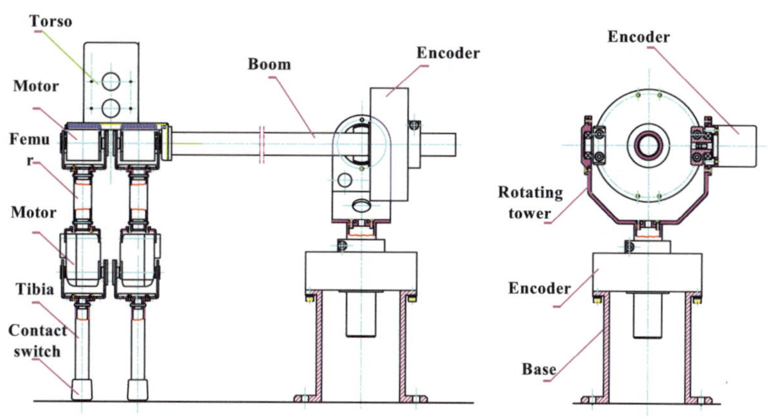

Fig. 11. Mechanical structure of THR-I.

Encoders are located between the boom and hip joint, and binary contact switches are located at the tip of the leg to detect whether or not a leg is in contact with the walking surface. There is no actuation at the stance leg tip. Hence, the robot is underactuated. It is assumed that walking consists of two successive phase: a single support phase and an instantaneous impact phase. Although this robot is simple, it captures the main difficulties: hybrid, static instability, and under-actuation. This model was also adopted in (Geng et al., 2006; Chevallereau et al., 2003).

To describe the shape of the biped, let $q_c = (q_1, q_2, q_3, q_4)'$ denote the configuration coordinates, and q_5 denote the absolute coordinate of the torso with respect to the coordinate frame as shown in Fig.12. The vector of the generalized coordinates of the biped robot is defined as $q = (q_1, q_2, q_3, q_4, q_5)'$. Let (x_{com}, y_{com}) denote the Cartesian coordinates of the center of mass. Torques u_i, $i = 1$ to 4, are applied between each connection of two links. Let σ denote the biped angular momentum around the pivot point of the stance leg. For the above choice of the coordinates in the support phase, σ has the following form (Chevallereau, 2004):

$$\sigma = -D_5(q_c)\dot{q} \tag{33}$$

where $D_5(q_c)$ is the fifth line of matrix $D(q_c)$.

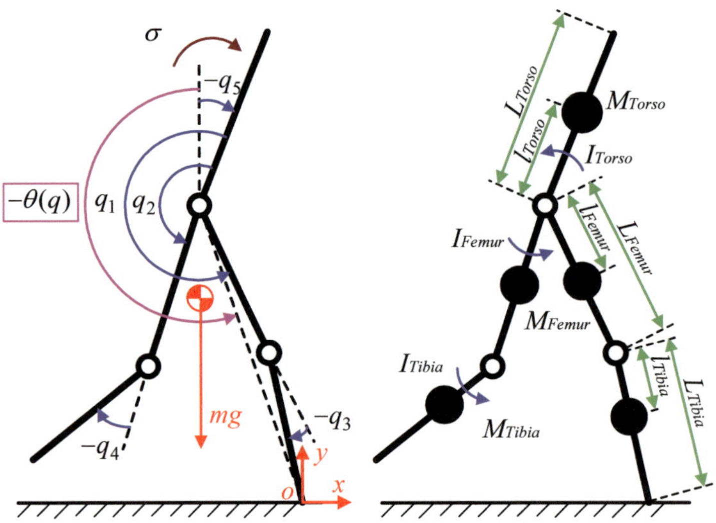

Fig. 12. Model of a 5-link THR-I biped robot

Since only the gravity affects the angular momentum around the pivot point, the angular momentum dynamics can therefore be written as

$$\dot{\sigma} = mg \cdot x_{com}(q) \tag{34}$$

Let q^-, q^+, $\dot{q}^-$, and $\dot{q}^+$ denote the pre- and post-impact generalized positions and generalized velocities, respectively. The superscript "–" and "+" will denote quantities immediately before and after impact thereinafter.

We assumed that the impact is instantaneous and inelastic. After impact the former stance leg is lifted off immediately, and the legs swap roles, which can be written as the following transformation equation:

$$q^{+} = Rq^{-} \tag{35}$$

where $R = \begin{bmatrix} 0 & 1 & 0 & 0 & 0 \\ 1 & 0 & 0 & 0 & 0 \\ 0 & 0 & 0 & 1 & 0 \\ 0 & 0 & 1 & 0 & 0 \\ 0 & 0 & 0 & 0 & 1 \end{bmatrix}$ is a circular matrix describing the exchange of the support leg.

According to the conservation of momentum about impact point and no rebound nor slip at impact of swing leg tip, the map from $\dot{q}^{-}$ and $\dot{q}^{+}$ can be obtained respectively by

$$\dot{q}^{+} = \Delta_{\dot{q}}(q^{-})\dot{q}^{-} \tag{36}$$

where $\Delta_{\dot{q}}(q^{-})$ can be found in (Westervelt et al., 2003a).

5.2 Synthesizing periodic walking patterns based on section-map factor

Consider the following output function (Westervelt et al., 2003a):

$$y = h(q) := h_0(q) - h_d \circ \theta(q) \tag{37}$$

where $h_0(q)$ specifies the four actuated joints that are to be controlled and $h_d \circ \theta(q)$ specifies the desired evolution of these joints as a function of the monotonic quantity θ(q), as shown in Fig. 13. Driving y to zero will force $h_0(q)$ to track $h_d \circ \theta(q)$; thus the configuration of the robot is being controlled as a holonomic constraint parameterized by $\theta(q)$.

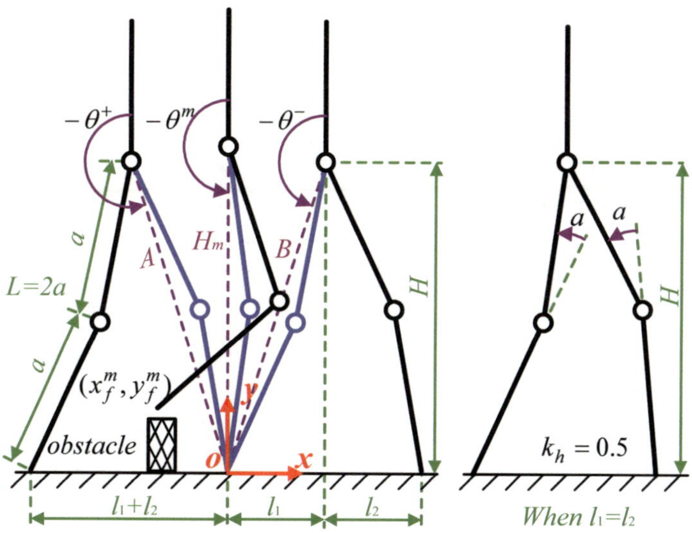

Fig. 13. Parameters of the walking pattern

Choosing

$$h_0(q) := H_0 qa \tag{38}$$

$$\theta(q) := cq \tag{39}$$

where $H_0 = \begin{bmatrix} I_{4\times4} & 0_{4\times1} \end{bmatrix}$, c = [-1 0 -1/2 0 -1].
To obtain a stable walking pattern, we propose a synthesizing method consisting of the following four steps:

1. parameterize position constraints $h_d(\theta)$ at all breakpoints;

Since one of the basic aspects of biped locomotion is to maintain a constant erect torso, we specify q5 = 0 in the whole walking cycle, as shown in Fig. 13.
To shape the impact posture, we define two normalized non-dimensional parameters:

$$k_s := (l_1 + l_2) / L \tag{40}$$

$$k_h := l_1 / (l_1 + l_2) \tag{41}$$

where k_s describes the magnitude of the stride relative to leg length, and k_h describes the ratio of the hip abscissa to the stride. k_s and k_h could take values from 0 to 1 during normal gaits.
Let H denote the height of hip joint at impact, and. H can be determined by the following equation:

$$H = L\sqrt{\cos^2(\alpha / 2) - k_s^2 / 4} \tag{42}$$

where α is the angle of the knee joint when $k_h = 0.5$.
To be compatible with the ground condition, it is necessary to specify several middle postures to describe the swing foot over rough terrain or in environments with obstacles. For simplicity, we select one middle posture q^m where $\theta_m = (\theta^+ + \theta^-)/2$. The height of the hip in the middle posture can be determined by $H_m = (A+B)/2$. We utilize the Cartesian coordinates of swing foot (x_f^m, y_f^m), to parameterize q^m. The robot can negotiate different obstacles on the ground by varying (x_f^m, y_f^m).
Since both impact postures and middle postures are determined, the configuration at all breakpoints can be written as

$$h_d(\theta) = \begin{cases} H_0 q^+(k_s, k_h), & \theta = \theta^+ \\ H_0 q^m(x_f^m, y_f^m), & \theta = \theta^m \\ H_0 q^-(k_s, k_h), & \theta = \theta^- \end{cases} \tag{43}$$

2. determine the derivative constraints $dh_d(\theta) / d\theta$ at impact postures;

Since we assume that the robot maintain a constant erect posture during the whole walking cycle, the following two conditions must be satisfied:

$$\dot{q}_5^- = 0 \tag{44}$$

$$\dot{q}_5^+ = [0_{1\times4} \quad 1]\Delta_{\dot{q}}\dot{q}^- = 0 \tag{45}$$

Observing from human walking, we find that human beings appear to hold his support knee joint and relative angle between two thighs intendedly just before impact, so we get the following equation:

$$\dot{q}_3^- = 0 \tag{46}$$

$$\dot{q}_2^- - \dot{q}_1^- = 0 \tag{47}$$

Let σ^- denote the angular momentum just before impact. According to (33), one can obtain

$$\sigma_- = -D_5(q^-)\dot{q}^- \tag{48}$$

According to equation (44) to (48), the generalized velocities $\dot{q}^-$ can be expressed as

$$\dot{q}^- = \Pi(q^-)\sigma^- \tag{49}$$

where $\Pi(q^-) = \begin{bmatrix} 0 \quad 0 \quad 0 \quad 0 \quad 1 \\ [0_{1\times4} \quad 1]\Delta_{\dot{q}} \\ 0 \quad 0 \quad 1 \quad 0 \quad 0 \\ -D_5(q^-) \\ -1 \quad 1 \quad 0 \quad 0 \quad 0 \end{bmatrix}^{-1} \cdot \begin{bmatrix} 0 \\ 0 \\ 0 \\ 1 \\ 0 \end{bmatrix}.$

Considering the equation (36), the generalized velocities $\dot{q}^+$ can also be uniquely determined. Considering (38), (39), and (49), the derivative constraints at impact postures can be written as

$$\frac{dh_d}{d\theta} = \begin{cases} H_0\Pi(q^-)/(c\Pi(q^-)), & \theta = \theta^+ \\ H_0\Delta_{\dot{q}}\Pi(q^-)/(c\Delta_{\dot{q}}\Pi(q^-)), & \theta = \theta^- \end{cases} \tag{50}$$

3. obtain the continuous trajectory $h_d(\theta)$ by interpolation;

To satisfy constraint (43), (50), and the continuity conditions of the first derivative and the second derivative at all breakpoints, $h_d(\theta)$ are characterized by two third-order polynomial expressions:

$$h_d(\theta) = \begin{cases} \sum_{i=0}^{3} M_i \cdot (\theta - \theta^+)^i, & \theta \in [\theta^+, \theta^m] \\ \sum_{i=0}^{3} N_i \cdot (\theta - \theta^m)^i, & \theta \in [\theta^m, \theta^-] \end{cases} \tag{51}$$

Thereby, we can obtain M_i and N_i by third-order spline interpolation. In this way, $h_d(\theta)$ is twice differentiable during the whole single support phase. When (x_f^m, y_f^m) and α are specified, the walking pattern can be determined by the two non-dimensional parameters k_s and k_h uniquely.

4. determine the parameters with a small section-map factor.
Considering (33), (36), (48), and (49), the section-map factor can be calculated as

$$\delta = \frac{\sigma^+}{\sigma^-} = \frac{-D_5(q^+)\dot{q}^+}{\sigma^-} = \frac{-D_5(q^+)\Delta_{\dot{q}}\dot{q}^-}{\sigma^-} = -D_5(q^+)\Delta_{\dot{q}}(q^-)\Pi(q^-) \tag{52}$$

According to (52), the section-map factor only depends on q at impact. Since there are only two parameters k_s and k_h, we can easily obtain a small section-map factor by exhaustive search computation (Fu et al., 2006).

5.3 Stable walking transition and its stability analysis

According to section-map stability criterion, the robot can achieve stable walking provided that all section-map factors is less than one. Fig. 14 shows the property of angular momentum during one-step transition.

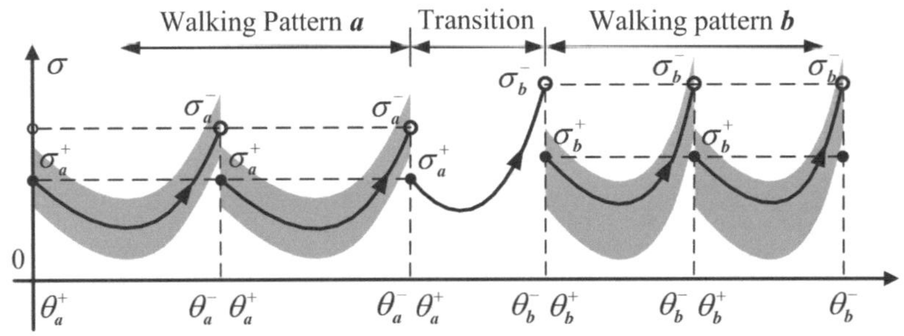

Fig. 14. Property of angular momentum during one-step transition

Moreover, each walking pattern has a domain of stable attraction, and we assume the domains before and after transition are respectively

$$a_{\min} < \sigma_a^- < a_{\max} \tag{53}$$

$$b_{\min} < \sigma_b^- < b_{\max} \tag{54}$$

Define one-step transition map $P_{a\to b} : \sigma_a^- \to \sigma_b^-$

$$P_{a\to b}(\sigma_a^-) = \sqrt{(\delta_a \cdot \sigma_a^-)^2 + 2\int_{\theta_a^+}^{\theta_b^-} I(\theta)J(\theta)d\theta} \tag{55}$$

To realize a stable one-step transition, the following two conditions must be satisfied:
The domain of attraction of walking pattern a can be steered into the domain of attraction of walking pattern b under transition map (Westervelt, 2003b), that is,

$$\{P_{a\to b}(\sigma_a^-) \mid a_{\min} < \sigma_a^- < a_{\max}\} \cap \{\sigma_b^- \mid b_{\min} < \sigma_b^- < b_{\max}\} \neq \varnothing \tag{56}$$

The walking with perturbation should be in the intersection set of domains:

$$\sigma_a^- \in \{\sigma_a^- \mid a_{\min} < \sigma_a^- < a_{\max}\} \cap \{P_{a\to b}^{-1}(\sigma_b^-) \mid b_{\min} < \sigma_b^- < b_{\max}\} \quad (57)$$

Since one-step transition map (55) is a monotonic increasing function, as shown in Fig. 15, the two stable transition conditions can be written as

$$\begin{cases} P_{a\to b}(a_{\max}) > b_{\min} \\ P_{a\to b}(a_{\min}) < b_{\max} \end{cases} \quad (58)$$

$$\min\{a_{\min}, P_{a\to b}^{-1}(b_{\min})\} < \sigma_a^- < \max\{a_{\max}, P_{a\to b}^{-1}(b_{\max})\} \quad (59)$$

Fig. 16 shows the property of angular momentum during multi-step transition, and we assume the domains before and after transition are respectively

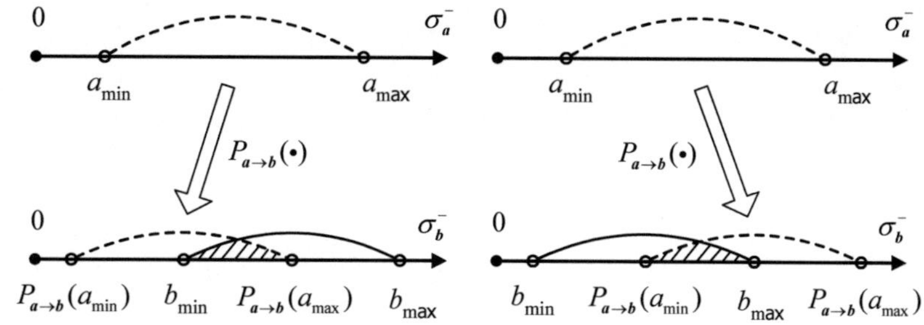

Fig. 15. Phase portraits during one-step transition

$$a_{\min} < \sigma_a^- < a_{\max} \quad (60)$$

$$d_{\min} < \sigma_d^- < d_{\max} \quad (61)$$

Define multi-step transition map $P_{a\to d}: \sigma_a^- \to \sigma_d^-$, to realize a stable multi-step transition, the following two conditions must be satisfied:

$$\{P_{a\to d}(\sigma_a^-) \mid a_{\min} < \sigma_a^- < a_{\max}\} \cap \{\sigma_d^- \mid d_{\min} < \sigma_d^- < d_{\max}\} \neq \varnothing \quad (62)$$

$$\sigma_a^- \in \{\sigma_a^- \mid a_{\min} < \sigma_a^- < a_{\max}\} \cap \{P_{a\to d}^{-1}(\sigma_d^-) \mid d_{\min} < \sigma_d^- < d_{\max}\} \quad (63)$$

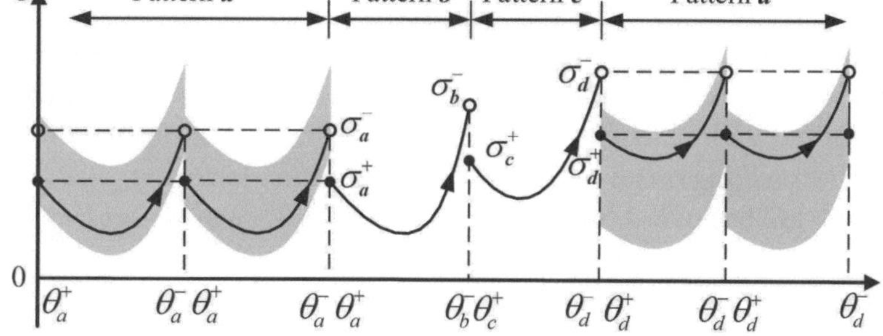

Fig. 16. Property of angular momentum during multi-step transition

6. Walking eperiments

6.1 Stable walking experiment

This section provides several experimental results toward checking the section-map stability criterion.

In the first experiment, THR-I was controlled to walk on a flat floor with a section-map factor $\delta = 0.89$. The experiment lasted more than 120s and THR-I took approximately 600 steps, which indicates the walking period is 0.2s per step. Fig. 17 gives video frames of THR-I taking four steps for a typical walking motion. Fig. 18 is the real joint angles versus time during walking. Fig. 19 are the section-map factors calculated from encoders during walking.

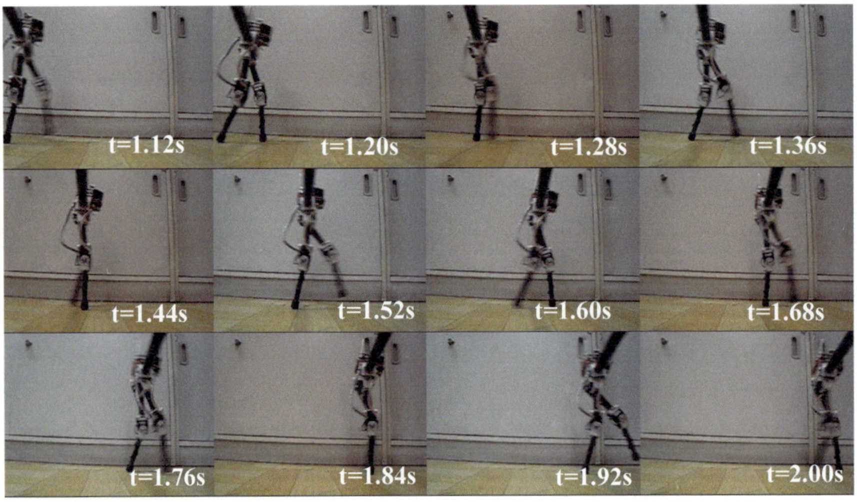

Fig. 17. Video frames of THR-I taking four consecutive steps with $\delta = 0.89$. The robot is walking at 0.20 s per step

6.2 Unstable walking experiment

The second experiment demonstrated the walking result with a section-map factor $\delta = 1.20$, which indicates the corresponding biped walking is unstable stable. Fig. 20 shows the desired and real values of holonomic constraints, from which we can observe that the walking pattern can not be imposed on the robot. Fig. 21 is the corresponding snapshot of the walking experiment, from which one can see the robot falls forward finally.

For periodic forward walking, the minimum of the angular momentum around the pivot point during a walking cycle should be positive; otherwise the robot has no enough energy to achieve a step and will fall backward. Fig. 22 is the desired and real values of holonomic constraints during walking on level ground with the section factor $\sigma_{\min} < 0$, which indicates that the robot will fall backward finally. Fig. 23 is the snapshot of walking experiment.

6.3 Stable walking transition experiment

The fourth experiment demonstrated the walking transition. Fig. 24 is simulation results of phase portraits during one-step transition with a 5% error from limit cycles before transition. The section-map factor before transition is 0.86, and after transition is 0.89; therefore, the walking is stable. Fig. 25 gives video frames of THR-I walking from 0.2s/step to 0.3s/step.

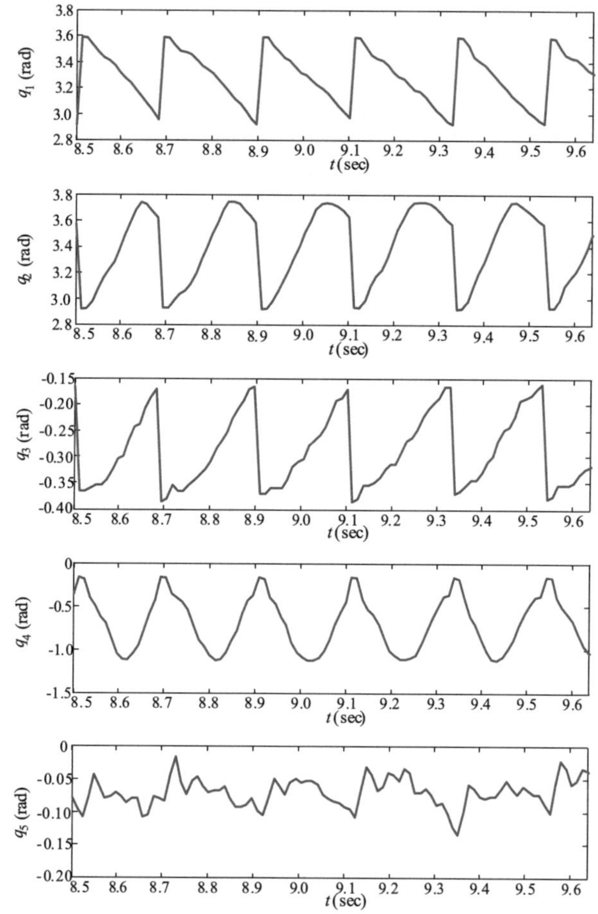

Fig. 18. Real joint angles of THR-I with $\delta = 0.89$

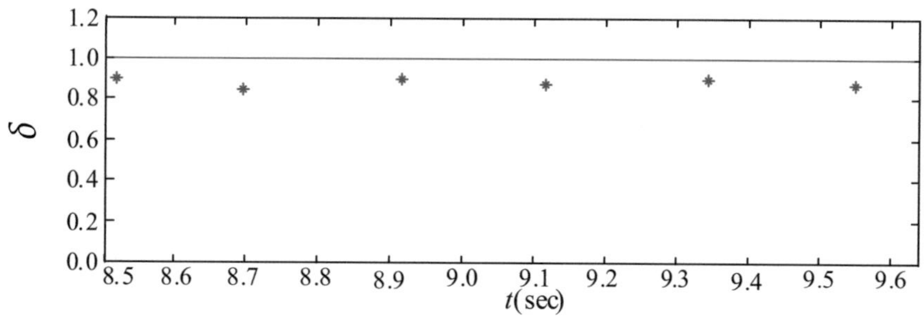

Fig. 19. Section-map factor estimated by rotary encoders during walking. All section-map factors are.less than one

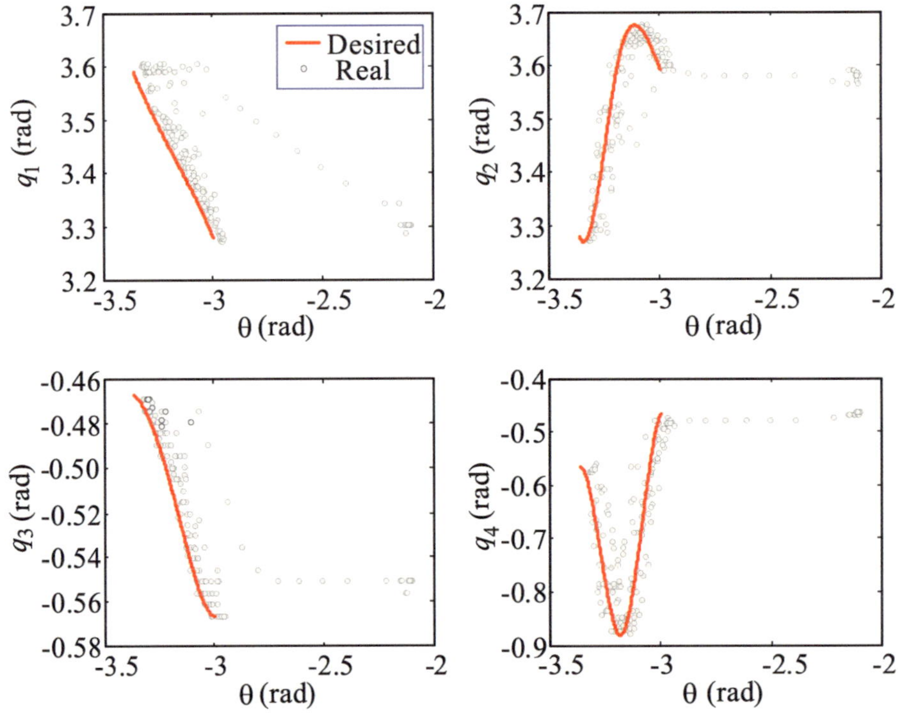

Fig. 20. Desired and real values of holonomic constraints with $\delta = 1.2$

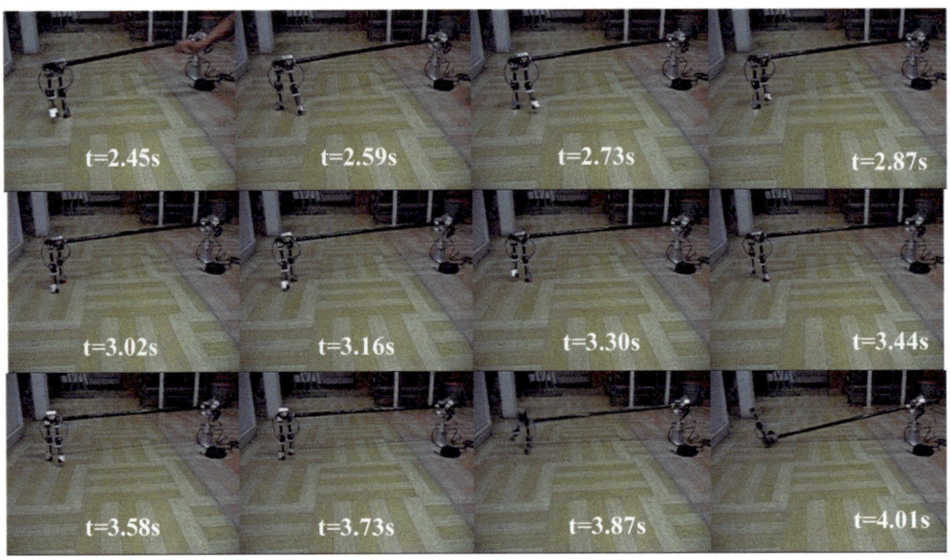

Fig. 21. Video frames of biped walking experiment with $\delta = 1.2$

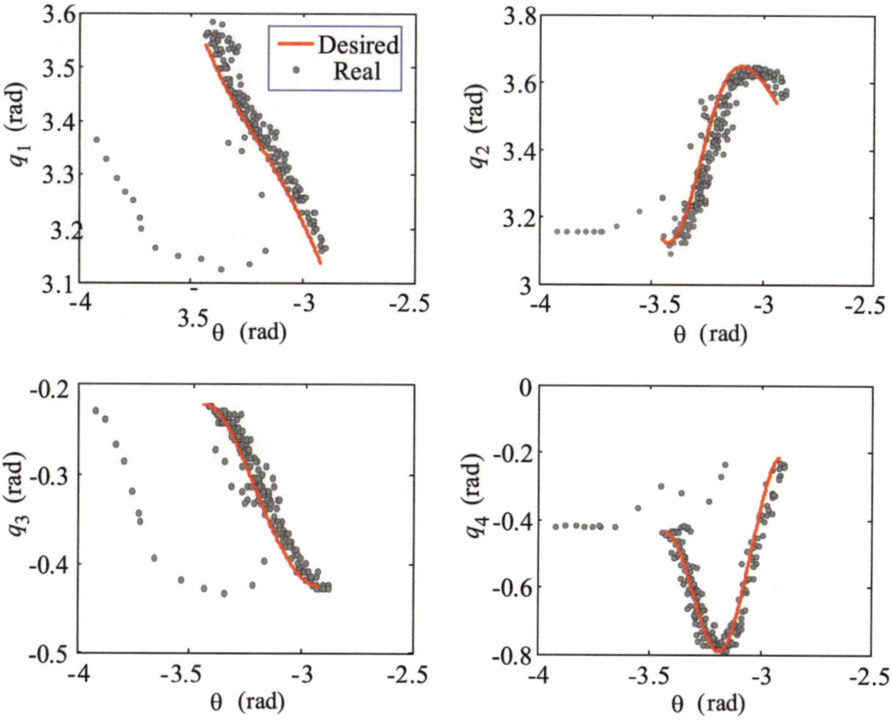

Fig. 22. Desired and real values of holonomic constraints with $\sigma_{\min} < 0$

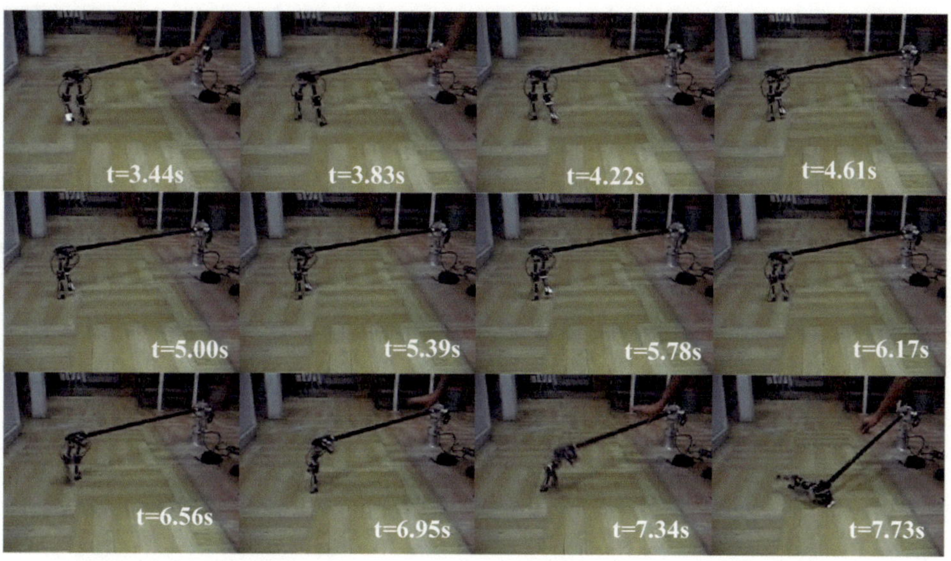

Fig. 23. Video frames of biped walking experiment with $\sigma_{\min} < 0$

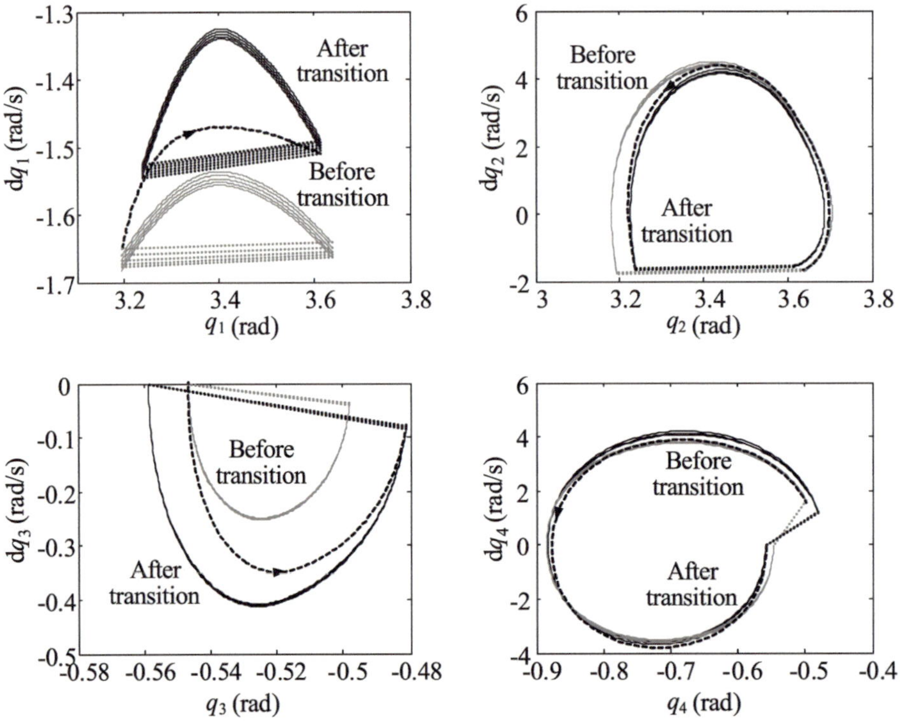

Fig. 24. Simulation results of phase portraits during one-step transition

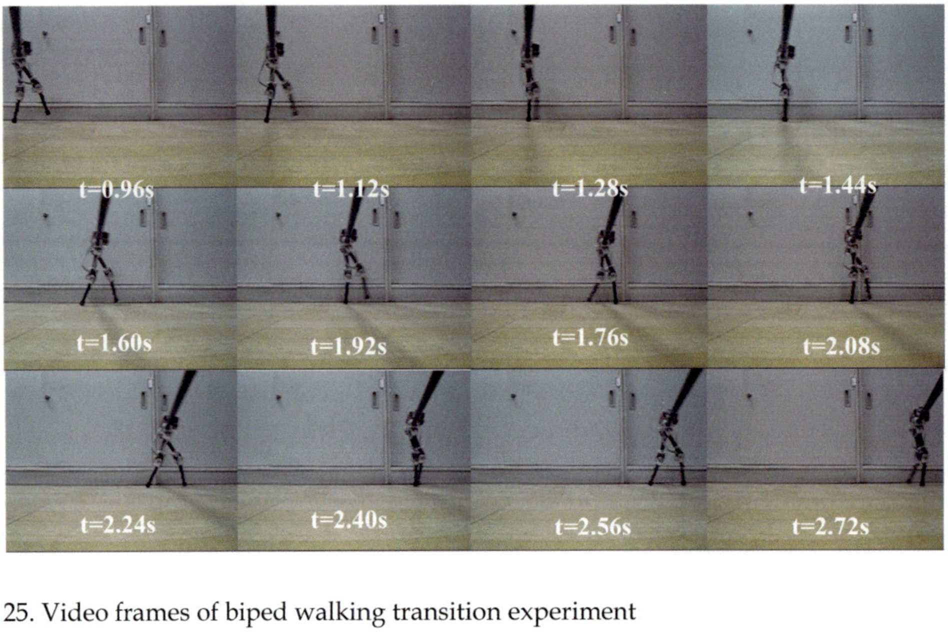

Fig. 25. Video frames of biped walking transition experiment

7. Conclusion

When publications are referred in the text, enclose the author's name and the date of publication within the brackets. For one author, use author's surname and the date (Arkin, 2004). For two authors, give both names & the year (Mataric & Brooks, 1999). For three or more authors, use the first author, plus „et al.", and the date (Siegwart et al., 2006). If giving a list of reference, separate them using semicolons.
In this study, we focused on a coherent stability criterion and its application methods for biped walking. The main results of this chapter are summarized as follows:

1. An overall mathematical modelling method for biped walking is proposed based on dimension-variant hybrid automata. This method expresses the overall biped walking model as an 8-tuple and can reflect all kinds of continuous and discrete properties of biped walking, which makes it possible to study stability and design control strategy for biped locomotion from a global point of view.
2. A rigorous mathematical definition of biped walking stability is presented by combining the character of biped locomotion with the notion of classical stability from the view of hybrid trajectory. It is pointed out that the model in the task space is a length-varying and inertia-varying inverted pendulum, and the analytic form of the inverted pendulum model is derived. This makes it possible to study stability of biped walking in a low-dimension task space.
3. It is pointed out that, under some assumption, stability of the hybrid trajectory is equivalent to that of the section sequence at switch section in the task space of biped walking. Based on this result, section-map stability criterion is presented. This criterion is applicable not only to dynamic walking which ZMP criterion can not solve, but also to non-periodic walking which Poincare return map criterion can not solve.
4. By the proposed criterion, a synthesizing method for walking patterns based on section-map factor is presented. The effectiveness of this method is confirmed by a biped robot THR-I, which can walk with a relative speed of 2 leg lengths per second. This robot is one of the few biped machines which can walk so fast and stable (Geng et al., 2006).

Since the sagittal plane dynamics of biped walking are almost decoupled from those in the frontal plane (Furusho & Sano, 1990; Kuo, 1999), this chapter is only concentrated on stability issue in sagittal plane. The future work is to extend this method to the frontal plane to produce stable, dynamic three-dimensional walking.

8. References

Canudas-de-Wit, C. (2004). On the concept of virtual constraints as a tool for walking robot control and balancing. *Annual Reviews in Control,* Vol. 28, No. 2, 157-166

Chevallereau, C.; Abba, G.; Aoustin, Y. ; Plestan, F. ; Westervelt, E. R.; Canudas-de-wit, C.; & Grizzle, J. W. (2003). RABBIT: A testbed for advanced control theory. *IEEE Contr. Syst. Mag.,* Vol. 23, No. 5, 57-79

Chevallereau, C.; Formal'sky, A. & Djoudi, D. (2004). Tracking of a joint path for the walking of an underactuated biped. *Robotica,* Vol. 22, No. 1, 15-28

Fu, C.; Shuai, M.; Huang, Y.; Wang, J. & Chen, K. (2006). Parametric walking patterns and optimum atlases for underactuated biped robots, *Proc. IEEE Int. Conf. Intell. Robots Syst.,* 342-347

Furusho, J. & Sano, A. (1990). Sensor-based control of a nine-link biped. *Int. J. Robot. Res.*, Vol. 9, No. 2, 83-98

Geng, T.; Porr, B. & Worgotter, F. (2006). Fast biped walking with a sensor-driven neuronal controller and real-time online learning. *Int. J. Robot. Res.*, Vol. 25, No. 3, 243-259

Goswami, A.; Espiau, B. & Keramane, A. (1996). Limit cycles and their stability in a passive bipedal gait," in Proc. IEEE Int. Conf. Robot. Autom., 246-251

Goswami, A. & Kallem, V. (2004). Rate of change angular momentum and balance maintenance of biped robots, *Proc. IEEE Int. Conf. Robot. Autom.*, 3785-3790

Grizzle, J. W.; Abba, G. & Plestan, F. (2001). Asymptotically stable walking for biped robots: Analysis via systems with impulse effects. *IEEE Trans. Autom. Contr.*, Vol. 46, No. 1, 51-64

Guckenheimer, J. & Holmes, P. (1985). *Nonlinear Oscillations, Dynamical Systems, and Bifurcations of Vector Fields*. New York: Springer-Verlag

Hirai, K.; Hirose, M.; Haikawa, Y. & Takenaka, T. (1998). The development of Honda humanoid robot, *Proc. IEEE Int. Conf. Robot. Autom.*, 1321-1326

Huang, Q. ; Yokoi, K.; Kajita, S.; Kaneko, K.; Arai, H.; Koyachi, N. & Tanie, K. (2001). Planning walking patterns for a biped robot. *IEEE Trans. Robot. Autom.*, Vol. 17, No.3, 280-289

Huang, Q. & Nakamura, Y. (2005). Sensory reflex control for humanoid walking. *IEEE Trans. Robot. Autom.*, Vol. 21, No. 5, 977-984

Hurmuzlu, Y. &Moskowitz, G. Bipedal locomotion stabilized by impact and switching. (1993). *Dyn. Stab. Syst.*, Vol. 60, No. 2, 331-344

Kajita, S.; Matsumoto, O. & Saigo, M. (2001). Real-time 3-D walking pattern generation for a biped robot with telescopic legs, Proc. *IEEE Int. Conf. Robot. Autom.*, 2299-2306.

Kuo, A. D. (1999). Stabilization of lateral motion in passive dynamic walking. *Int. J. Robot. Res.*, Vol. 18, No. 9, 917-930

Kuo, A. D. (2002). Energetics of actively powered locomotion using the simplest walking model. *Journal of Biomechanical Engineering*, Vol. 124, 113-120

Lim, H.; Kaneshima, & Takanishi, Y. (2002). A. On-line walking pattern generation for biped humanoid robot with trunk, *Proc. IEEE Int. Conf. Robot. Autom.*, 3111-3116

McGeer, T. (1990). Passive walking with knees, *Proc. IEEE Int. Conf. Robot. Autom.*, 1990, 1640-1645

Nishiwaki, K.; Kagami, S.; Kuniyoshi, Y.; Inaba, M. &Inoue, H. (2002). Online generation of humanoid walking motion based on fast generation method of motion pattern that follows desired ZMP, *Proc. IEEE Int. Conf. Intell. Robots Syst.*, 2684-2689

Popovic, M. & Englehart, A. (2004). Angular momentum primitives for human walking: biomechanics and control, *Proc. IEEE Int. Conf. Intell. Robots Syst.*, 1658-1691

Pratt, J. & Tedrake, R. (2005). Velocity based stability margins for fast bipedal walking, *Proc. 1st Ruperto Carola Symp. Fast Motions in Biom. and Robot.*, 1-27

Raibert, M. H. (1986). *Legged robots that balance*. MIT Press, Cambridge, MA

Takanishi, A.; Ishida, M.; Yamazaki, Y. &Kato, I. (1985). The realization of dynamic walking robot WL-10RD, *Proc. Int. Conf. Adv. Robot.*, 459-466

Vukobratovic, M. & Juricic, D. (1969). Contribution to the synthesis of biped gait. *IEEE Trans, Biomed. Eng.*, Vol. BME-16, No. 1, 1-6

Westervelt, E. R.; Grizzle, J. W. & Koditschek, D. (2003a). Hybrid zero dynamics of planar biped walkers. *IEEE Trans. Autom. Contr.*, Vol. 48, No. 1, 42-56

Westervelt, E. R.; Grizzle, J. W. & Canudas-de-Wit, C. (2003b). Switching and PI control of walking motions of planar biped walkers. *IEEE Trans. Autom. Contr.*, Vol. 48, No. 2, 308-312

Part 2

Dynamics, and Gait Generation

9

Dynamical Analysis of a Biped Locomotion CPG Modelled by Means of Oscillators

Armando Carlos de Pina Filho and Max Suell Dutra
Federal University of Rio de Janeiro – UFRJ
Brazil

1. Introduction

The study of mechanisms like mechanical members intends not only to build autonomous robots, but also to help in the rehabilitation of human being. The study of locomotion to take part in this context, and has been intensively studied since the second half of 20th century. An ample vision of the state of the technique up to 1990 can be found in works as Raibert (1986) and Vukobratovic et al. (1990).

Year after year, from technological advances, based on theoretical and experimental researches, the man tries to copy or to imitate some systems of the human body. It is the case, for example, of the central pattern generator (CPG), responsible for the production of rhythmic movements. Modelling of this CPG can be made by means of coupled oscillators, and this system generates patterns similar to human CPG, becoming possible the human gait simulation. There are some significant works about the locomotion of vertebrates controlled by central pattern generators: Grillner (1985), and Pearson (1993).

From a model of two-dimensional locomotor, oscillators with integer relation of frequency can be used for simulating the behaviour of the hip angle and of the knees angles. Each oscillator has its own parameters and the link to the other oscillators is made through coupling terms. We intend to evaluate a system with coupled van der Pol oscillators. Some previous works about CPGs using nonlinear oscillators, applied in the human gait simulation, can be seen in Bay & Hemami (1987), Zielinska (1996), Dutra et al. (2003), Pina Filho (2005), and Pina Filho (2008).

The objective of this work is to analyze the dynamics of this coupled oscillators system by means of bifurcation diagrams and Poincaré maps. From the analysis and graphs generated in MATLAB, it was possible to evaluate some characteristics of the system, such as: sensitivity to the initial conditions, presence of strange attractors and other phenomena of the chaos.

2. Central Pattern Generator - CPG

The first indications that the spinal marrow could contain the basic nervous system necessary to generate locomotion date back to the early 20th century. According to Mackay-Lyons (2002), the existence of nets of nervous cells that produce specific rhythmic movements for a great number of vertebrates is something unquestionable. Nervous nets in the spinal marrow are capable of producing rhythmic movements, such as: swimming, jumping, and walking, when isolated from the brain and sensorial entrances. These

specialized nervous systems are known as nervous oscillators or central pattern generators (CPGs). The human locomotion is controlled, in part, by a central pattern generator, which is evidenced in the works as Calancie et al. (1994), and Dimitrijevic et al. (1998).
The choice of an appropriate pattern of locomotion depends on the combination of a central programming and sensorial data, as well as of the instruction for one determined motor condition. This information determines the way of organisation of the muscular synergy, which is planned for adequate multiple conditions of posture and gait (Horak & Nashner, 1986).
Figure 1 presents a scheme of the control system of the human locomotion, controlled by the central nervous system, which the central pattern generator supplies a series of pattern curves for each part of the locomotor. This information is passed to the muscles by means of a network of motoneurons, and the conjoined muscular activity performs the locomotion. Sensorial information about the conditions of the environment or some disturbance are supplied as feedback of the system, providing a fast action proceeding from the central pattern generator, which adapts the gait to the new situation.

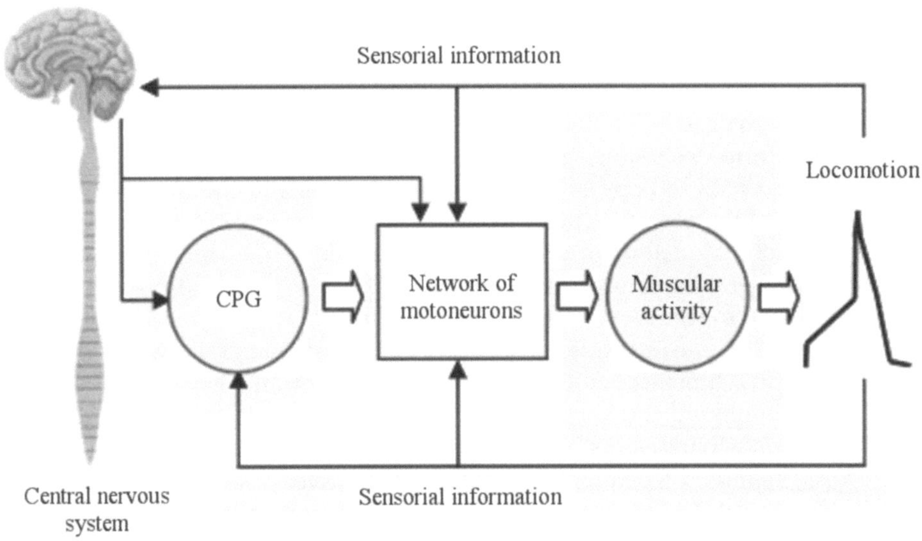

Fig. 1. Control system of the human locomotion.

Despite the people not walk in completely identical way, some characteristics in the gait can be considered universal, and these similar points serve as base for description of patterns of the kinematics, dynamics and muscular activity in the locomotion.
In the study to be presented here, the greater interest is related to the patterns of the hip and knee angles. From the knowledge of these patterns of behaviour, the simulation of the central pattern generator using the system of coupled oscillators becomes possible.
Considering the anatomical planes of movement, we need to know the behaviour of hip and knee in sagittal plane. Figure 2 presents the movements of flexion and extension of the articulation of hip and knee in sagittal plane.
According Pina Filho et al. (2006), figures 3 and 4 present the graphs of angular displacement and phase space of the hip in the sagittal plane, related to the movements of flexion and extension, while figures 5 and 6 present the graphs of angular displacement and phase space of the knee, related to the movements of flexion and extension.

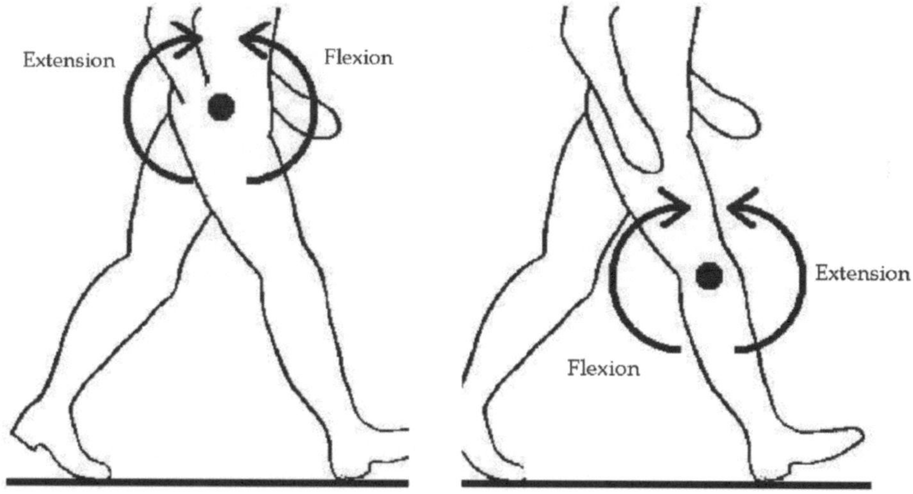

Fig. 2. Movements of the hip and knee: flexion and extension.

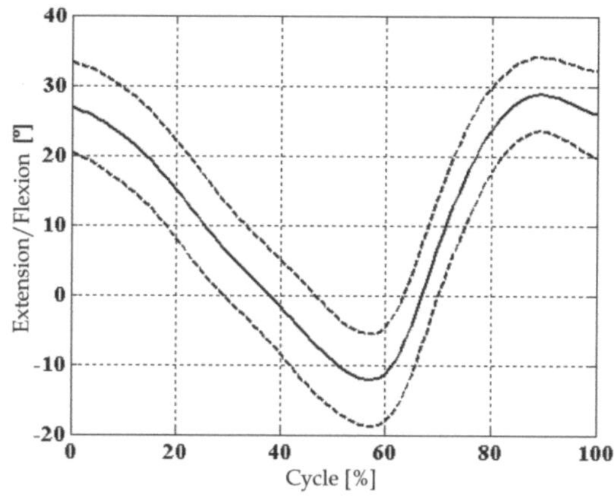

Fig. 3. Angular displacement of the hip in the sagittal plane (mean ± deviation).

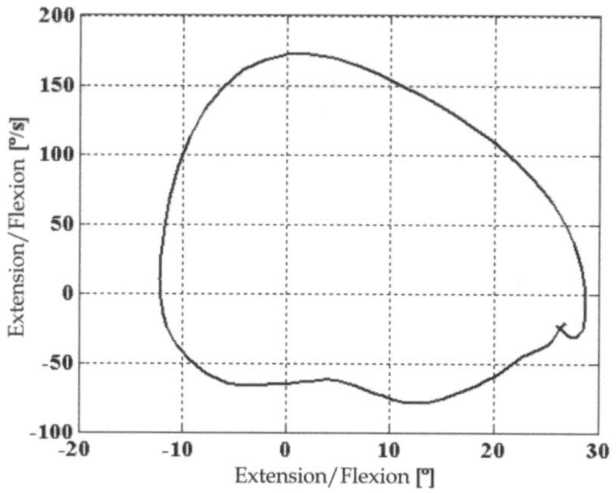

Fig. 4. Phase space of the hip in the sagittal plane.

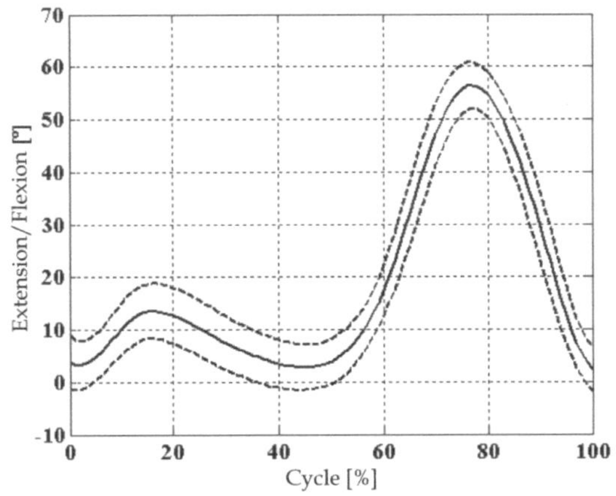

Fig. 5. Angular displacement of the knee in the sagittal plane (mean ± deviation).

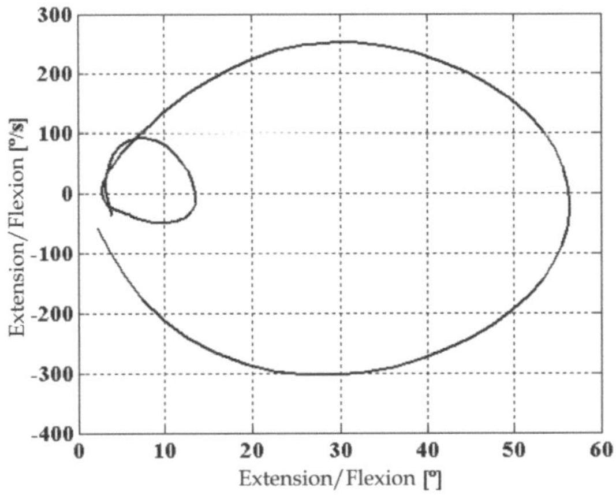

Fig. 6. Phase space of the knee in the sagittal plane.

3. Biped locomotor model

Before a model of CPG can be applied to a physical system, the desired characteristics of the system must be completely determined, such as: the movement of the leg or another rhythmic behaviour of the locomotor. Some works with description of the rhythmic movement of animals include Eberhart (1976), Winter (1983) and McMahon (1984), this last one presenting an ample study about the particularities of the human locomotion. To specify the model to be studied is important to know some concepts related to the bipedal gait, such as the determinants of gait.

The modelling of natural biped locomotion is made more feasible by reducing the number of degrees of freedom, since there are more than 200 degrees of freedom involved in legged locomotion. According to Saunders et al. (1953), the most important determinants of gait are: 1) the compass gait that is performed with stiff legs like an inverted pendulum. The pathway of the centre of gravity is a series of arcs; 2) pelvic rotation about a vertical axis. The influence of this determinant flattens the arcs of the pathway of the centre of gravity; 3)

pelvic tilt, the effects on the non-weight-bearing side further flatten the arc of translation of the centre of gravity; 4) knee flexion of the stance leg. The effects of this determinant combined with pelvic rotation and pelvic tilt achieve minimal vertical displacement of the centre of gravity; 5) plantar flexion of the stance ankle. The effects of the arcs of foot and knee rotation smooth out the abrupt inflexions at the intersection of the arcs of translation of the centre of gravity; 6) lateral displacement of the pelvis.
Figure 7 presents a 3D model with 15 degrees of freedom, and the six determinants of gait. The kinematical analysis, using the characteristic pair of joints method is presented in Saunders et al. (1953).

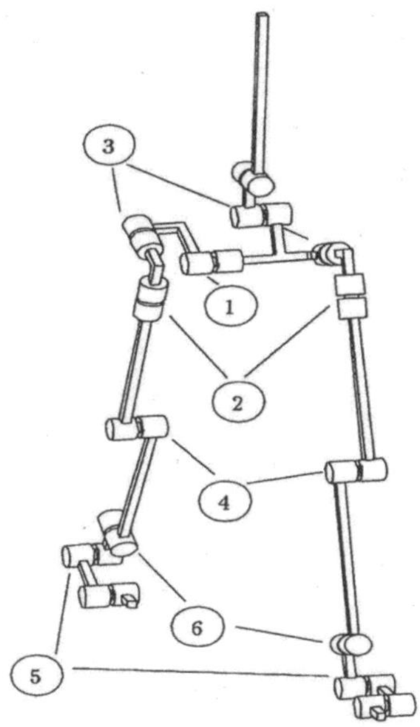

Fig. 7. Three-dimensional model with the six determinants of gait.

In order to simplify the investigation, a 2D model that performs motions parallel only to the sagittal plane will be considered. This model, showed in Fig. 8, characterises the three most important determinants of gait, determinants 1 (the compass gait), 4 (knee flexion of the stance leg), and 5 (plantar flexion of the stance ankle). The model does not take into account the motion of the joints necessary for the lateral displacement of the pelvis, for the pelvic rotation, and for the pelvic tilt.
Figure 8 presents too the angles and lengths of the model, where: ℓ_s is the length of foot responsible for the support (toes), ℓ_p is the length of foot that raises up the ground (sole), ℓ_t is the length of tibia, and ℓ_f is the length of femur. The angle of the hip θ_4 and the angles of the knees θ_3 and θ_5 will be determined by a coupled oscillators system, representing the CPG, while the other angles are calculated by the kinematical analysis of the mechanism. In this work we will not present details of this analysis, which can be seen in Pina Filho (2005).
This model must be capable to show clearly the phenomena occurred in the course of the motion, and works with the hypothesis of the rigid body, where the natural structural movements of the skin and muscles, as well as bone deformities, are disregarded. The

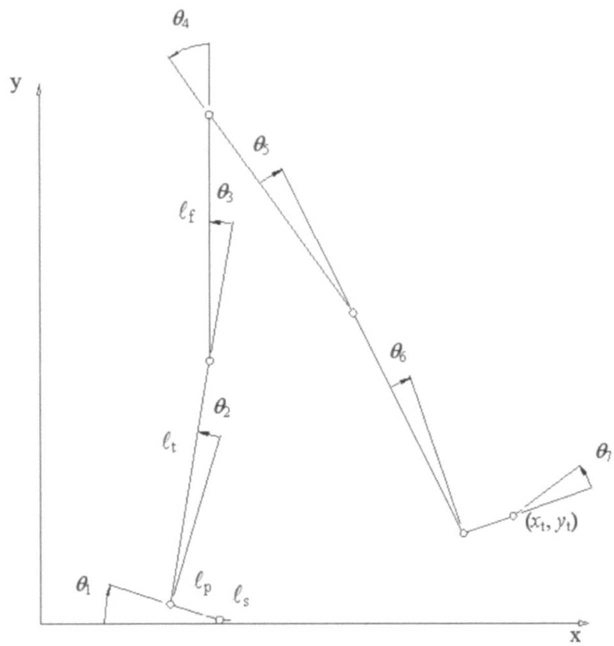

Fig. 8. Two-dimensional model with the determinants of gait, angles and lengths.

locomotion cycle can be divided in two intervals: double support phase, with the two feet on the ground, and single support phase, with only one foot touching the ground, and one of the legs performs a balance movement (the extremity of the support leg is assumed as not sliding).

From this model, we can now to study the CPG, simulated by means of nonlinear oscillators, which can be used in control systems of locomotion, providing the approach trajectories of the legs. The CPG is composed of a set of oscillators, where each oscillator, with own amplitude, frequency and parameters, generates angular signals of reference for the movement of the legs, as we will see in the next section.

4. Modelling of the oscillators system

Coupled oscillators systems have been extensively used in studies of physiological and biochemical modelling. Since the years of 1960, many researchers have studied the case of coupling between two oscillators, because this study is the basis to understand the phenomenon in a great number of coupled oscillators. One of the types of oscillators that can be used in coupled systems is the auto-excited ones, which have a stable limit cycle without external forces. The van der Pol oscillator is an example of this type of oscillator, and it will be used in this work. Then, considering a system of n coupled van der Pol oscillators, from van der Pol equation:

$$\ddot{x} - \varepsilon\left(1 - p\left(x - x_0\right)^2\right)\dot{x} + \Omega^2\left(x - x_0\right) = 0 \qquad \varepsilon, p \geq 0 \tag{1}$$

where ε, p and Ω correspond to the parameters of the oscillator, and adding coupling terms that relate the oscillators velocities, we have:

$$\ddot{\theta}_i - \varepsilon_i\left[1 - p_i(\theta_i - \theta_{io})^2\right]\dot{\theta}_i + \Omega_i^2(\theta_i - \theta_{io}) - \sum_{j=1}^{n} c_{i,j}\left(\dot{\theta}_i - \dot{\theta}_j\right) = 0 \qquad i = 1, 2, \dots, n \tag{2}$$

which represents coupling between oscillators with the same frequency, where θ corresponds to the system degrees of freedom. In the case of coupling between oscillators with integer relation of frequency, the equation would be:

$$\ddot{\theta}_h - \varepsilon_h\left[1 - p_h(\theta_h - \theta_{ho})^2\right]\dot{\theta}_h + \Omega_h^2(\theta_h - \theta_{ho}) - \sum_{i=1}^{m} c_{h,i}\left[\dot{\theta}_i(\theta_i - \theta_{io})\right] - \sum_{k=1}^{n} c_{h,k}\left(\dot{\theta}_h - \dot{\theta}_k\right) = 0 \tag{3}$$

where $c_{h,i}[\dot{\theta}_i(\theta_i - \theta_{io})]$ is responsible for the coupling between oscillators with different frequencies, while $c_{h,k}\left(\dot{\theta}_h - \dot{\theta}_k\right)$, also seen in Eq. (2), effects the coupling between oscillators with the same frequency. Both terms were defined by Dutra (1995).

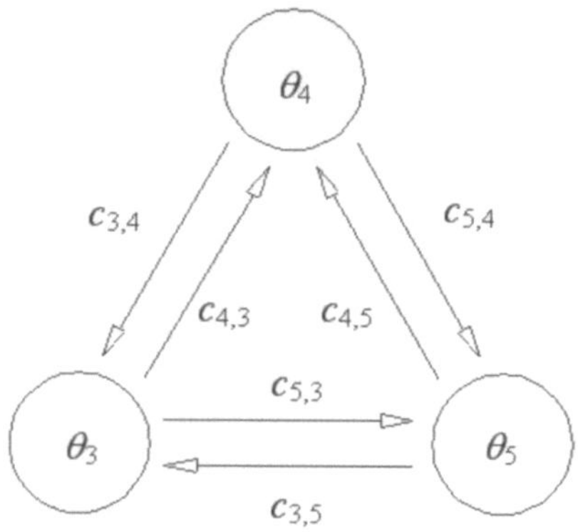

Fig. 9. Structure of coupling oscillators.

Experimental studies of human locomotion (Braune & Fischer, 1987) and Fourier analysis of these data (Dutra, 1995) show that the movements of θ_3, θ_4 and θ_5 (see Fig. 8) can be described very precisely by their fundamental harmonic, whether the biped in single or double support phase.

To generate the angles θ_3, θ_4 and θ_5 as a periodic attractor of a nonlinear net, three coupled van der Pol oscillators were used. These oscillators are mutually coupled by terms that determine the influence of one oscillator on the others (Fig. 9).

Applying Eq. (2) and (3) to the proposed problem, knowing that the frequency of θ_3 and θ_5 (knee angles) is double of θ_4 (hip angle), we have the following equations:

$$\ddot{\theta}_3 - \varepsilon_3[1 - p_3(\theta_3 - \theta_{3o})^2]\,\dot{\theta}_3 + \Omega_3^2(\theta_3 - \theta_{3o}) - c_{3,4}\left[\dot{\theta}_4(\theta_4 - \theta_{4o})\right] - c_{3,5}\left(\dot{\theta}_3 - \dot{\theta}_5\right) = 0 \tag{4}$$

$$\ddot{\theta}_4 - \varepsilon_4[1 - p_4(\theta_4 - \theta_{4o})^2]\,\dot{\theta}_4 + \Omega_4^2(\theta_4 - \theta_{4o}) - c_{4,3}\left[\dot{\theta}_3(\theta_3 - \theta_{3o})\right] - c_{4,5}\left[\dot{\theta}_5(\theta_5 - \theta_{5o})\right] = 0 \tag{5}$$

$$\ddot{\theta}_5 - \varepsilon_5[1 - p_5(\theta_5 - \theta_{5o})^2]\,\dot{\theta}_5 + \Omega_5^2(\theta_5 - \theta_{5o}) - c_{5,4}\left[\dot{\theta}_4(\theta_4 - \theta_{4o})\right] - c_{5,3}\left(\dot{\theta}_5 - \dot{\theta}_3\right) = 0 \tag{6}$$

From Eq. (4)-(6), using the parameters shown in Table 1 together with values supplied by Pina Filho (2005), the graphs were generated in MATLAB as shown in Fig. 10 and 11, which present, respectively, the behaviour of θ_3, θ_4 and θ_5 as a function of time and stable limit cycles of oscillators.

$c_{3,4}$	$c_{4,3}$	$c_{3,5}$	$c_{5,3}$	$c_{4,5}$	$c_{5,4}$	ε_3	ε_4	ε_5
0.001	0.001	0.1	0.1	0.001	0.001	0.01	0.1	0.01

Table 1. Parameters of van der Pol oscillators.

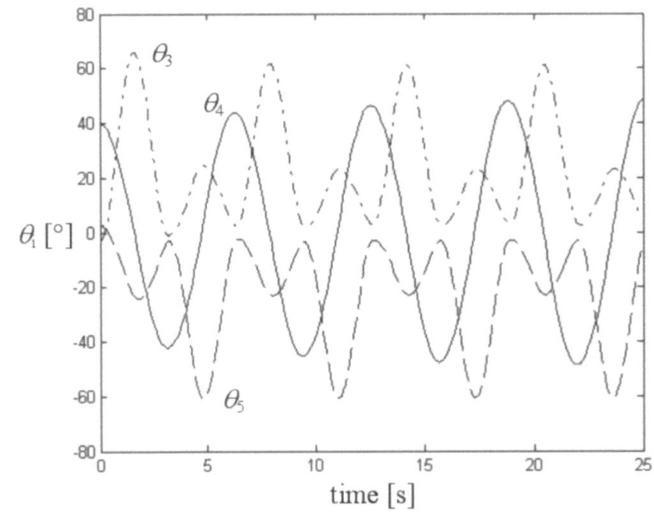

Fig. 10. Angles as a function of time.

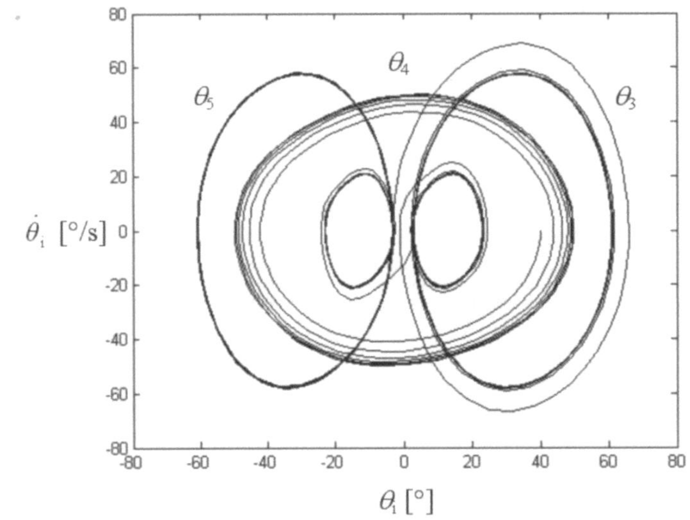

Fig. 11. Trajectories in the phase space.

Comparing Fig. 10 and 11 with the experimental results presented in Section 2 (Fig. 3, 4, 5, 6), it is verified that the coupling system supplies similar results, what confirms the possibility of use of mutually coupled van der Pol oscillators in the modelling of the CPG.

Figure 12 shows, with a stick figure, the gait with a step length of 0.63 m. Figure 13 shows the gait with a step length of 0.38 m. Dimensions used in the model can be seen in Table 2.

Thumb	Foot	Leg (below the knee)	Thigh
0.03 m	0.11 m	0.37 m	0.37 m

Table 2. Model dimensions.

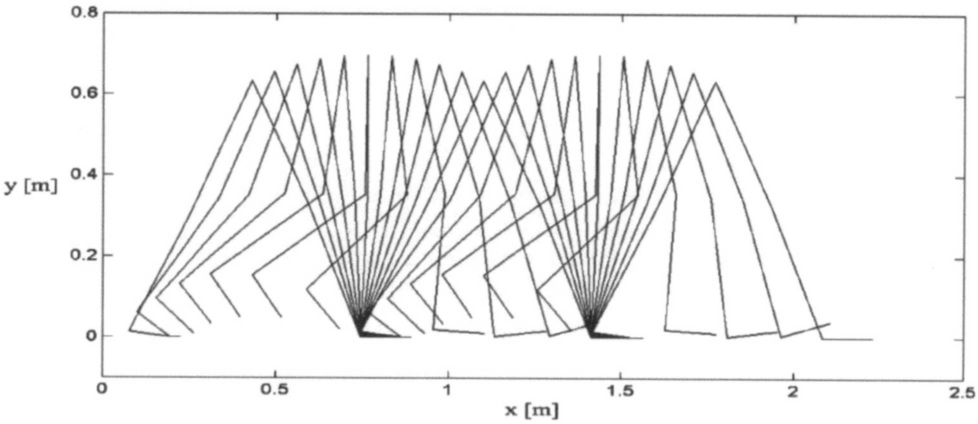

Fig. 12. Stick figure showing the gait with a step length of 0.63 m.

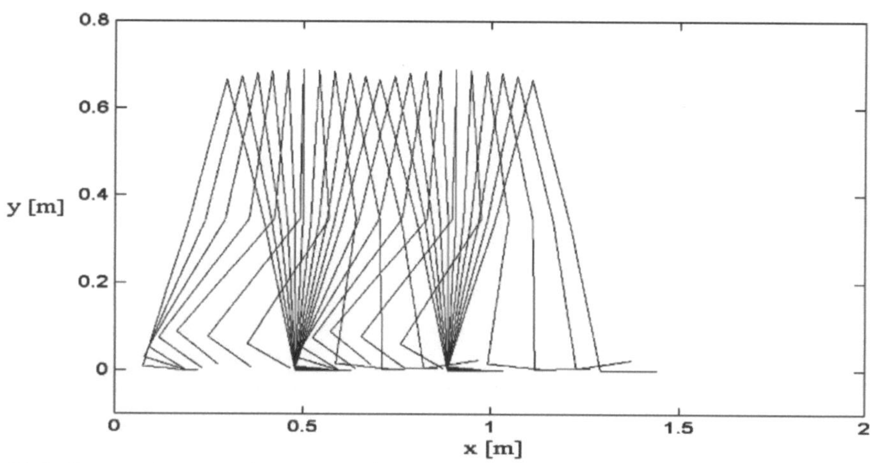

Fig. 13. Stick figure showing the gait with a step length of 0.38 m.

5. Dynamical analysis of the oscillators system

The nonlinear dynamical analysis of the system presented here requires the definition of some usual concepts. Usually, for some values of parameters, the system behaviour is periodic, and for other values the behaviour is chaotic. A periodic system returns to its state after a finite time t. The trajectory in the phase space is represented by a closed curve. The chaotic system presents trajectories of non-closed orbits that are generated by the solution of a deterministic set of ordinary differential equations.

Two conditions must be satisfied to make possible that a system presents chaotic behaviour: the equations of motion must include a nonlinear term; and the system must have at least three independent dynamic variables. The main consequence associated with the chaos is the sensitivity to the initial conditions. In chaotic systems, a small change in the initial conditions results in a drastic change in the system behaviour. More details about the Chaos theory and its characteristics can be found in many works, such as: Strogatz (1994) and Baker & Gollub (1996).

The existence of bifurcation in a system is related with the existence of chaos. In all chaotic system, it is possible to observe the bifurcation phenomenon, however, not all system that presents bifurcation necessarily presents a chaotic response. The influence of some parameter in the system response can be identified by means of bifurcation diagrams, which present the stroboscopic distribution of the system response from slow variation of a parameter (Thompsom & Stewart, 1986). This method was applied here, which implies to simulate different parameter values that we want to analyze, evaluating the response in Poincaré maps.

The Poincaré map consists in the reduction of continuous systems in time (flows) in discrete systems in time (maps). Then, a Poincaré map allows that system dynamics to be represented in a space with lesser dimension than original system, reducing a n-dimensional space for $n-1$ dimensions. The Poincaré map is obtained from the phase space diagram by observing this "stroboscopically", i.e., sample points in the phase space in regular intervals.

Then, considering different values for the parameters ε_3, ε_4 and ε_5, the tests have been performed using MATLAB to generate the bifurcation diagrams and Poincaré maps. In principle, keeping values of $\varepsilon_4 = 0.1$ and $\varepsilon_5 = 0.01$, the value of ε_3 was varied from 0 to 2. Other values of the system have been kept. Figure 14 presents the bifurcation diagram showing the behaviour of knee oscillator θ_3 with variation of parameter ε_3, which represents the damping term related with this oscillator.

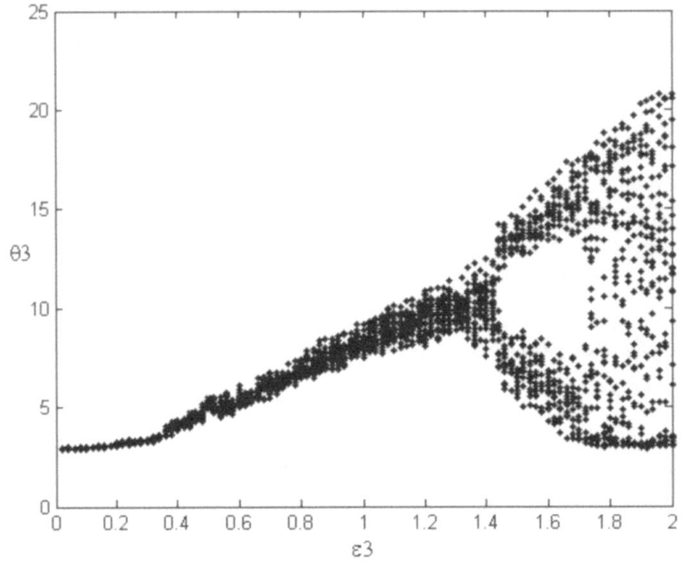

Fig. 14. Bifurcation diagram for θ_3 with variation of ε_3.

The diagram of Fig. 14 does not represent the bifurcation as simple curves, which normally happens in dynamical analysis of a system, but with a cloud of points. Considering the complexity of coupled oscillators system, this fact can be explained by relation between coupling terms or by quasiperiodic response of the system.

According to Santos et al. (2004), a great variation between coupling terms, with one of them approaching to zero, makes the system presents practically a unidirectional coupling, and consequently the response in bifurcation diagram is represented by a cloud of points, characterizing not only the presence of periodic and chaotic orbits, as also pseudo-trajectories. More details about this subject can be seen in Grebogi et al. (2002).

In relation to system behaviour, with small values of damping term, below 0.1, the system presents a periodic response (Fig. 15). With the increase of damping term, the system starts to present a quasiperiodic response and later chaotic response, as presented in Fig. 16 and 17, respectively.

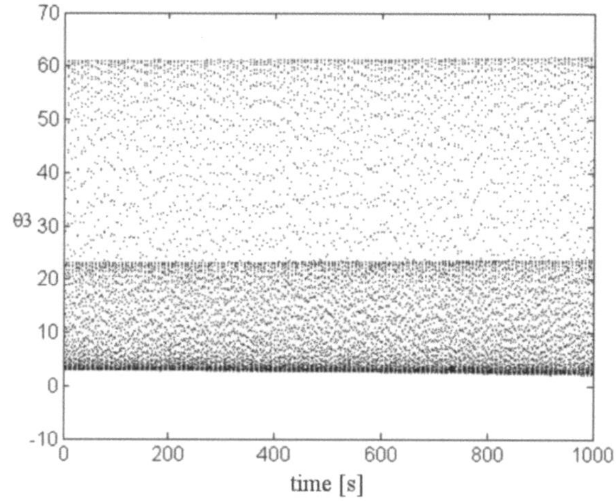

Fig. 15. Periodic response: $\varepsilon_3 = 0.01$.

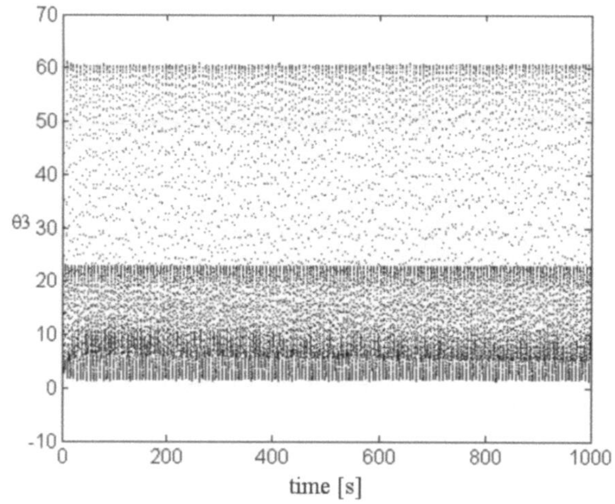

Fig. 16. Quasiperiodic response: $\varepsilon_3 = 1$.

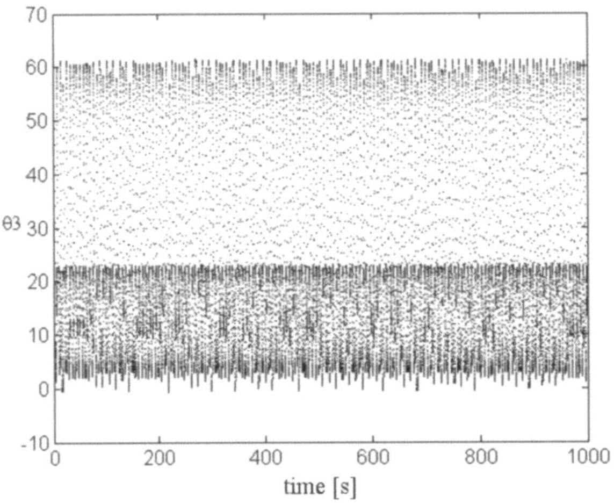

Fig. 17. Chaotic response: $\varepsilon_3 = 2$.

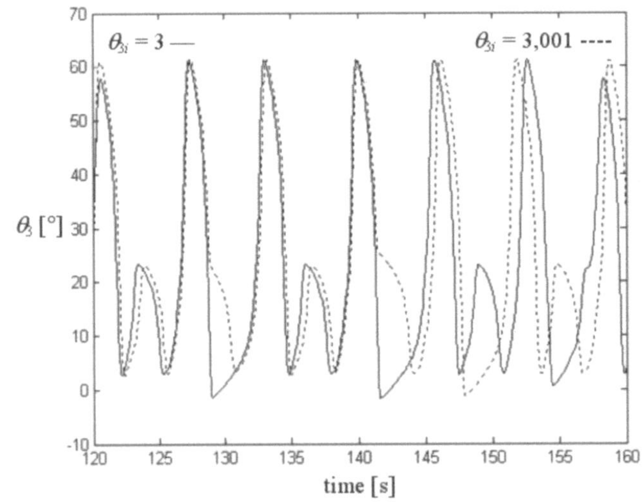

Fig. 18. Sensitivity to the initial conditions in chaotic response.

From Fig. 14 and 17, we observed the configuration of chaotic regime when $\varepsilon_3 = 2$. More details about transition between quasiperiodic and chaotic response are presented by Yoshinaga & Kawakami (1994), Yang (2000) and Pazó et al. (2001).

Sensitivity to the initial conditions can be verified considering two simulations with different conditions, for example, with $\varepsilon_3 = 3$ (chaotic regime), choosing initial values for the angles: $\theta_3 = 3°$, $\theta_4 = 50°$, $\theta_5 = -3°$, and changing $\theta_3 = 3.001°$, we observed the influence of initial conditions in the system response (Fig. 18).

Another interesting point of the chaos analysis is the presence of strange attractors, which can be observed in Poincaré map. In dissipative systems the Poincaré map presents a set of points disposed in an organized form, with a preferential region in phase space that attracts the states of dynamical system. Figure 19 showing the strange attractor generated in the analysis of knee oscillator θ_3.

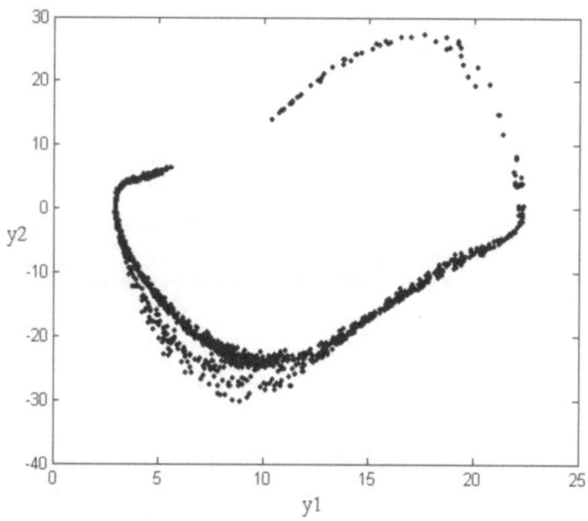

Fig. 19. Strange attractor for θ_3.

Considering the coupling oscillators, the degree of influence between them is defined by the coupling term. Then, a change of oscillator parameters must influence the behaviour of other oscillators. Figure 20 presents the bifurcation diagram showing the behaviour of knee oscillator θ_5 with variation of parameter ε_3.

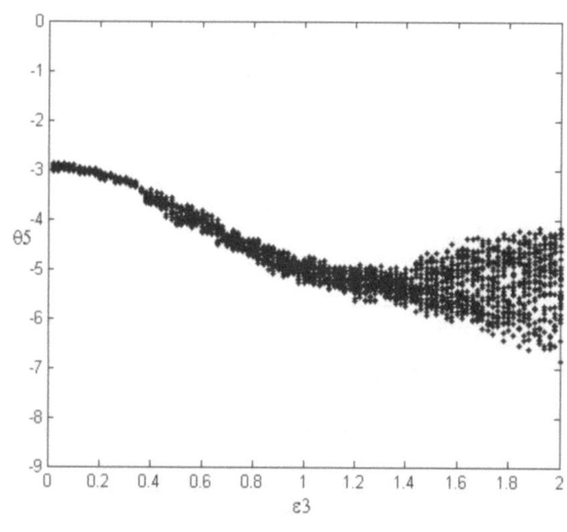

Fig. 20. Bifurcation diagram for θ_5 with variation of ε_3.

In relation to the hip angle, the influence of knee oscillator θ_3 on the hip is small, therefore the behaviour of θ_4 does not show many alterations. This occurs due to small value adopted for the coupling term between the oscillators ($c_{34} = c_{43} = 0.001$). In relation to the knees, the coupling term is greater ($c_{35} = c_{53} = 0.1$), configuring a more significant influence.

Similarly to analysis of ε_3, the system response can be analyzed by varying the values of ε_4 (from 0 to 2) and keeping the other values fixed. Figure 21 presents the bifurcation diagram showing the behaviour of hip oscillator θ_4 with variation of parameter ε_4, which represents the damping term related with this oscillator. Figure 22 showing the strange attractor generated in the analysis of this oscillator.

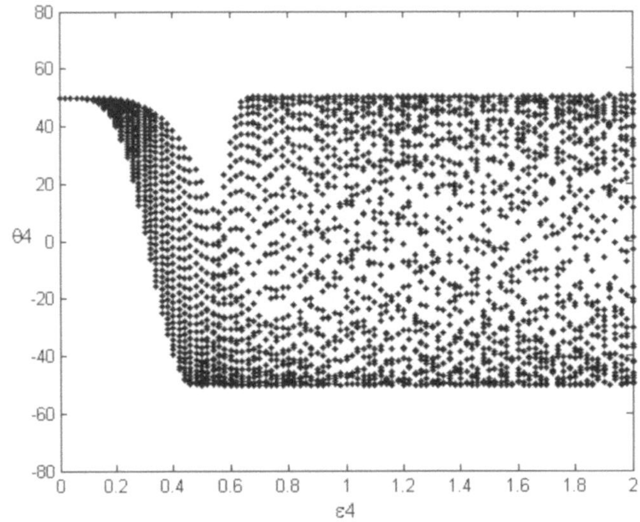

Fig. 21. Bifurcation diagram for θ_4 with variation of ε_4.

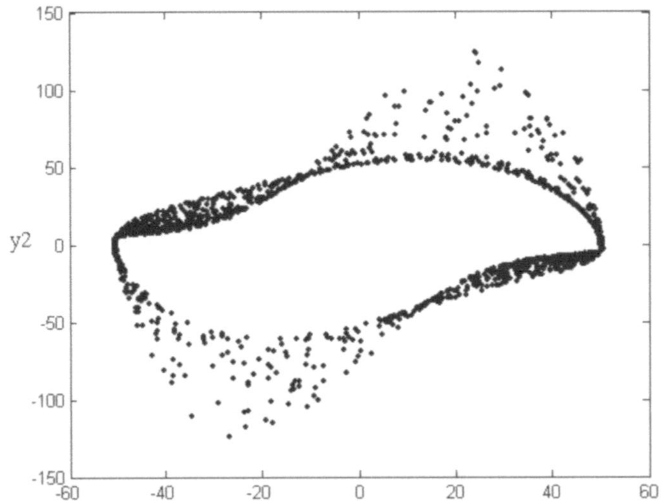

Fig. 22. Strange attractor for θ_4.

As seen previously in the analysis of ε_3, the influence of hip on the knees is small, then a variation of ε_4 does not cause great changes in θ_3 and θ_5.

Finally, the system response can be analyzed by varying the values of ε_5 (from 0 to 2) and keeping the other system values fixed. Figure 23 presents the bifurcation diagram showing the behaviour of knee oscillator θ_5 with variation of the parameter ε_5, which represents the

damping term related with this oscillator. Figure 24 showing the strange attractor generated in the analysis of this oscillator.
Figure 25 presents the bifurcation diagram showing the behaviour of knee oscillator θ_3 with variation of the parameter ε_5. In relation to the hip, the knee oscillator θ_5 presents small influence on the hip angle θ_4.

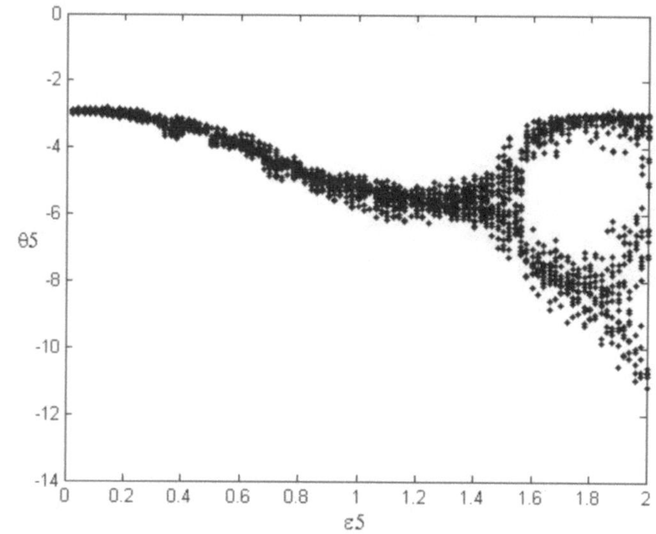

Fig. 23. Bifurcation diagram for θ_5 with variation of ε_5.

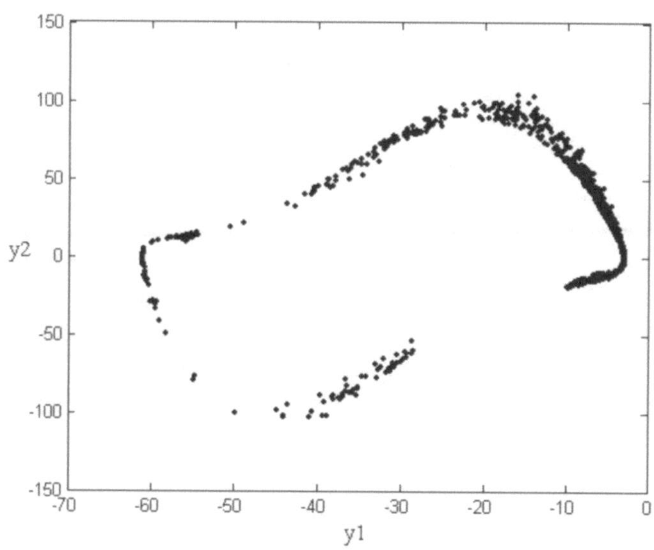

Fig. 24. Strange attractor for θ_5.

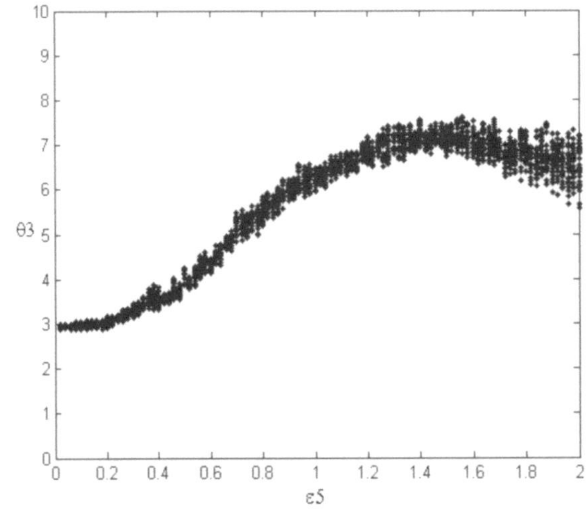

Fig. 25. Bifurcation diagram for θ_3 with variation of ε_5.

6. Conclusion

In this chapter, we present the study of a biped locomotor with a CPG formed by a system of coupled van der Pol oscillators. A biped locomotor model with three of the six most important determinants of human gait was used in the analyses. After the modelling of the oscillators system, a dynamical analysis was performed to verify the performance of the system, in particular, aspects related to the chaos. From presented results and discussion, we come to the following conclusions: the use of mutually coupled nonlinear oscillators of van der Pol can represent an excellent way to generate locomotion pattern signals, allowing its application for the control of a biped by the synchronization and coordination of the legs, once the choice of parameters is correct, which must be made from the data supplied by the analysis of bifurcation and chaos. Through the dynamical analysis it was possible to evidence a weak point of coupling systems. The influence of the knee oscillators on the hip, and vice versa, is very small, what can harm the functionality of the system. The solution for this problem seems immediate: to increase the value of the coupling term between the hip and knees. However, this can make the system unstable. Then, it is necessary a more refined study of the problem, which will be made in future works, as well as a study of the behaviour of the ankles, and simulation of the hip and knees in the other anatomical planes, increasing the network of coupled oscillators, and consequently simulating with more details the human locomotion CPG. This study has great application in the project of autonomous robots and in the rehabilitation technology, not only in the project of prosthesis and orthesis, but also in the searching of procedures that help to recuperate motor functions of human beings.

7. Acknowledgment

The authors would like to express their gratitude to InTech - Open Access Publisher, for the invitation to participate in this book project.

8. References

Baker, L.G. & Gollub, J.P. (1996). *Chaotic Dynamics, an Introduction*, 2nd edition, Cambridge University Press

Bay, J.S. & Hemami, H. (1987). Modelling of a neural pattern generator with coupled nonlinear oscillators. *IEEE Trans. Biomed. Eng.* 34, pp. 297-306

Braune, W. & Fischer, O. (1987). *The Human Gait*, Springer Verlag, New York

Calancie, B.; Needham-Shropshire, B.; Jacobs, P.; Willer, K.; Zych, G. & Green, B.A. (1994). Involuntary stepping after chronic spinal cord injury. Evidence for a central rhythm generator for locomotion in man. *Brain* 117(Pt 5), pp. 1143-1159

Dimitrijevic, M.R.; Gerasimenko, Y. & Pinter, M.M. (1998). Evidence for a spinal central pattern generator in humans. *Annals of the New York Academy of Sciences* 860, pp. 360-376

Dutra, M.S. (1995). *Bewegungskoordination und Steuerung einer zweibeinigen Gehmaschine*, Shaker Verlag, Germany

Dutra, M.S.; Pina Filho, A.C.de & Romano, V.F. (2003). Modelling of a Bipedal Locomotor Using Coupled Nonlinear Oscillators of Van der Pol. *Biological Cybernetics* 88(4), pp. 286-292

Eberhart, H.D. (1976). Physical principles for locomotion, In: *Neural Control of Locomotion*, R. M. Herman et al., Plenum, USA

Grebogi, C.; Poon, L.; Sauer, T.; Yorke, J.A. & Auerbach, D. (2002). *Handbook of Dynamical Systems*, Vol. 2, Chap. 7, Elsevier, Amsterdam

Grillner, S. (1985). Neurobiological bases of rhythmic motor acts in vertebrates. *Science* 228, pp. 143-149

Horak, F.B. & Nashner, L.M. (1986). Central programming of postural movements adaptation to altered support-surface configurations. *Journal of Neurophysiology* 55, pp. 1369-1381

Mackay-Lyons, M. (2002). Central Pattern Generation of Locomotion: A Review of the Evidence. *Physical Therapy* 82(1)

McMahon, T.A. (1984). *Muscles, Reflexes and Locomotion*, Princeton University Press

Pazó, D.; Sánchez, E. & Matías, M.A. (2001). Transition to High-Dimensional Chaos Through Quasiperiodic Motion. *International Journal of Bifurcation and Chaos*, Vol. 11, No. 10, pp. 2683-2688

Pearson, K.G. (1993). Common principles of motor control in vertebrates and invertebrates. *Annu. Rev. Neurosci.* 16, pp. 265-297

Pina Filho, A.C.de (2005). *Study of Mutually Coupled Nonlinear Oscillators Applied in the Locomotion of a Bipedal Robot*, D.Sc. Thesis, PEM/COPPE/UFRJ, Brazil

Pina Filho, A.C.de (2008). Simulating the Hip and Knee Behaviour of a Biped by Means of Nonlinear Oscillators. *The Open Cybernetics & Systemics Journal*, Vol. 2, pp. 185-191

Pina Filho, A.C.de; Dutra, M.S. & Raptopoulos, L.S.C. (2006). Modelling of Bipedal Robots Using Coupled Nonlinear Oscillators, In: *Mobile Robots, Toward New Applications*, Aleksandar Lazinica, (Ed.), Pro Literatur Verlag, Germany/ARS, Austria, pp. 57-78

Raibert, M.H. (1986). Legged Robots, *Communications of the ACM* 29(6), pp. 499-514

Santos, A.M.dos; Lopes, S.R. & Viana, R.L. (2004). Rhythm Synchronization and Chaotic Modulation of Coupled van der Pol Oscillators in a Model for the Heartbeat. *Physica A* 338, pp. 335-355

Saunders, J.B.; Inman, V. & Eberhart, H.D. (1953). The Major Determinants in Normal and Pathological Gait. *J. Bone Jt. Surgery* 35A

Strogatz, S. (1994). *Nonlinear Dynamics and Chaos*, Addison-Wesley

Thompsom, J.M.T. & Stewart, H.B. (1986). *Nonlinear Dynamics and Chaos*, John Wiley & Sons, Chichester

Vukobratovic, M.; Borovac, B.; Surla, D. & Stovic, D. (1990). *Biped Locomotion*, Springer Verlag, Berlin

Winter, D. (1983). Biomechanical motor patterns in normal walking. *J. Motor Behav.* 15(4), pp. 302-330

Yang, J. (2000). Quasiperiodicity and transition to chaos. *Physical Review E* 61, pp. 6521-6526

Yoshinaga, T. & Kawakami, H. (1994). Bifurcations of Quasi-periodic Responses in Coupled van der Pol Oscillators with External Force. *IEICE Transactions on Fundamentals of Electronics, Communications and Computer Sciences*, Vol. E77-A, No. 11, pp. 1783-1787

Zielinska, T. (1996). Coupled oscillators utilised as gait rhythm generators of a two-legged walking machine. *Biological Cybernetics* 74, pp. 263-273

10

Some Results on the Study of Kneed Gait Biped

Zhenze Liu[1], Yantao Tian[1] and Changjiu Zhou[2]
[1]College of Communication Engineering, Jilin University
[2]Advanced Robotic Intelligent Control Center, School of Electrical Electronic Engineering, Singapore Polytechnic
[1]China
[2]Singapore

1. Introduction

The notion of obtaining passive gaits, powered only by gravity, was pioneered by McGeer[1], who thought that ,we can perhaps learn about the stability and control of walking by studying un-powered, uncontrolled models.

Some results with McGeer's passive dynamic models of human locomotion suggest that human body parameters such as mass distribution or limb lengths may have more influence on the existence and quality of gait than is generally recognized. The question has been subsequently studied by many other researchers-such as Collins, Garcia and Goswami.

Human locomotion is typically described as having a periodic movement pattern and stable passive gaits were found for both planar and non-planar bipeds on shallow downhill slopes. And the existence of passive limit cycles(periodic behavior) has important implications for the design of walking robots. Some basic definition about the limit cycle has been induced[2][3][6], discrete events, such as contact with the ground , can act to trap the evolving system state within a constrained region of the state space. Therefore, even when the underlying continuous dynamics are unstable, discrete events may induce a stable limit set and limit cycles are often created in this way.

Here the paper will take great interest in the model Goswami presented 1997 and will describe the model geometry, its dynamic parameters, and its governing equations during the swing stage and the transition stage. In addition, a typical walk cycle of the passive robot on a inclined plain with the help of a phase diagram will be discussed, this motion can continue indefinitely due to a delicate balance between the robot's kinetic energy and potential energy. The discussion about the intricate energy transition and also the mutual influence between the swing leg and stance leg will help us to be better aware of the passivity gait of this kind of compass-like biped robot, besides, some further control ideas will be educed based on this very character thus lead to systematic control design. In spite of this, the paper also present some applicable control strategies on the gait biped to improve its gaits and present some new idea of anti-phase synchronization.

The results of gait biped concluded above can also be extended to the model of three dimensional phase and some useful research results will shed light on new discovery of this terrific field of the gait biped.

2. The compass gait biped model

2.1 The model description and assumption

The paper will follow the model that Goswami presented in 1997, so called the compass gait biped, shown as Figure 1, is equivalent to a double pendulum with point masse m_H and m concentrated at the hip and legs respectively. The leg-length is l, which is divided into two parts: a and b, a is the distance from the leg-tip to the position of m and b is the distance from m to the hip center m_H. The support angle θ_s and nonsupport angle θ_{ns} determine the configuration of the compass gait. The angle was made by the biped leg with the vertical (counterclockwise positive). 2α is the total angle between the legs, which is defined as the "inter-leg angle", and in addition is formed during the instant when both legs are touching the ground. The slope of the ground with the horizontal is denoted by the angle ϕ

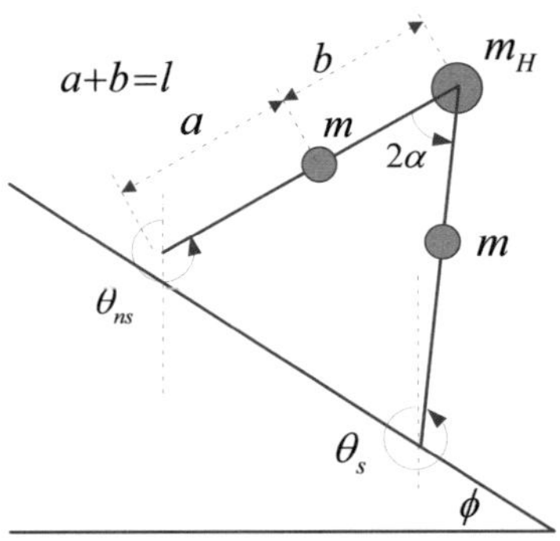

Fig. 1. Model of a compass gait biped robot on a slope

The model has been made by the following assumptions: the total mass of the robot $m_C = 2m + m_H$ is constant and equal to 20kg. For the sake of simplifying the model, all masses are considered point-masses and the legs are identical with each leg having telescopically retractable knee joint with a mass-less lower leg(shank), this retractable knee joint which is called prismatic-joint knee and is the imaginary concoction, the function of it is to address the conceptual problem of foot-clearance common to all knee-less planar bipeds. The gait consists of swing stage and an instantaneous transition stage: during the swing stage the robot behaves exactly like an inverted planar double pendulum with its support point being analogous to the point of suspension of the pendulum. During the transition stage the support is transferred from one leg to the other. The robot is assumed to move on a horizontal or inclined plane surface. The impact of the swing leg with the ground is assumed to be inelastic and without sliding[4]. This implies that during the instantaneous transition stage the robot configuration remains un-changed, and the angular momentum of the robot about the impacting foot as well as the angular momentum of the pre-impact support leg about the hip are conserved. These conservation laws lead to a discontinuous change in robot velocity.

2.2 Dynamics of the swing stage

The dynamic equations of the swing stage are similar to the well-known double pendulum equations. Since the legs of the robot are assumed identical, the equations are similar regardless of the support leg considered.

They have the following form

$$M(q)\ddot{q}+C(q,\dot{q})\dot{q}+g(q)=Bu \tag{1}$$

where $q=\begin{bmatrix}q_1\\q_2\end{bmatrix}=\begin{bmatrix}\theta_{ns}\\\theta_s\end{bmatrix}$; $B=\begin{bmatrix}-1 & 0\\1 & 1\end{bmatrix}$, and the vector u represents independent torques at the hip and ankle, which are assumed to be identically zero in the case of passive biped .

The matrices $M(q)$,$C(q,\dot{q})$, and $g(q)$ are given as

$$M(q)=\begin{bmatrix}mb^2 & -mlb\cos(q_2-q_1)\\ -mlb\cos(q_2-q_1) & (m_H+m)l^2+ma^2\end{bmatrix}$$

$$C(q,\dot{q})=\begin{bmatrix}0 & mlb\sin(q_2-q_1)\dot{q}_2\\ mlb\sin(q_1-q_2)\dot{q}_1 & 0\end{bmatrix}$$

$$g(q)=\begin{bmatrix}mb\sin(q_1)\\ -(m_Hl+ma+ml)\sin(q_2)\end{bmatrix}$$

The parameters used for our simulations are $a=b$ =0.5m, $l=a+b$, m_H =2 m =10kg. Since no dissipation takes place during swing stage, thus the total mechanical energy E of the robot is conserved during this stage.

2.3 Transition equations

The algebraic transition equations relate the robot's states just before and just after its collision with the ground. The support and the non-support legs switch during transition. The pre-impact and post-impact configurations of the robot can be simply related by

$$\theta^+ = J\theta^- \tag{2}$$

With

$$J=\begin{pmatrix}0 & 1\\1 & 0\end{pmatrix} \tag{3}$$

The matrix J exchanges the support and the swing leg angles for the upcoming swing stage. The pre-impact and post-impact variables are identified respectively with the superscripts – and +. The conservation of angular momentum principle applied to the robot gives us the following equation

$$Q^-(\alpha)\dot{\theta}^- = Q^+(\alpha)\dot{\theta}^+$$

From which we can write the joint-velocity relationship

$$\dot{\theta}^{+} = (Q^{+}(\alpha))^{-1} Q^{-}(\alpha)\dot{\theta}^{-} = H(\alpha)\dot{\theta}^{-}$$

where

$$Q^{-}(\alpha) = \begin{bmatrix} (m_H l^2 + 2ml^2)\cos(2\alpha) - mab \\ \cdots - 2mbl\cos(2\alpha) & -mab \\ -mab & 0 \end{bmatrix} \qquad Q^{+}(\alpha) = \begin{bmatrix} mb^2 - mbl\cos(2\alpha) & (ml^2 + ma^2 + m_H l^2) - mbl\cos(2\alpha) \\ mb^2 & -mbl\cos(2\alpha) \end{bmatrix}$$

With

$$\alpha = \frac{q_1 - q_2}{2}$$

The complete state vector q before and after impact can thus be written as

$$q^{+} = \dot{W}(\alpha)q^{-}$$

With

$$\dot{W}(\alpha) = \begin{pmatrix} J & 0 \\ 0 & H(\alpha) \end{pmatrix}$$

Moreover, it follows with the robot geometry during the transfer

$$\theta_{ns}^{-} + \theta_{s}^{-} = -2\varphi \qquad (\text{or } \theta_{ns}^{+} + \theta_{s}^{+} = -2\varphi\ )$$

$$\theta_{ns}^{-} - \theta_{s}^{-} = 2\alpha \qquad (\text{ or } \theta_{s}^{+} - \theta_{ns}^{+} = 2\alpha\ )$$

where +and - correspond to the instants just after and before the change of support, respectively.

The assumption that the angular momentum of the robot is conserved during the transition doesn't explicitly indicate how the mechanical energy of the robot changes during this stage. we will present a detailed explanation in the following section on the fact that through the transition stage, the change in mechanical energy is always negative.

3. Characteristics of steady passive compass gaits

3.1 Description of a typical limit cycle

Due to the hybrid nature[5] of the governing equations, it is impossible to utilize the traditional tool developed to aid the study of this nonlinear systems. McGeer has proposed an idea of linearizing the swing-stage equations of the robot about an equilibrium state, thus making it possible to explicitly integrate these equations. Next the transition equations are concatenated and the conditions for the existence of a periodic solution of this coupled system is found. To study the stability of this periodic solution, a second linearization about the periodic solution is necessary. The problem with this approach is that the linear solution is valid only within a narrow region around the point of linearization.

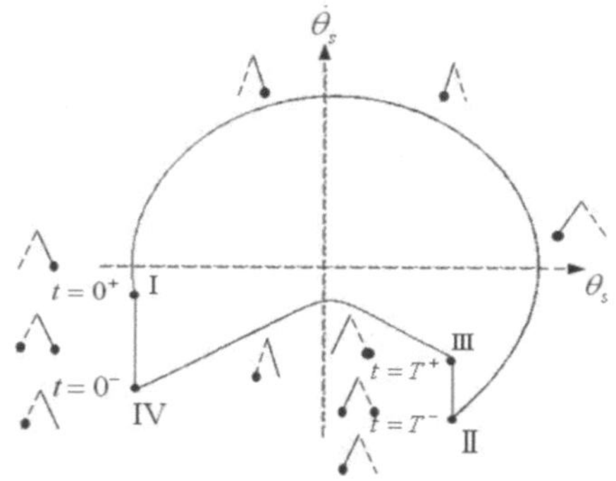

Fig. 2. Phase portrait of a periodic walk. This figure corresponds to only one leg of the biped, one cycle in the figure corresponds to two steps of the robot. In the figure we have indicated some of the time stamps important in the dynamic evolution of the biped. The configuration of the biped has been shown with small stick diagrams. In these diagrams, one leg is dotted, the other leg is solid, and a black dot at the foot indicates the supporting leg.

Figure 2 just presents the sketch of a phase-space limit cycle of a symmetric gait of the robot on a three degree slope.

Follow the phase trajectory at the instant marked I, corresponding to time $t = 0^+$, when the rear leg just loses contact with the ground and becomes the swing leg. The phase trajectory evolves in the clockwise sense in the diagram with the cycle from I - II, depicts the swing leg suspended as a simple pendulum from the moving point-hip, at the same time, stance leg "hinged" at the point of support as an inverted simple pendulum. While the swing leg will cross the velocity axis at a positive velocity, the biped is in the vertical configuration. During the process, only the stance leg contacts with the ground, please recall that we have the assumption that there is no slipping at the stance leg ground contact. Instant II corresponds to time $t = T^-$, when the swing leg is about to touch the ground. The impact between the swing leg and the ground occurs at the instant $t = T$, we observe a velocity jump from II - III due to the impact. In order to simplify the model, we assume that the time during the impact is instantaneous, which means there is an impulse force acting on the biped. Due to this presumption, constant angle momentum is possible and the decrease of the kinetics can be explained by the jump in velocity and inelastic property. At instant III, $t = T^+$, the swing leg becomes the support leg and executes the process of III - IV , and it corresponds to the motion of the support "hinged" at the point of support as an inverted simple pendulum. From IV — I , thus $t = 0^- - t = 0^+$, the velocity jump appears for another time due to the impact between the current swing leg with the ground, similar to the process II - III, and then the cyclic trajectory is a limit cycle. For the stable gaits, it will attract and absorb all nearby trajectories that enter its attractive basin. This property will be useful for the further control strategy design.

Simulation trials reveal that the passive compass gait robot can walk down a slope with a steady gait. For a given robot, one and only one stable gait on a given slope exists, which symbolize the periodicity of the humanoid gaits, if we can make full use of this property, we may find some idea on controlling the robots by maintaining the stability of limit cycle through the idea of adding some torque or only adjust the parameter of the system. Moreover, the initial value of the passive walking must correspond with an energy value , for the lost energy during the process of collision should conform to some regulations between the gravity and kinemics. To a certain slope, the limit cycle is the only, so the state point adjoin to the limit cycle can also converge to the limit cycle. The non-linear system possesses the property of being sensitive to the initial value, so the analytical procedure to find this limit cycle still remain a challenge.

3.2 The energy analysis in passive gait

Figures 3 depict the variation graph of kinetic energy, potential energy and total mechanical energy corresponding to the limit cycle of certain three slope respectively. Seen from figures, we can clearly specify the whole biped gaits of the robot, the kinetic energy (KE) and the potential energy (PE) have a complex variation process just not as we have expected before. KE just experiences an asymmetry periodic process. At instant T, a sharp downwards jump exists because of the inelastic impact of the legs and the ground thus causing the loss of the kinetic energy dramatically, we can clearly see from Figure 3 that the reduction of the energy is irregular just due to the inertial kinetic energy compensation of the stance leg, the enhancement of kinetic energy is partly compensated by gravity, the detailed message of the variation of the energy and conversion can be informed in figures. While PE just experiences a contrary process. During the swing stage, gravity and only gravity acts on the robot, so the whole mechanical energy of the system will keep constant. At instant $t = T$, mechanical energy will also have a downward jump, this variation value will be equal to the kinetics'. The potential energy decreases continuously during the whole process, we can tell from figures that some coupling phenomenon exists between the swing leg and the stance leg, similar to that 2-dof mechanical configuration. Mutual influence between the two legs can be observed indirectly by figures and also will help us in realizing this complex hybrid system.

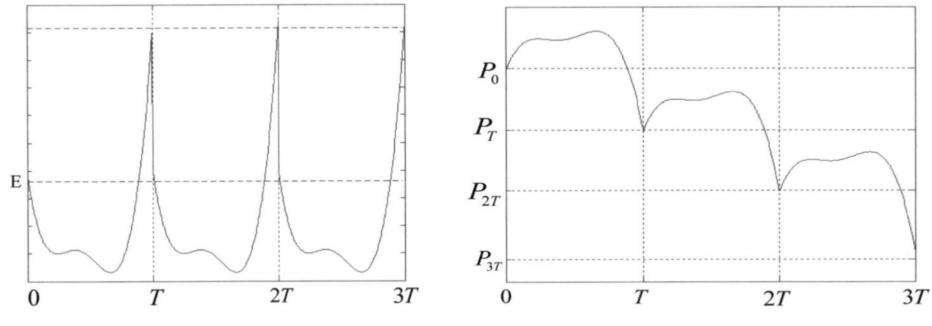

Fig. 3. The kinetic energy and potential energy variation graph corres-ponding to the limit cycle of certain three slope

Seen from Figures, we present the variation and comparison graph. And some important points t_0,t_1,t_2,t_3,t_4 have been selected out to explain the whole biped gaits, they represent the instant corresponding to different culmination points during the steady gaits period. During the whole walking course, KE and PE curve just go along with the direction $t_0 - t_1 - t_2 - t_3 - t_4$, amid of it, $t_0 = 0$, $t_4 = T$. The graph can tell us some details about the particular energy variation of the whole steady robot gaits.

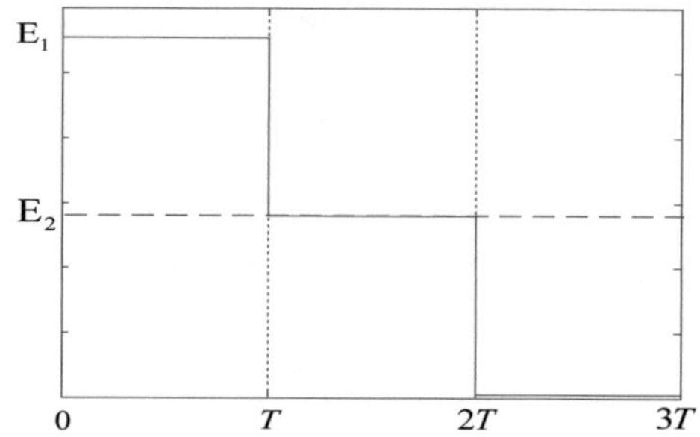

Fig. 4. The total mechanical energy variation graph corresponding to the limit cycle of certain three slope

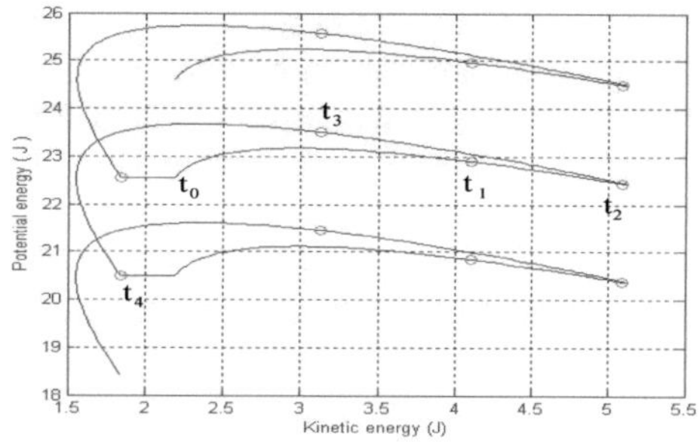

Fig. 5. The variation and comparison graph of KE and PE conversion of the swing leg during steady robot gaits of certain 3 degree slope

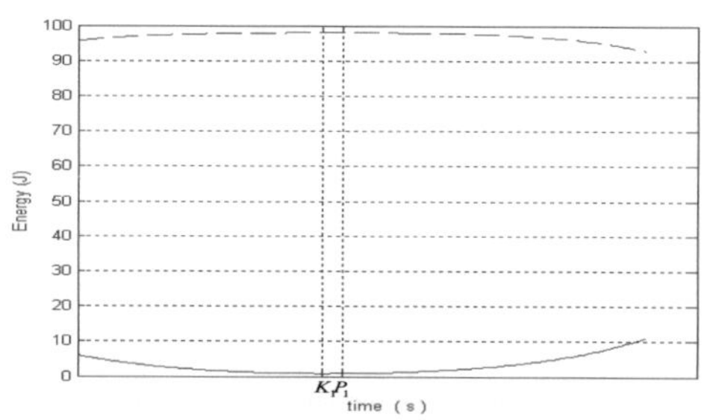

Fig. 6. The graph on the two nearest existed culmination values of KE and PE corresponding to the course of energy conversion of the swing leg within one gait cycle

Seen from Figures we address the culmination values corresponding to the course of energy conversion. There are three culmination points of KE and two culmination points of PE within one steady gait period. For Figure 11, we can set the culmination potential value of the swing leg as P_1, P_2, P_3 from the left to right, and also set two culmination kinetic value of the swing leg as K_1, K_2 with the same sequence above within one gait cycle, the maximum and minimum of the energy can be observed. The figure just search out the culmination point of the two nearest point as k and p, the potential energy culmination point just drop behind the kinetic energy culmination point even they are adjacent while

they are not the same point as we had thought before. The reason will be well explained in the following section.
Theoretical speaking, the mechanic energy should keep constant during the swing stage corresponding to the stable walking limit cycle. Virtually, total mechanic energy will decrease incessantly during the swing stage. While the reduction of the magnitude can be omitted comparing with total energy magnitude. The reason that we mention this problem is that due to the complexity of the non-linear system, it is necessary to make some adjustments sometimes in order to get the better results when considering the control strategy of the system.
The phenomena called "rub ground" will exist during the swing stage, this phenomena just happens at the time before the superposition of two legs and ends just at the instant of the superposition of two legs. The height between the swing leg and the ground will be negative when the swing leg swings from the start to the vertical position by simulation results corresponding to steady robot gaits of certain 3 degree slope, the maximum of the height will reach -0.0033m. Why this phenomenon exist and how to avoid this state which we intuitively sense unrealistic? We can solve this problem by some technique methods such as the assumption discussed in the second part of the paper- the introduction of a purely imaginary concoction so called prismatic joint knee. The prismatic joint is assumed to retract the lower leg to clear the ground, and the retraction of the lower leg is assumed mass-less, it will not affect the robot dynamics and the swing leg returns to its original length l at transition. The assumption is very necessary for the existence of the limit cycle and many properties of the bipedal gaits can be observed directly and also will guide us in some directions: for example, what is the relationship between the point of intersection of two legs and the height between the swing leg and the ground? For real gaits of the robot, we can modify the value of the graph to keep tracing the steady gaits of the robot.

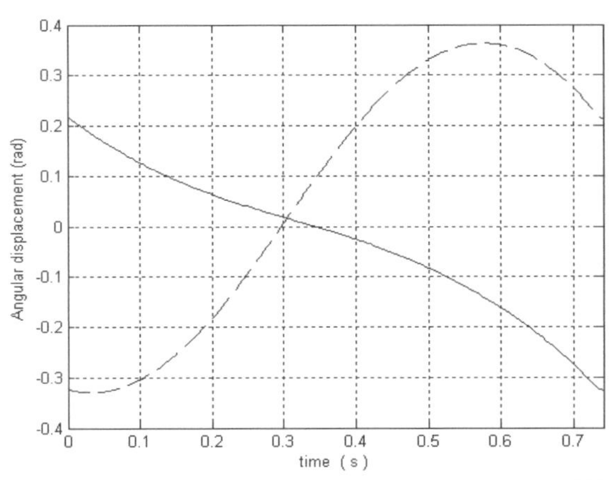

Fig. 7. The relationship graph of angle position between the swing leg and the stance leg

'---' just represents the swing leg and '—' just represents the stance leg, the same with the following figures.
Seen from Figure 15, the angle position curve of the swing leg is much more approach to sine wave, while the stance leg has a comparative big difference with the swing leg. This can be explained that the stance leg experiences a compelled motion, with the action force

coming from the swing leg as well as gravity. And in addition, the coupling degree of the two legs vary at different instant.
The angle position will be more than zero when two legs are in the state of superposition, which means the joint of superposition lies in the left side of the vertical direction. When the swing leg becomes straight, the angle of the stance leg will be positive, and the stance leg the same.
Here the focus of the work is a relative further study of the passive gait of a compass-like, planar, biped robot on inclined slopes, an analysis about the distribution of the energy and also the conversion law between the swing leg and the stance leg during the process of the steady robot gaits, have been discussed in the paper. Phase-position property corresponds to the limit cycle, the coupling properties between two legs, the existence of the culmination points which produced in the course of the conversion of KE and PE are also the topic of the research. To a certain slope angle ϕ, one and only one stable limit cycle exists.
The research of the paper will have positive significance in getting better aware of the law and global property to biped gaits of the robot. The model we adopt here is an ideal position, how to induce or modify a more realistic model for biped gaits, and how to enlarge the initial value attraction region of the limit cycle as well as how to apply the efficient control on the robot combined with its own property with the least energy possible will guide our further research direction.

4. Some simple control laws

The existence of passive gaits in simple bipeds is interesting and may help to explain the efficiency of human locomotion. In particular, the sensitivity to initial conditions and ground slope must first be emphasized [8]. In spite of this, robustness to external disturbances and parameter uncertainty must be investigated. In the paper, we address simple control law for the compass gait biped by tracking a given mechanical energy of the robot with the torque added on the hip and ankle respectively.

4.1 The idea of control law tracking passive energy level

As the robot walks down on a slope, its support point also shifts downward at every touchdown, the kinetic energy will increase accordingly as it loses gravitational potential energy. In a steady walk, at the end of each step by the impact, the amount of kinetic energy will absorb the loss of the gravitational potential energy. This character presents us an idea on control passive biped robot, if, at every touchdown we reset our potential energy reference line to the point of touchdown, then the total energy of the robot appears constant regardless of its downward descent. We name the characteristic energy of the passive limit cycle on a given slope as "reference energy", the function of it is to drive the robot toward it thus attain to a mobile balance.
The approach assumes that we have already identified the passive limit cycle for a given slope and the advantage of it is that it is able to generate gaits which don't exist for the un-powered robot. In addition, at the same time, only those neighborhoods of the passive gait can function well by this control law. The total mechanical energy E of the robot can be expressed as $E = 0.5\dot{\theta}^T M\dot{\theta} + PE$. The power input to the system is the time rate of change of the total energy, $\dot{E} = \dot{\theta}^T B\, u$ for a passive cycle $u = 0$, and the reference energy of the limit cycle is

$E^* = E(\dot{\theta}^*, \theta^*)$. Here we use a simple damper control law of the form $Bu = -\beta\dot{\theta}$. So the power input to the system is therefore $\dot{E} = -\dot{\theta}^T \beta\dot{\theta}$. For a positive definite β, the quantity $-\dot{\theta}^T \beta\dot{\theta} < 0$, which means that the robot's kinetic energy decreases monotonically. In order to simplify the choice, we can further specify that the control law should bring the current level of the robot to the reference energy level at an exponential rate.
Three ways of the control law[9] will be implemented: by means of the two actuators acting independently or them acting together on the hip or in the supporting leg at the point of support, the latter will be also called as "support ankle torque". The paper will have a discussion about the latter two control strategies.

4.2 Control with hip torque

We propose a control law of the following form based on the idea presenting abov then after the calculation we get

$$u_H = -\frac{\lambda(E - E^*)}{\dot{\theta}_s - \dot{\theta}_{ns}} \tag{4}$$

λ is a parameter influencing the degree about the rate of convergent to the reference energy level. At the state of $\dot{\theta}_s - \dot{\theta}_{ns} = 0$, the control law will have a singularity, to solve this problem, the common idea is to set the control to zero whenever $\| \dot{\theta}_s - \dot{\theta}_{ns} \| < \varepsilon$.
To the passive limit cycle on a 3° slope, we make some active phase cycle superimposed on it, and from the picture, we let the starting position, denote as A, lie outside the basin of the attraction of the passive limit cycle. In this situation the passive robot would have fallen down soon, while the control law will lead the gait of the robot go back into the state of limit cycle and thus keep the periodic state. We can come to the conclusion from the control that the basin of attraction of the passive limit cycle has been enlarged and this will have a realistic sense in the application of the further study.

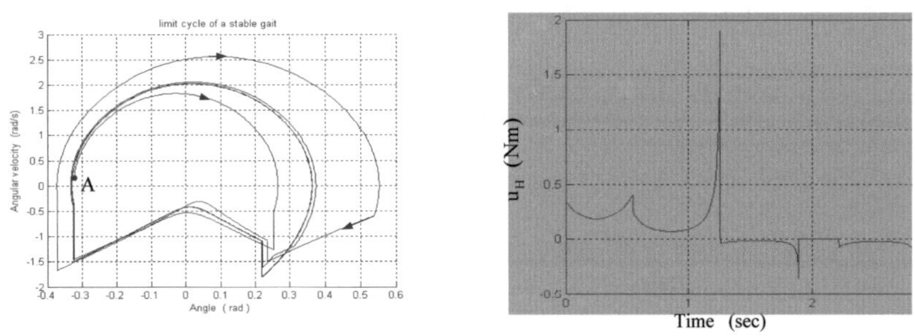

Fig. 8. Active stabilization of a limit cycle. Here we show the performance of the energy tracking law for a robot walking down a 3° slope. The system driven only by a hip torque seeks and returns to the passive cycle of the robot. The initial condition is denoted as point A, lying outside the basin of attraction of the passive limit cycle. Through the control strategy, the system has been brought back to the limit cycle. The right will be the graph of the variation graph on the control added on the hip

Seen from Figure, we conclude that the system will run away from the limit cycle without the added control. While added with control, the system will make a 3-4 gaits adjustment to converge to the original limit cycle and then keep its stable gait state, which proves the validity of the control. Figure 4 just depicts the variation on total energy based on the hip control condition, we can clearly see that the total energy will fluctuate within a transitory process and then go into a constant value which corresponding to the energy of the limit cycle. From Figure, we can explicitly be aware of the detailed variation on the control torque act of the hip and of the every gait state. To seek for the deep relationship between the variations of added torque, the variation regulation of the energy control and also the property of the limit cycle will be useful.

Then, we observe the process with the starting position lying inside the basin of attraction of the passive limit cycle of certain slope(here 3°).

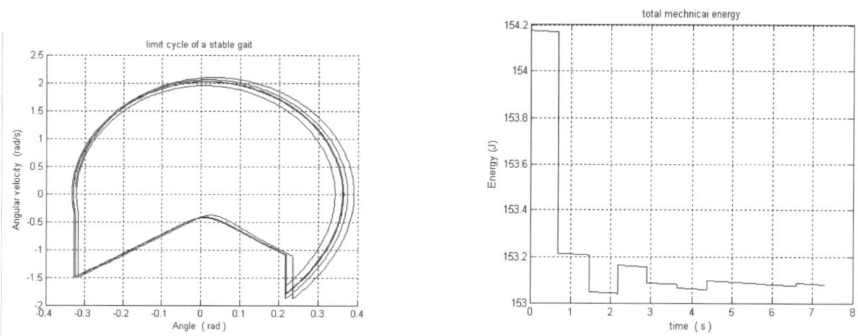

Fig. 9. Limit cycle corresponding to the certain three slope with the initial condition lying in the limit cycle The variation on total energy corresponding to the condition as Figure has demonstrated.

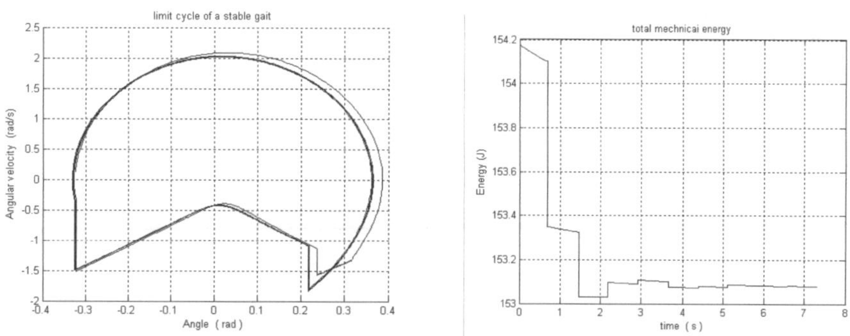

Fig. 10. Limit cycle with hip control corresponding to the condition as Figure And the variation on total energy corresponding to the condition as Figure 8 has demonstrated.

Figures 9 and 10 just show us something about the limit cycle corresponding the certain three slope with the initial condition lying in the limit cycle and also the variation of total mechanical energy corresponding that condition.

Figures 9 and 10 just show us the whole process with control added on the hip. Seen from the picture, we may safely find that time consuming in going into the limit cycle has been improved a lot evidently, with only two gaits the gait will converge to its stable period with control comparing at least 4 gaits to the same limit cycle by its own convergence. The variation on the total mechanical energy just describes the whole process of the biped gaits. In another words, with the control on the hip we may efficiently improve the quality of the convergence of the passive limit cycle.

Next, we'd meant to make unnatural limit cycles to track the certain target mechanical energy denoted by E^{tar}, which is different from the reference energy corresponding to that slope.

By the use of hip control, we can successfully produce new gaits, while it is very interest for us to see that the consequence in tracking the specified target energy can not match the very exact result that we expect, it will converge to the energy cycle which is adjacent to the target energy cycle. That is to say, the control strategy can help us to track any appointed target energy in some degree and will guide us to get better aware of the property of the passive gait control.

$$E^{tar} = 156J$$

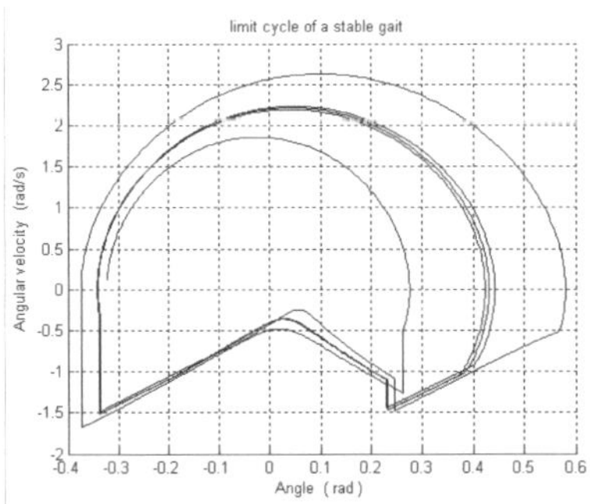

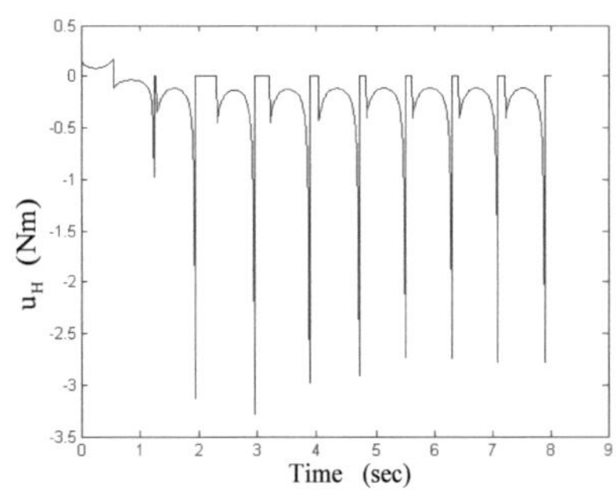

Fig. 11. Target energy tracking control and also the variation graph by using hip actuator on the control added on the hip

The target energy we'd like to track	The final attained energy with hip control
$E^{tar} = 154J$	$E^{final} = 153.1348J$
$E^{tar} = 156J$	$E^{final} = 153.1520J$
$E^{tar} = 158J$	$E^{final} = 153.1480J$
$E^{tar} = 160J$	$E^{final} = 153.1088J$

Table 1. The relationship between active biped gait for different target energies and the energy level at which the robot converged at the end.

Seen from Table 1, we find that no cycle with an energy level E^{tar} less than that corresponding to the passive cycle could be generated

4.3 Control with ankle torque

We will implement the same control law employing only the support ankle torque following the above hip control, and then we have:

$$u_s = -\frac{\lambda(E - E^*)}{\dot{\theta}_s} \qquad (5)$$

With the same procedure depicted as section 3.2, to the passive limit cycle on a 3° slope, we make some active phase cycle superimposed on it, we let the starting position, denote as A lie outside the basin of the attraction of the passive limit cycle, and in this situation the passive robot would have fallen down soon. While the control law will make the gait go into the state of limit cycle and thus enlarge the basin of attraction of the passive limit cycle.

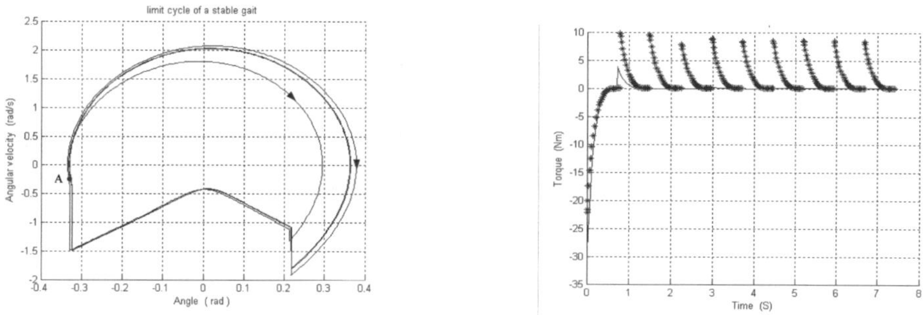

Fig. 12. Active stabilization of a limit cycle. Here we show the performance of the energy tracking law for a robot walking down a 3° slope. The system driven only by an ankle torque on the stance leg seeks and returns to the passive cycle of the robot. The initial condition is denoted as point A, lying outside the basin of attraction of the passive limit cycle. Through the control strategy, the system has been brought back to the limit cycle. And the right one will be the Energy tracking control using support ankle actuation. Ten steps of the robot are presented here. The support ankle alternates between the left and the right ankle. The black dot represents the energy Etar=152.6J and the real line represents the energy Etar=153.08J.

Figure 12 reveals the evolution of ankle torque. The repeated peaks in the control torque correspond to the time instants of foot touchdown. The zero of the time axis in the figure represents the beginning of a swing stage. Seen from this Figure, control is active from the beginning and as the robot's energy reaches the reference energy, the control becomes zero, clearly, foot touchdown has caused a sudden change in the angular velocity and also the system energy. In reality, arbitrarily large torques can't be applied as it may cause the robot foot to roll on the ground or maybe leave the ground.

We can come to a conclusion from Figure 16 that the target energy that we appoint ahead must lie in a relatively narrow region in order not to run away from the stable periodic state due to the property of non-linear system. We have found during the study that if we choose the target energy a little farther away from the reference energy, with only ankle control, the target energy that we expected can't be tracked successfully. Seen from Figure, the total energy will never converge to a constant value as we expect. Figure12 below just address the region about limit cycle of a stable gait with ankle control in certain slope.

For the study to the ankle torque control and the hip torque control, we come to the result that the main difference between them, is that for the hip control discussed above, we can effectively converge to any target energy (within a limit), which has much larger region than the ankle control. The reason will be explained as the ankle control is capable of more directly affecting the overall dynamics of the robot. Whether controlled at the hip or at the ankle, the control law will enlarge the basin of attraction of the limit cycle.

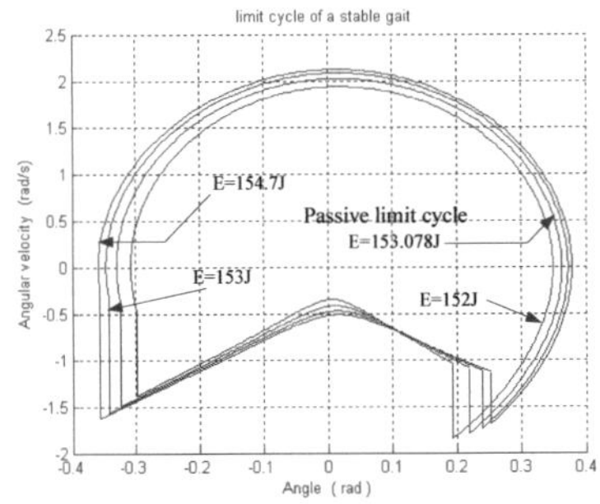

Fig. 13. Some region about limit cycle of a stable gait with ankle control in certain slope

In addition, we will pay more attention to the study about two actuators added at the hip and ankle together in the future work and it will be useful to identify the boundary of the basin of attraction and to determine the favorable initial conditions. In spite of all stated above, we should also know that the robot's behavior is heavily influenced by the impact model which is not the only available impact model, how to model some new realistic foot/ground impact models possessing such a manner that reasonable perturbations of the model parameters don't dramatically change the gait, should be considered.

5. The complicated idea of controlling the gait biped with energy based control slope invariance law

5.1 Controlled symmetry and slope invariance

The idea that passive limit cycles can be made slope invariant by a control that compensates the gravitational torques acting on the biped has been proposed by Mark. Spong.

The result just relies on some symmetry properties in the Lagrangian dynamics of robots with respect to rotations of inertial frame. A group action of $SO(n)$ has been defined to change the ground slope with respect to the inertial frame on Q, for $n=2$ in the planar case, this group action takes a particularly simple form as

$$A = \begin{bmatrix} \cos\varphi & -\sin\varphi \\ \sin\varphi & \cos\varphi \end{bmatrix} \in SO(2) \tag{6}$$

The group action, $\phi_A : Q \to Q$ is given by

$$\phi_A(q) = (q_1 + \varphi, q_2 + \varphi) \tag{7}$$

The so called lifted action on TQ is

$$(\phi_A(q), T_q\phi_A(\dot{q})) = (\phi_A(q), \dot{q}) \tag{8}$$

The kinetic energy and impact equations are invariant under this group action and if $q(t)$, $\dot{q}_i(t)$ is a solution trajectory of (1), with $u=0$ then $\phi_A(q)$, $\dot{q}$ is a solution of

$$M(q)\ddot{q}+C(q,\dot{q})\dot{q}+g(\phi_A(q))=0 \tag{9}$$

Via the control

$$u=B^{-1}(g(q)-g(\phi_A(q))) \tag{10}$$

The limit cycle of Figure can be reproduced on any ground slope via active control that effectively cancels the gravity vector that corresponds to the current slope.

5.2 Energy based control to the gravity compensation control

Using this gravity compensation control of the previous section. We let

$$u=B^{-1}(g(q)-g(\phi_A(q))+\bar{u}$$

So that (1) becomes

$$M(q)\ddot{q}+C(q,\dot{q})\dot{q}+g(\phi_A(q))=B\bar{u}$$

The design of the additional term $\bar{u}$ is to increase the robustness to slope variations. Set S as a storage function

$$S=\frac{1}{2}(E-E_{ref})^2 \tag{11}$$

E is the total (kinetic and potential) energy

$$E=\frac{1}{2}\dot{q}^T M(q)\dot{q}+V(q)$$

And E_{ref} is the constant energy of the biped along the limit cycle trajectory of the system corresponding to a fixed ground slope. A simple calculation shows that

$$\begin{aligned}\dot{S}&=(E-E_{ref})\dot{E}\\&=(E-E_{ref})\dot{q}^T B\bar{u}\end{aligned}$$

Where the second equality comes from the usual passivity or skew-symmetry property of rigid robots. Based on the above deduction, we design the following control scheme

$$\bar{u}=-kB^{-1}(E-E_{ref})\dot{q} \tag{12}$$

Where k is a scalar gain, and we can easily get the results as

$$\dot{S}=-2k\left\|\dot{q}\right\|^2 S\leq 0 \tag{13}$$

So the function S works just as a Lyapunov function. Thus the total energy of the biped will thus converge exponentially toward the reference between impacts. At the impact the storage function will exhibit a jump discontinuity. It follows from standard results in hybrid system theory that, if less than its value at the previous jump, then the total energy will converge asymptotically to the reference energy as $t \to \infty$.

5.3 Some results with energy tracking

Figures in this section will show that the addition of the total energy shaping control $\bar{u}$ results in both an increase in the basin of attraction of the limit cycle and increased convergence to the limit cycle. This has important consequences for robustness to external disturbances as well as uncertainty and variations in the ground slope.

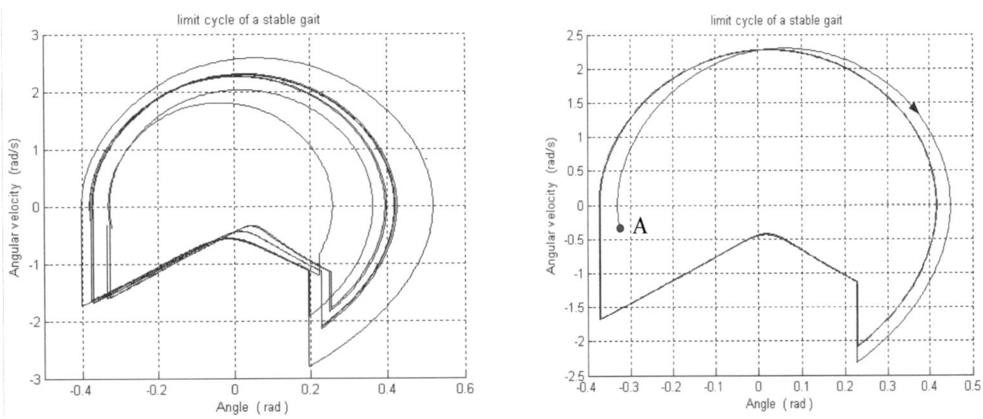

Fig. 14. Convergence to the limit cycle (a)without total energy control (b)with total energy control

With total energy control, the biped trajectory converges to the limit cycle in one to three steps depending on the initial conditions whereas without the total energy control convergence is much slower, on the order of ten to twelve steps just as Figure 3 shows. And Figure just presents the variation process with the energy explanation variation graph. The whole detailed convergence process will be identified in the figure. The value of the storage function, S, shown in Figure 14, will decrease at each step, the implication of this is that the trajectory after each step moves closer to the limit cycle on which the energy equals the reference energy.

Seen from figure 14 (a), the initial condition lies outside the region to the limit cycle, and the robot will fall down with asymmetry gaits under this condition. By the idea of control, the trajectory of the gait will be brought back to the stable limit cycle only within few steps, this proves that, with control, an increase occur in the basin of attraction of the limit cycle.

The convergence speed and convergence efficiency of the control will be influenced by scalar k in great degree. With the simulation, a result comes out that it is not right for k to be the larger the better, virtually this feedback coefficient will possess a more complex variation during the whole control process. That is to say, there exists an "optimal" choice as well as appropriate value for k beyond which the stability is degraded, the choose value of this k will be preferable important for the control of the limit cycle.

5.4 Slope variation

As a further illustration, the performance of the system when the slope exhibits a sudden change will be presented. The control input is determined by the local slope, which is the ground slope at the stance leg. The local slope can be determined by the two-point contact condition which occurs at the moment of contact of the swing foot with the ground thus for a discrete slope change. Figure 8(a) shows that, without the total energy control, the robot is not able to maintain a stable gait. Figure 8(b) shows that, with the addition of the total energy based control $\bar{u}$, the biped successfully makes the transition between slopes. During the course of simulation, we can come to the conclusion that E_{ref} will be the decisive factor to the control, and it must correspond to certain angle ϕ, otherwise the control strategy will be out of function.

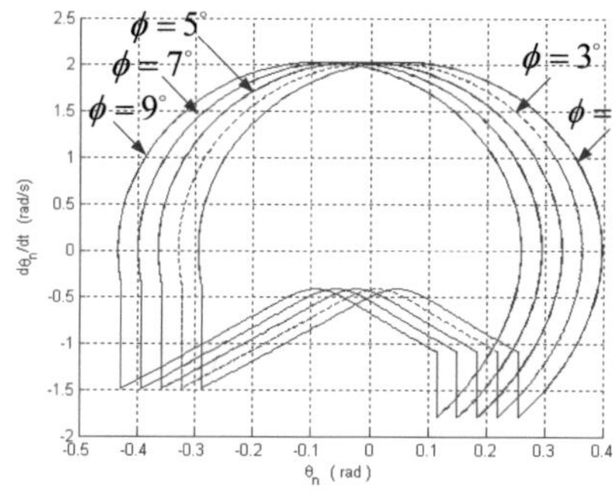

Fig. 15. The limit cycle corresponding to different slope angle using the control $\bar{u}$ with $\phi_0 = 3^\circ$

Figure 15 just addresses the detail for the limit cycle corresponding to different slope angle using the control $\bar{u}$ with the initial $\phi_0 = 3^\circ$. The control idea is thus to make the robot vary at different limit cycle to keep stable walking gaits when facing different suddenly slope change and in addition this control idea is shown to be effective in generating new stable walking gaits for biped robots. Definitely the total energy control increases the basin of attraction but there are still limits to the range of slope variation as well as disturbances that the biped can tolerate. Increasing the basin of attraction further would improve the applicability of these passivity based ideas.

6. Control of average progression speed with two actuators

Energy based control slope invariance law just discussed above works pretty well in some occasion while the law neglects the truth that actually the speed of walking gaits should be considered in some degree in passive walking when creating steady gaits. In order to solve the problem, so called average progression speed control strategy which Goswami has proposed will help us to establish the relationship between the average speed of progression and the target energy to improve the robot performance.

This control strategy for the robot is on the basis of the principle that the total energy of robot appeared constant regardless of its downward descent. The control law tries to drive the robot toward the reference energy corresponding to the energy of the limit cycle on given slope. The assumption is that, for the given slope, a passive limit cycle exists and have

already been identified. Although this may appear extremely constraining at first, while the advantage is that gaits can be generated which don't exist for the un-powered robot.
The total mechanical energy E of the robot can be expressed as $E = 0.5\dot{\theta}^{T} M \dot{\theta} + PE$. The power input to the system is the time rate of change of the total energy, $\dot{E} = \dot{\theta}^{T} S u$, for a passive cycle $u = 0$ and the reference energy $E^{*} = E(\dot{\theta}^{*}, \theta^{*})$. A simple damper control law of the form $S u = -\beta \dot{\theta}$ is used here. The power input to the system is therefore $\dot{E} = -\dot{\theta}^{T} \beta \dot{\theta}$. For a positive definite β the quantity $-\dot{\theta}^{T} \beta \dot{\theta}$ <0, which means that the robot's kinetic energy decreases monotonically. In order to simplify the choice, specify that the control law should attempt to bring the current level of the robot to the reference energy level at an exponential rate.
The hip actuator and the actuator in the supporting leg at the support of leg are available at any instant. This section will have a study on the performance of the control law with both actuators.

6.1 Control of two actuators

The idea on the control of average progression speed will be discussed as the following. The average speed per step is given by $v = \dfrac{2l \sin \alpha}{T}$, the kth step target energy is E_k^{tar}, which is equal to that the $k-1^{th}$ step with an added term. And this target energy is proportional to the error in speed. E_k^{tar} is expressed as

$$E_k^{tar} = E_{k-1}^{tar} + \eta (v^{tar} - v_{k-1}) \tag{14}$$

η is a weighting factor between energy and speed. A simplification is obtained by imposing that the hip torque be proportional to the ankle torque with a proportionality constant of μ, thus $\mu = [1 \quad \mu]^{T} \mu_H$, while the relationship of the actuator between hip torque and ankle torque can be assigned at any rate that we expect.

$$\mu_H = \begin{cases} \dfrac{-\lambda(E - E_k^{tar})}{\dot{\theta}_s(1+\mu) - \dot{\theta}_n} & \text{if } \left\| \dot{\theta}_s(1+\mu) \neq \dot{\theta}_n \right\| \succ \varepsilon \\ 0 & \text{otherwise} \end{cases}$$

With the implementation of the control law corresponding to two actuators, some satisfactory control results have been acquired and the strategy has been proved to be valid in tracking reference energy considering the influence of speed.
Figures 16 will show us some detailed message about the process that through the two actuators control in tracking the limit cycle. It will just take the robot about 30-40 gaits to walk into the limit cycle that we appoint. The collision with the ground is avoided by means of the retraction of the mass-less shank of the swing leg. In general, if the inclination of the upward slope is increased, the robot tends to lengthen the step length in order to maintain the specified speed. The same control law can be easily extended to control the robot on a terrain with a series of plane surfaces with changing slopes.
As shown in Figures 17, the desired speed is reached for a large range of values of λ. As the target speed is less than that corresponding to the passive limit cycle, the robot tries to lengthen its step length and the step period to maintain a constant average speed.

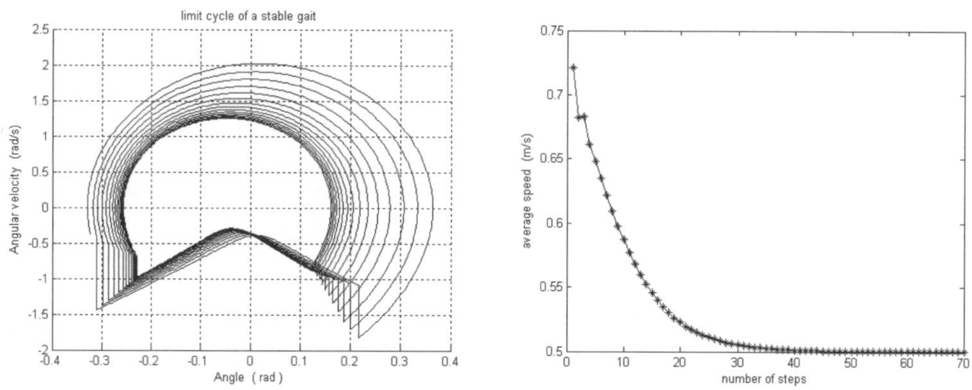

Fig. 16. Phase plane representation of the energy tracking control with two actuators added on the hip and the ankle together. The right will be the average speed and the number of steps corresponding to the two actuators control

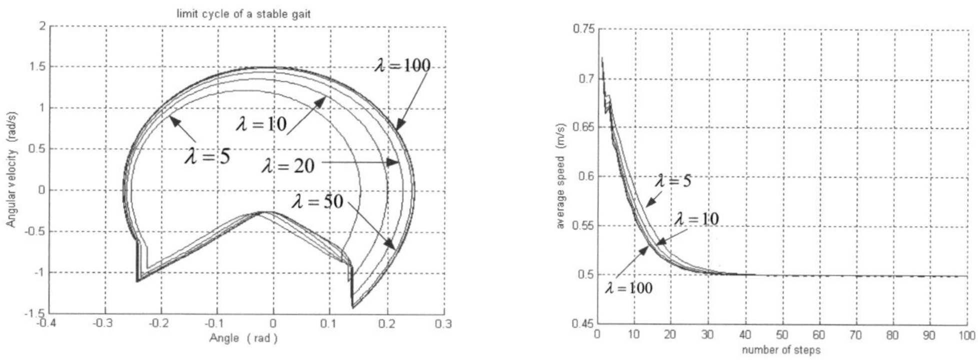

Fig. 17. Phase plane representation of the energy tracking control with respect to the different parameter λ using two actuators. The average speed and the number of steps with respect to the different parameter λ using two actuators

6.2 Some discussion about parameter variation

With the control, we find that the average progression speed control strategy can works pretty well in solving some more difficult walking gait with the appropriate parameter variation. Control of the average speed with two actuators ensures the convergence to an active cycle for a reasonably specified speed. The control law has been studied in detail by changing one parameter at a time while holding the others fixed. The parameters concerned are λ, μ, η , E_0^{tar} . The following simulations are carried with for the following parameter as

$$\phi = 0.0524, \ \alpha = 0.2710, \ E_{ref} = 153J,$$
$$\lambda = 5, \ \mu = 5, \ \eta = 2, \ V_{tar} = 5m/s.$$

λ	The desired average speed is reached for a large range of values of λ
μ	For higher values of μ, it will cause a bifurcation leading to asymmetric or 2-periodic gaits. In such a gait the average speed oscillates around the target speed, the amplitude of this oscillation increases with μ
η	Will slightly affects the speed of convergence to the cycle. And when it is zero the target energy is not updated at every step so for a target energy equal to the reference energy the robot converges to the passive limit cycle.
E_0^{tar}	Affects the rate of convergence to the target speed. The target energy is modified at every step and we can't predict a priority to what final energy the robot will converge.

Table 1. Effect of λ, μ, η, E_0^{tar}

μ	$\alpha(^\circ)$	$v(m/s)$	$T(s)$	$E_{final}(J)$
5	9.9240	0.5	0.6890	148.9767
8	10.5769	0.5	0.7339	149.2529
10	10.8739 and 10.4443	0.5222 and 0.4778	0.7222 and 0.7584	149.8433 and 149.9025

Table 2. Effect of μ on the control performance. The table corresponds to simulations on a 3° slope with parameters: $\lambda = 5$, $E^{tar} = 153\,J$, $\eta = 5$

The most curious effect of μ is that for higher values, it will cause a bifurcation leading to asymmetry or 2-periodic gaits just shown as the data of Table 2. In such a gait, the average speed oscillates around the target speed, and the amplitude of this oscillation increase with μ. And figure 15 will show us the limit cycle under the so called 2-periodic gaits state. At this time, the gaits just locate in the limit cycle and is about to get away from this stable state if some slight disturbances working on the gaits. In addition, when getting out of this state, the walking gait of the robot will go into chaos and then slip down. Figure 16 is the graph of the average speed and the number of steps with 2-periodic gaits state. With different initial energy value, the graph of average speed and the number of steps will be different due to the sensitivity of chaos.

Furthermore, some more attention should be paid to the work of how to identify the boundary of the basin of attraction and how to determine the favorable initial conditions effectively. In spite of all those stated in the paper, there still exists some other problems such as the robot's behavior is heavily influenced by the impact model which the paper proposed is not the only available impact model. How to model some new realistic foot/ground impact models possessing such a manner that reasonable perturbations of the model parameters don't dramatically change the gait, should be considered.

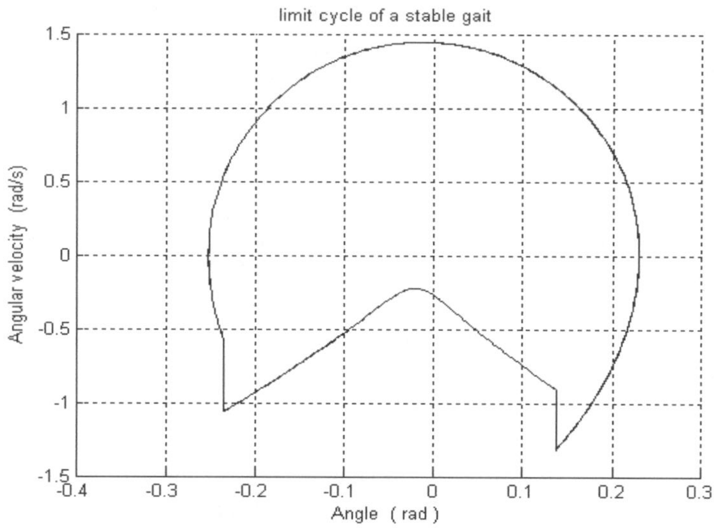

Fig. 18. Phase plane representation of the energy tracking control using two actuators with 2-periodic gaits state.

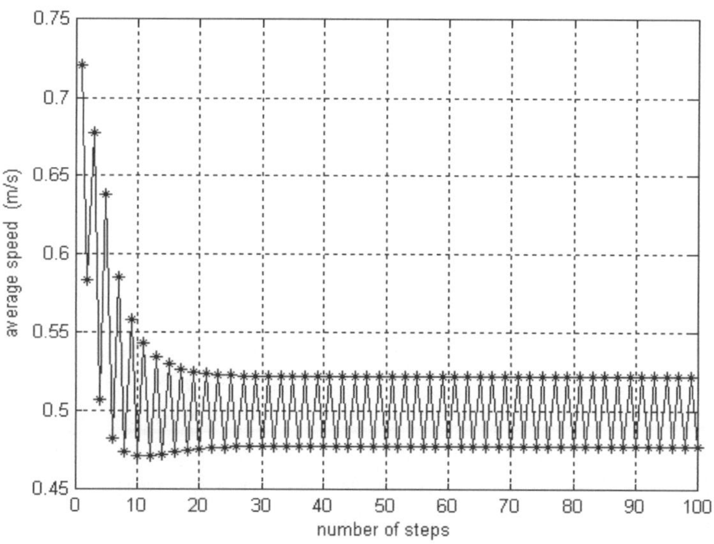

Fig. 19. The average speed and the number of steps corresponding to the two actuators control with 2-periodic gaits state

7. Influence of robot parameters on the gait

This section presents the effects of continuous change of the parameters ϕ, μ and β on the gait of our compass-like biped robot. First we discuss the limitations of al linear model in predicting the robot's long term behavior. Next we point out the general features of the symmetric gaits of the robot-this section mainly consists of a graphical presentation. When one of the parameters exceeds a certain limiting value, we observe bifurcation of the dynamics which we discuss subsequently. Finally we focus on the features of chaotic behavior of the robot gait.

7.1 Symmetric gait

This section presents the evolution of pertinent gait descriptors as functions of the three parameters during the symmetric gait regime of the robot. As opposed to a parameters which can be directly altered , a gait descriptors is an observed(measurable or computable) quantity which cannot be modified directly but is indirectly influenced by the parameters. The gait descriptors that appear the most meaningful to us for this study are the state variables q, the half inter-leg angle at touchdown α ,the step period T ,the average speed of progression v ,the total mechanical energy of the robot E, and the loss of mechanical energy ΔE due to impact.

The evolution of the gait descriptors is presented in the form of so-called bifurcation. Figs.5(a) to 5(f), 6.(a) to 6.(f) and 7(a) to 7(f). present the evolution of the gait descriptors that appear the most meaningful to us for this study are the state descriptors T , α , $\dot{\theta}_S$ (at the beginning of a step), v , E and $\frac{\Delta E}{E}$ as functions ,respectively ,of the parameters ϕ and μ but decreases with β .The results show that both the step period and the step length of the robot .The overall behavior of the robot can be summarized qualitatively as follows:

	T	L	E	V
$\phi \nearrow$	$\nearrow$	$\nearrow$	$\nearrow$	$\nearrow$
$\mu \nearrow$	$\nearrow$	$\nearrow$	$\nearrow$	$\nearrow$
$\beta \nearrow$	$\nearrow$	$\nearrow$	$\searrow$	$\searrow$

Some interpretations are in order here. Let us consider the evolution of total mechanical energy E of the robot in response to parameter changes. As the ground slope ϕ increases the potential energy PE of the robot available per step slightly increases. The kinetic energy KE ,being roughly proportional to $\|\theta\|^2$,increases also, see Fig.5(c). As a consequence the total energy E, Fig.5(e). An increase in β results in a lowering of the center of mass of the robot, which lowers PE available per step and increases the step period . The latter results in a decreases in the average velocity of the robot (Fig.7 (d)). The increase in KE caused by the small increase in the $\dot{\theta}_s$ cannot compensate for the decreases in PE and consequently lowers E. Conversely, an increase in μ , which results in raising the center of mass of the robot , increase E.

It is interesting to look at the effect of a parameter change on the evolution of entire limit cycles as shown in Fig.s5(g), 6(g) and 7(g). In response to an increase in ϕ the limit cycle expands along both axes, see Fig.5(g), implying an increase in the range of joint angle and joint velocity .The limit cycles are compensated along the joint velocity axis for an increase in the parameters μ and β (Figs .6(g) and 7(g)). A shorter reach of the limit cycle along the joint velocity axis means a smaller maximum joint velocity but dose not necessarily mean a slower robot . We see in Fig6 (d) that an increase in μ is associated with an increase in the average speed of progression v .

7.2 Chaotic bifurcation

7.2.1 Period-doubling bifurcation

We noticed in Figs.5 and 6 that for the range of variations of the parameters considered in this study an increase in ϕ and β cause a bifurcation in all the gait descriptors. Bifurcation was also observed for higher values of μ especially when coupled with higher values of ϕ (Fig.7).

As a consequence of the period-doubling bifurcation the limit cycle becomes 2-periodic and the robot gait becomes asymmetric with a shorter step and a longer step. The occurrence of bifurcation is shown in Figs.5,6,7 by the emergence of two branches in the curves, each associated with one of two dissimilar steps and describing its characteristic variables. Since bifurcation involves the state of the system and since all the gait descriptors, in turn, depend on the robot states, the occurrence of bifurcation is simultaneously manifested in all the gait descriptors.

On further increasing the parameters , the robot gait may experience a further period-doubling, giving rise to a 4-periodic limit cycle . This phenonmenon , repeated ad infinitum, is called a period doubling cascade and is recognized as one of the possible routes leading to chaos. Regardless of the parameter considered, we observe that the successive period doubling occur after progressively smaller intervals of parameter variation. This is expected in view of general results on period doubling casacades.

Period doubling cascades leading to chaotic behavior have already been observed for passive planar hopping robots which possess a smaller dimension than that of the compass. 2^n-periodic gaits, termed as "limping gaits," were observed and analyzed for hopping robots.

In Fig.9 we introduce a novel way of capturing the behavior of the biped during a period doubling cascade ensuring from the parameter ϕ (other parameters are kept constant at $\mu = 2, \beta = 1$). The figure plots the first plots the first return map of θ_{ns}. For a 1-periodic robot gait θ_{ns} is the same in every step. This gait is therefore represented by a point on the 45° line.

As we change the ground slope, this point moves along the 45° line from the right-hand top corner of Fig.9, as indicated by the arrow.

The first period doubling occurs at $\phi = 4.38^{\circ}$ when the gait turns 2-periodic and is therefore represented by 2 points. Just after the first bifurcation the 2 representative points differ only slightly from that of the 1-periodic gait from which they originate. The two steps are therefore very similar to the steps of the steps of the symmetric gaits. On further changes , in the parameter the two representative points move away from the 45° line along the two branches shown by dotted lines in Fig.9. It follows that one step length is slightly longer and the other slightly shorter than those of the corresponding symmetric gait. As we increase the slope the longer step is further elongated and the shorter step further shortened.

This continues until a second period doubling occurs at $\phi = 4.93^{\circ}$ when each branch gives rise to two sub-branches. In this 4-periodic gait the 4 different steps are visited in the same order with a longer step always followed by a shorter step. The last clearly identifiable bifurcation occurs when $\phi = 5.02°$ as the robot gait becomes 8-periodic.

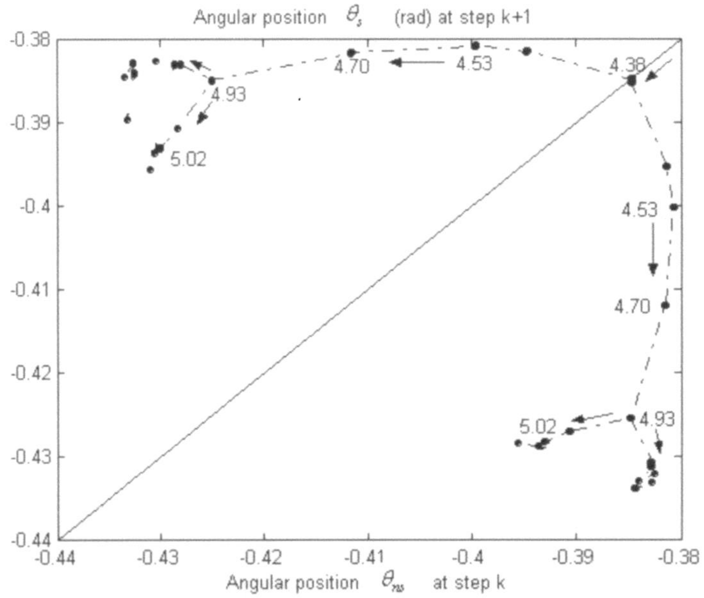

Fig. 20. First-return map of 2^n-periodic steady gaits

The period doubling cascade may also be observed using phase plane diagrams. The phase plane diagram for a symmetric gait. which is a single-loop closed trajectory repeated after two robot steps. During one step the considered leg is in the swing stage and during the following one, it is in the support stage. Since the gait is symmetric, the robot legs are indistinguishable and the phase plane cycles of the two legs are identical.

In case of a 2-periodic gait, since all state variables are identical after every two steps, the phase plane limit cycle associated with one leg is still a single-loop closed trajectory repeated after two robot steps, see Fig.21(a). However, since the gait is asymmetric, the limit cycles associated with the legs are no longer identical.

In case of 2^n-periodic gaits, all the state variables repeat themselves after every 2^n steps. The phase plane diagram associated with one leg is therefore a 2^{n-1}-loop closed trajectory repeated after every 2^n steps, distinguishable from the phase diagram of the other leg. The visual inspection of the phase plane diagrams of the 4-periodic and the 8-periodic gaits (Fig. 21(b) and 21(c), respectively) correctly indicates that they resulted from the bifurcation of respectively the preceding 2-periodic and the 4-periodic gaits.

7.2.2 Chaotic gaits

The chaotic gait is an extreme case of the asymmetric gait and is characterized by a complete disappearance of order in a system. During a chaotic gait on a given slope, the states, and consequently the gait descriptors, of the biped robot never completely repeat themselves. Chaotic gaits are represented in the bifurcation diagrams by a continuous distribution of points. We explicitly show this on Figs.22(a) and 22(b) and omit them in the other bifurcation diagrams for the sake of clarity.

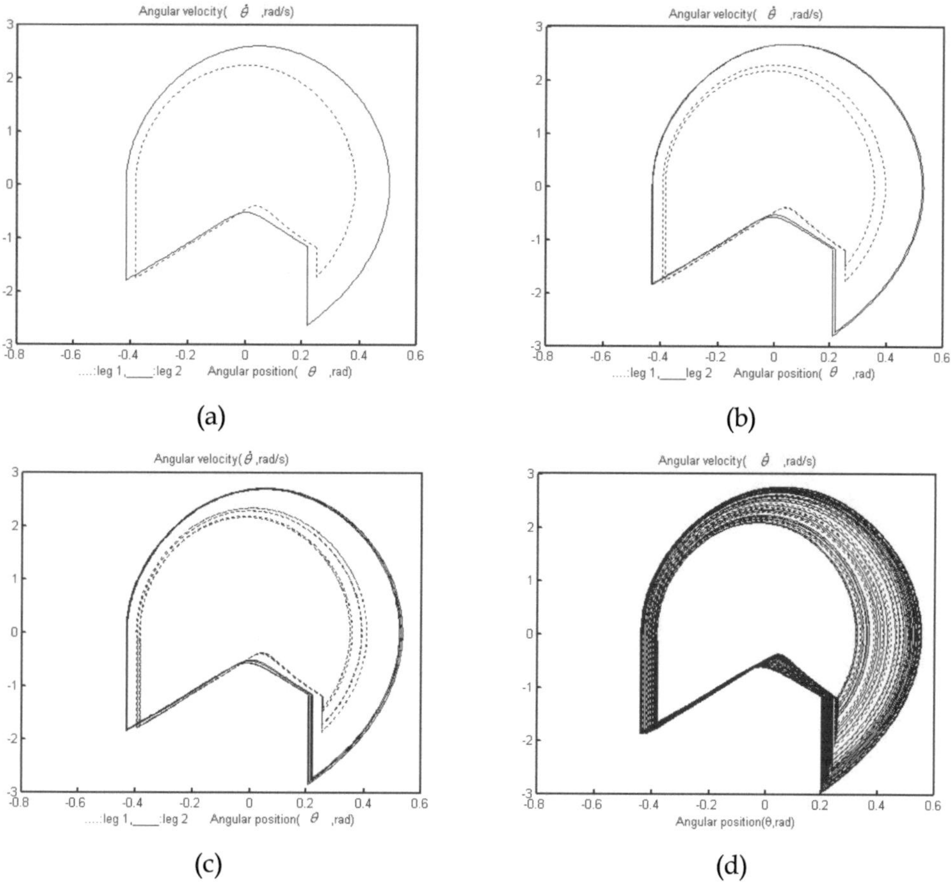

Fig. 21. Phase plane limit cycles of a) 2-periodic steady gait($\phi = 4.38^{\circ}$),b)4-periodic steady gait($\phi = 4.93^{\circ}$), c) 8-periodic steady gait($\phi = 5.02°$) d) chaotic gait associated with one leg, 100 robot steps, ($\phi = 5.2^{\circ}$). For all the 4 subplots $\mu = 2, \beta = 1$.

The gradual progression of the robot gait to the chaotic regime is well depicted in the first return maps of $\theta_{ns,k+1} = f(\theta_{ns,k})$ shown in Figs. 11(a) to 11(d). When $\phi = 5.02^{\circ}$, the gait is 8-periodic and its first return map consist of 8 points. At $\phi = 5.05^{\circ}$,the first return map still consists of 8 distinguishable clusters of points(Fig.11(a)). Through multiple period doubling bifurcation this 8-periodic gait gives rise to a 2^n -periodic gait with a large n . This gait will still preserved and θ_{ns} is still always followed by a small one. The same property is still preserved, since a large θ_{ns} is still always followed by a small one. The same property still holds for $\phi = 5.13^{\circ}$, but in this case the first return map appears as a continuum of points(Fig.11(c)). We are therefore very close to the "broad-band frequency" characteristic typical of chaotic behavior. Finally, when $\phi = 5.21^{\circ}$, we observe that predictability and periodicity have been completely destroyed, since a large θ_{ns} can be followed by another large one. The layered structure of the strange attractor can also be guessed from the first return map.

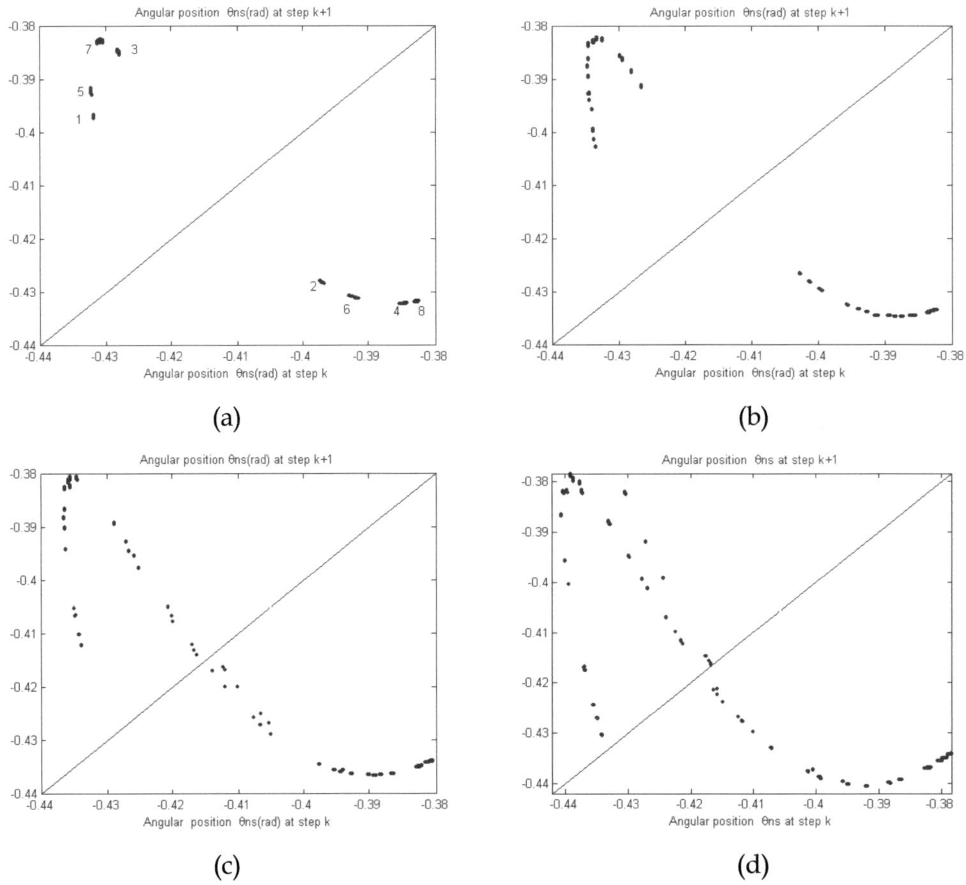

Fig. 22. First return map of θ_{ns}: a) 2^n-periodic gait, n large($\phi = 5.05°$), b) 2^n-periodic gait, n very large($\phi = 5.10°$), c) approaching chaotic gait ($\phi = 5.13°$), d) chaotic gait ($\phi = 5.21°$).For all the 4 subplots $\mu = 2, \beta = 1$.

7.2.3 Local stability of the limit cycle

One way to investigate the orbital stability of a limit cycle is by means of studying the stability of its fixed point in the Poincare map. As a natural choice studying of the Poincare section of the compass biped we take the condition that the swing leg of the robot touches the ground. For two successive touchdowns of the same leg the states of the robot can be related as

$$x_k = F(x_{k+1}) \tag{14}$$

Where $x = [\theta_{ns}, \theta_s, \dot{\theta}_{ns}, \dot{\theta}_s]^T$ is the 4-component state vector of the robot.

For a cyclic phase trajectory the first return map is fixed point of the mapping. On a cyclic trajectory, therefore , $x_k = x_{k+1}$ and we can write, $x^* = F(x^*)$ .For a small perturbation Δx^* around the limit cycle the nonlinear mapping function F can be expressed in terms of Taylor series expansion as

$$F(x^* + \Delta x^*) \approx F(x^*) + (\nabla F)\Delta x^* \tag{15}$$

Where ∇F is the gradient of F with respect to the states. Since x^* is a cyclic solution ,we can rewrite Eq.2as

$$F(x^* + \Delta x^*) \approx x^* + (\nabla F)\Delta x^* \tag{16}$$

The mapping F is stable if the first return map of a perturbed state is closer to the fixed point. This property can be viewed as the contraction of the phase eigenvalues of ∇F at the fixed point x^* are strictly less than one. From Eq.3 we write $(\nabla F)\Delta x^* \approx F(x^* + \Delta x^*) - x^*$ where $F(x^* + \Delta x^*)$ is the first return map of the perturb one state $x^* + \Delta x^*$. As it is not practical to analytically calculate perturb one state at a time by a small amount and observe its first return map. Repeating this procedure at least four times (once for each of the four states) we obtain an equation of the form

$$(\nabla F)\tau = \Psi \tag{17}$$

When ϕ increases from 0.0524 to 0.0824, the variation of eigenvalues are as follows:

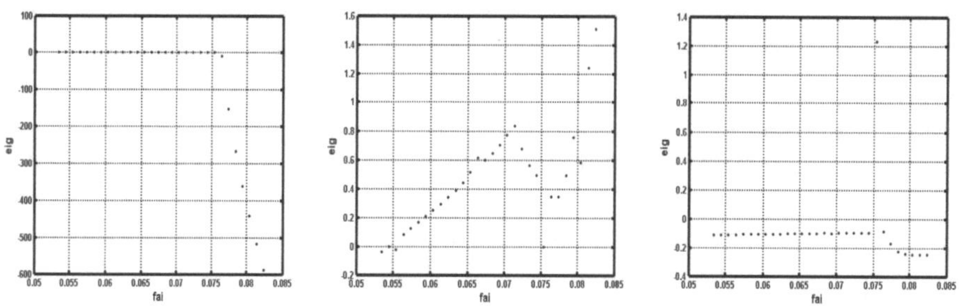

When $\phi = 0.0908$, the limit cycle is as follows:

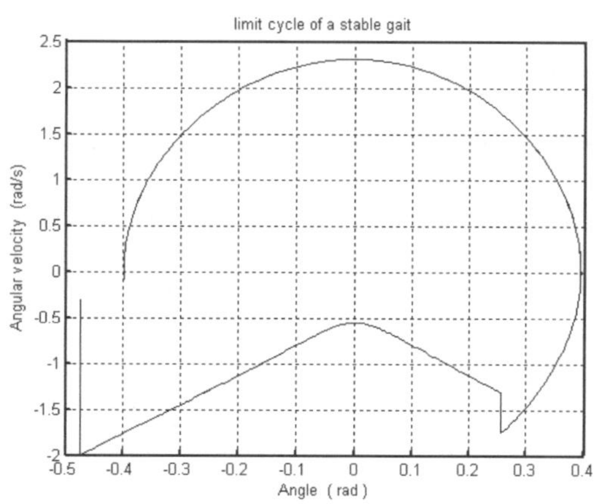

When μ increases, the variation of the first, the second and the fourth eigenvalues are as follows:

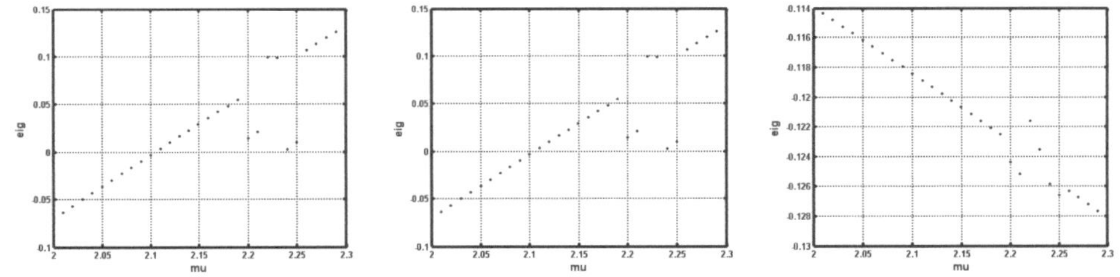

When β increases, the variation of the first, the second and the fourth eigenvalues are as follows:

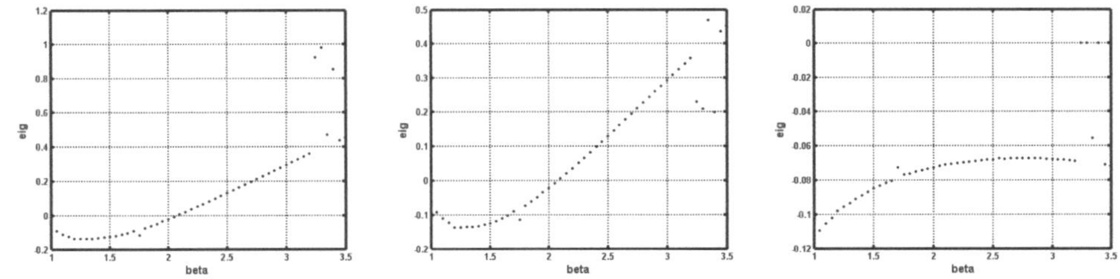

When μ and β increase, the real parts of the first and the second vary from negative to positive. It shows that the system varies from stable to unstable.

7.2.4 Chaotic control laws

Here we introduce a simple control law which was inspired by the passive energy characteristics of the compass model. As the robot walks down on a slope its support point also shifts downward at every touchdown . As it loses gravitational potential energy in this way its kinetic energy increases accordingly. In a steady walk this is exactly the amount of kinetic energy that is to be absorbed at the end of each step by the impact. If, at every touchdown we reset our potential energy reference line to the point of touchdown, the total energy of the robot appears constant regardless of its downward descent. We formulate a control strategy for the robot based on this principle. The control law, aware of this characteristic energy of the passive limit cycle, called the *reference energy* in this section, of the robot on a given slope tries to drive the robot toward it.

8. The introduction of anti-phase synchronization

Observing from the human gait biped, symmetry is an important indicator of healthy gait[6]. The presence and nature of asymmetry in gait can be a useful diagnostic tool for the clinicians. Symmetry can be measured through the use of so many kinetics variables such as acceleration, force, moment, energy, power, step period and step length. Is it possible to apply this obviously symmetry property of healthy gait in human walking into the design of the robot's gait and explicitly explain the efficiency of human and animal locomotion more in detail will be a new challenge. Some new control strategy of "anti-phase synchronization" has been presented here to reduce the complexity such as the property of nonlinearity and strong coupling of this hybrid dynamic system. To the best of our knowledge, it is the first time to introduce the concept of synchronization to explain and control the motion of passive biped theoretically.

For a perfectly symmetric gait a properly synchronized twin trajectories from corresponding joints should be identical. Through the control and the reduced presumption of the collision model, the strong coupling between two legs has been successfully erased. A controller which is able to solve the synchronization problem in such a way that the pendulum reaches the desired level of energy and they move synchronously in opposite directions has been presented and in addition the construction of new Lyapunov function and simulation results prove the validity of the strategy. The method stated in the paper is helpful to practical application of the design of the robot's gait.
The paper is organized as follows. First we formulate the problem statement. Next we analyze the behavior of compass-like biped. Then the symmetry property in gait biped and the possibility to the application of the anti-phase synchronization have been discussed in III as well as the problem of erasing the coupling between two legs. The main contribution of V is the construction of Lypunov function, the proposed controller and also the local stability analysis. Simulation results stated in IV just verify the effectiveness of the proposed method. The conclusions and future work are formulated in the final part.

8.1 Some symmetry property in gait biped

One of the most important properties in steady gait biped is that there exists some kind of symmetry with the variation of angle position and angle velocity.
The comparison relationship on the angular position of the two legs during steady periodic gait cycle has been presented in Fig. 3. It is obviously that angular positions of the two legs are asymmetry. The gait biped walking works as a double pendulum, while the stance leg has a comparative big difference with the swing leg. This can be explained that the stance leg experiences a relative compelled motion with the action force coming from the swing leg as well as from gravity. And in addition, the coupling degree of the two legs varies at different instant.
Presume the intersection point between two legs in Fig. 3 and the middle point with the two culmination value within one cycle of the swing leg as the symmetry point respectively, we get the asymmetry degree figures about two legs.
Observing from Fig. 4(a), different hip mass will correspond to the result that the larger the hip mass is, the higher the symmetry degree is. That is to say, the coupling effect between two legs will be influenced by the hip mass in great degree. With Fig. 4(b), different μ will correspond to the different error about the angular position and angular velocity. The conclusion is very important for it will help in modifying the gait biped model when choosing the parameter and adjusting the gait biped cycle. As stated above, Figures 4(a) and 4(b) just provide us a kind of symmetry. While for the coupling of the two legs, it is impossible to construct the same ideal sub-systems to fulfil traditional master-slave synchronization corresponding to the dynamic system of the robot. The most direct way is to erase the coupling of the two legs and construct the subsystem of the swing leg and stance leg respectively, observing their position and the angle velocity relationship.

8.2 Anti-phase synchronization

In order to make the system to reach a kind of synchronization, the complex dynamic equation should be simplified as: 1. erase the coupling of two legs. 2. construct the new collision model between the swing leg and the ground. 3. add the control with the least energy consumption at appropriate instant as impulse force to imitate the behaviour of human gait. The paper will pay great attention to the solution of the first problem.

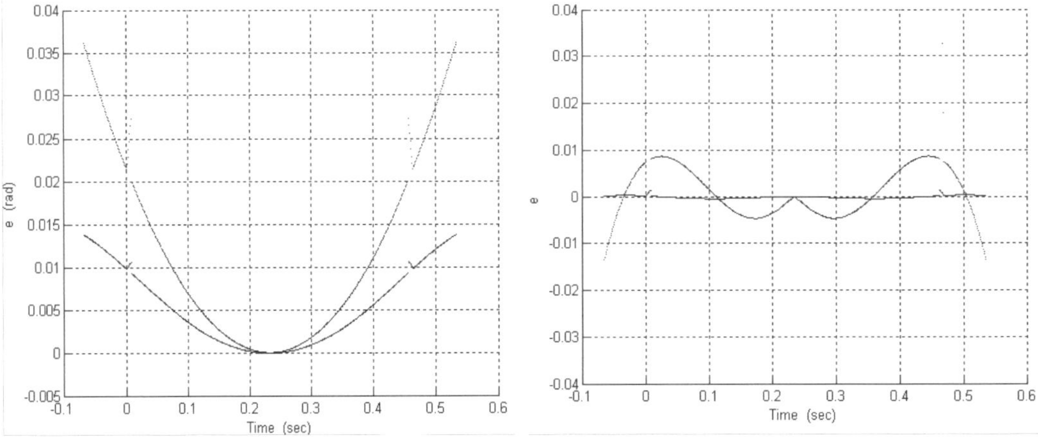

(a) (b)

Fig. 23. The asymmetry degree Figures about two legs with different symmetry point (a) inter-section point between legs (b) centre value with two culmination within one cycle of the swing leg

By applying the idea of "inverse dynamic control"[10], make both the gravity torque act on the robot and the added control torque be equivalent to a gravity action, then a closed loop linear system with the same effect on the robot can be obtained. The advantage of the idea is that it can help adjust the gait distance and also the period corresponding to the forward varied velocity at any instant. Through the control, the swing leg acts as the single pendulum and the stance leg works as the inverted pendulum, the dynamic property of the two legs are the same except the analysis of the equilibrium point and the stability.

In addition, a reasonable presumption can be provided that there exists no collision with the swing leg and the ground with respect to this kind of pendulum walking. That means with the algebra constraint added, during the cycle of gait biped, the tip of the swing leg slides with the ground all the time and no friction will be considered when the robot moving forward. The construction of the new collision model can be solved by the consideration of knees which is not the topic of this paper. It is reasonable for us to eliminate the impact of collision model here for the impact can be solved by some idea such as time delay set and other counteract equipment when designing the real robot. Under this condition, the phase graph of the gait will be a perfect circle, at the end of each gait cycle, the velocity of the leg will be set zero and at the same time preceding the velocity conversion.

8.3 Erasing the coupling

There exists strong coupling action between θns and θs when analysing the dynamic equation. Erasing the coupling and construct the same sub-system with the idea of "inverse dynamic control", then obtain a closed loop linear system. For the non-linear equations (2) of our biped, as the stance leg is about to leave the ground, the anti-phase control is induced to the equation with the form

$$u = B^{-1}(M(\dot{q})a + C(q,\dot{q})\dot{q} + g(q)) \tag{18}$$

Reduces the system to the decoupled double integrator system

$$\ddot{q} = a \tag{19}$$

Joint angles can then be controlled independently using a control law

$$a = -K_p q - K_d \dot{q} + r \tag{20}$$

Where K_p and K_d are diagonal matrices with elements consisting of position and velocity gains, respectively. For a given desired trajectory

$$t(q^d(t), \dot{q}^d(t)) \tag{21}$$

We can choose the input $r(t)$ as

$$r = \ddot{q}^d(t) + K_p q^d(t) + K_d \dot{q}^d(t) \tag{22}$$

The desired trajectory can be obtained as cubic trajectory as shown in [11] if the initial and final states of the trajectory are known. Thus a kind of synchronization can get with the walking trajectory and the given trajectory.
Simulation demonstrates that the rule for the swing leg is similar with the simple pendulum, during the process of anti-phase synchronization control, keep the dynamic state of the swing leg and make the stance leg act with the same rule, then

$$a = \begin{bmatrix} -\frac{g}{kl}\sin(\theta_{ns} + \phi) \\ -\frac{g}{kl}\sin(\theta_s + \phi) \end{bmatrix} \tag{23}$$

k is the parameter representing the mass centre of the pendulum. With the same parameter described in section 2, the distance between the hip and the mass centre is b, the distance between the foot and the mass centre is a, where $k = \frac{\beta}{\beta + 1}$, $\beta = \frac{b}{a}$.
The dynamic equation of the robot can be divided into two independent parts, and both of them possess the same expression as

$$\ddot{\theta}_{ns} + \frac{g}{kl}\sin(\theta_{ns} + \phi) = 0 \tag{24}$$

$$\ddot{\theta}_s + \frac{g}{kl}\sin(\theta_s + \phi) = 0 \tag{25}$$

8.4 Control synchronization

For global application of the synchronization method, introduce the new coordination

$$\theta_{ns}^t = \theta_{ns} + \phi$$

$$\theta_s^t = \theta_s + \phi$$

With the control strategy stated above, equations of the two legs have been given as (9) and (10) with the same dynamic control rule. Assume that both the legs possess the point mass m, and then the virtual mechanical energy of the robot is

$$V = \frac{1}{2} m(kl)^2 ({\theta_{ns}^{t}}^2 + {\dot{\theta}_{s}^{t}}^2) + mgkl(1 - \cos\theta_{ns}^{t}) \\ + mgkl(1 - \cos\theta_{s}^{t}) \tag{26}$$

Obviously $V \geq 0$, appoint V as the Lyapunov function of the system, for the collision has been avoided here, thus the system is conservative, then $\dot{V} = 0$. Seen from equation (11), the mechanical energy of the system is constant, this proves that the dynamic behaviour of the two legs can come to the state of anti-phase synchronization; expected ideal symmetry property appears here.
Synchronizing the two dynamic systems with the idea (12) so called mutual direction coupling synchronization.

$$\begin{cases} \dot{x} = f(x) + \bar{K}(\bar{x} - x) \\ \dot{\bar{x}} = f(\bar{x}) + \bar{K}(x - \bar{x}) \end{cases} \tag{27}$$

Where $\bar{K}(\bar{x} - x)$, $\bar{K}(x - \bar{x})$ is the mutual coupling synchronization control item of the two sub-systems. In addition, it will be adjusted different for the purpose of improving the synchronized precision and enhancing the synchronized velocity.
$K = diag(k_1, k_2, ..., k_n)$ is so called coupling length, where $n = 2$ here. Therefore the control objective can be formalized by the following relations

$$\lim_{t \to \infty}(\theta_{ns}^{t} + \theta_{s}^{t}) = 0 \tag{28}$$

$$\lim_{t \to \infty}(\dot{\theta}_{ns}^{t} + \dot{\theta}_{s}^{t}) = 0 \tag{29}$$

Synchronized time of the system should be considered here and two legs would come to the state of anti-phase synchronization within one cycle with the control. In order to fulfil the control of the gait biped, two main problems are discussed as follows: 1. stability with which the robot will not to slip forwards or downwards corresponding to the gait biped in a 2-dimension plain. 2. the gait should satisfy any given target velocity and gait distance as well. To reach this kind of target control goal, an applicable method is to control the amplitude of the swing leg with energy consumption consideration.
Add the new controller which is related to in system (9) and (10)

$$\ddot{q} = a + \bar{B}\bar{u} \tag{30}$$

Where

$$\bar{B} = \begin{bmatrix} 1 & 0 \\ -1 & 1 \end{bmatrix}$$

$$\bar{u} = \begin{bmatrix} -\gamma\,\dot{\theta}_{ns}^{t} H - \lambda_1 \sin(\theta_{ns}^{t} + \theta_{s}^{t}) - \lambda_2(\dot{\theta}_{ns}^{t} + \dot{\theta}_{s}^{t}) \\ -2\gamma\,\dot{\theta}_{ns}^{t} H - 2\lambda_1 \sin(\theta_{ns}^{t} + \theta_{s}^{t}) - 2\lambda_2(\dot{\theta}_{ns}^{t} + \dot{\theta}_{s}^{t}) \end{bmatrix} \tag{16}$$

Where $H = H(\theta_{ns}, \dot{\theta}_{ns}) + H(\theta_s, \dot{\theta}_s) - 2H^*$. λ_1, λ_2 is the ratio coefficient, which decides the converging speed of anti-phase synchronization; γ is positive gain coefficient. H represents Hamiltonian function of each pendulum-like leg and the designed controller (16) can swing the pendulum up to the desired energy level H^* in such a way that the pendulum-like two legs move in opposite directions.

Theorem: For any given controller presented as (16), if satisfies $\lambda_2 > 2H^*$, then the set $\theta_{ns}^{t} = -\theta_{s}^{t}$, $\dot{\theta}_{ns}^{t} = -\dot{\theta}_{s}^{t}$ is globally asymptotically stable with respect to the controlled system (15).

Proof: Construct the virtual Hamiltonian function of each pendulum-like leg

$$H(\dot{\theta}, \theta) = \frac{1}{2} m(kl)^2 \dot{\theta}^{t2} + mgkl(1 - \cos\theta^{t}) \tag{31}$$

The control objective can be formalized by the following relations

$$\lim_{t\to\infty} H(\theta(t), \dot{\theta}(t)) = H^* \qquad \theta = \theta_{ns}^{t}, \theta_{s}^{t} \tag{32}$$

The relation implies that the periods of oscillations of each pendulum are identical (frequency synchronization).

Using the control law (16), analyses the equations of the closed loop system with respect to the variable $x = \theta_{ns}^{t} + \theta_{s}^{t}$, then

$$m(kl)^2 \ddot{x} + \lambda_2 \dot{x} + \lambda_1 \sin x$$

$$= -\gamma[H(\theta_{ns}^{t}, \dot{\theta}_{ns}^{t}) + H(\theta_{s}^{t}, \dot{\theta}_{s}^{t}) + \lambda_2 - 2H^*]\dot{x} \tag{33}$$

Define a new Lyapunov function to the whole system

$$V = \frac{1}{2} m(kl)^2 \dot{x}^2 + \lambda_2(1 - \cos x)$$

$$+ mgkl(2 - \cos\theta_{ns}^{t} - \cos\theta_{s}^{t})$$

Obviously $V \geq 0$

$$\dot{V} = m(kl)^2 \dot{x}\ddot{x} + \lambda_2 \dot{x} \sin x$$

$$= -\gamma[H(\theta_{ns}^{t}, \dot{\theta}_{ns}^{t}) + H(\theta_{s}^{t}, \dot{\theta}_{s}^{t}) + \lambda_2 - 2H^*]\dot{x}^2$$

for $H(\theta_{ns}^{t}, \dot{\theta}_{ns}^{t}) \geq 0$, $H(\theta_{s}^{t}, \dot{\theta}_{s}^{t}) \geq 0$, if $\lambda_2 > 2H^*$, $\dot{V} \leq 0$. Then the set $x = 0$ is globally asymptotically stable.

From this observation one can make a few important conclusions. First, the uncontrolled system can exhibit synchronous behaviour. Clearly, it follows that the Hamiltonian of each pendulum tends to a common limit. However, due to energy dissipation the limit value depends on the initial conditions and particularly, if one initializes the pendulum from an identical point, the oscillations will decay. Therefore the uncontrolled system exhibits a behaviour which is very close to the desired one, and there is one thing to the controller-to maintain the energy level for each pendulum, and this problem will be useful for the further research.

As predicted by the theorem, there is a set of zero Lebesgue measure of exceptional initial conditions for which the control objective can not be achieved. For example, if one initiate the system at the point where $\theta_{ns}^{t} = \theta_{s}^{t} = 0$, $\dot{\theta}_{ns}^{t} = \dot{\theta}_{s}^{t} = 0$ the anti-phase synchronization control $\bar{u}$ based on the energy can't drive the system away from the zero condition, however from practical point of view, it is not difficult to modify the controller to handle this problem.

Presume the system has been in the condition of anti-phase synchronization, that is, the stable point of the closed loop system, the limit dynamics of each pendulum is given by the following equation

$$m(kl)^2\ddot{\theta}^t + mgkl\sin(\theta^t) = -2\gamma\left[H(\theta^t,\dot{\theta}^t) - H^*\right]$$

And therefore the control objective

$$\lim_{t\to\infty} H(\theta^t,\dot{\theta}^t) = H^*$$

is achieved.

8.5 Simulation and discussion

To verify the effectiveness of the proposed method, we conduct the following simulation results.

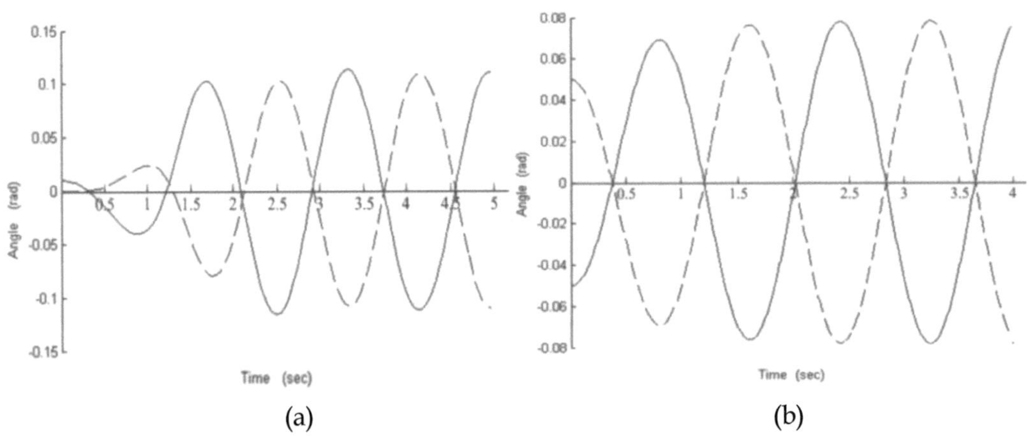

(a) (b)

Fig. 24. Anti-phase synchronization of two legs in the controlled system with different initial condition $(\theta_{ns}, \theta_s, \gamma)$ (a) (0.01, 0, 10) (b) (-0.05, 0.05, 10)

The limit cycle under anti-phase synchronization is given by Figure 5 and its Mechanical energy of the robot under anti-phase synchronization is just depicted by Figure 24. It is clearly that the anti-phase synchronization can enlarge the convergence region of the limit cycle and it appears the typical double-pendulum property.

Figures 24 present the results of the anti-phase synchronization based on the energy control. With the anti-phase synchronization control, the angular position and angular velocity of the two legs can reach the synchronization with the same magnitude and the opposite direction at any moment. Simulation results prove the validity of the control method. In addition, as simulation demonstrates, with different initial condition of the gait biped will be around zero as well as any given biped value and with the control it can converge to the state of anti-phase synchronization corresponding to any given target energy. The converging speed depends on the controlled parameter $\gamma,\lambda_1,\lambda_2$. Virtually the higher the value of the parameters are, the faster the converging speed is. The consequence is valid with the consideration of both the constraint of the manipulator and the appropriate maximized sustaining force added on the system. In practice, for not using the initial condition as it does in simulation, so the effect of the control will be better in controlling the practical robot.

In this work, we considered the problem of controlled synchronization of the decoupled two legs based on the compass-like biped model. Some new method so called "anti-phase synchronization" has been presented to explain the perfectly symmetric gait typical of healthy gait in human walking. The paper also proposed a useful controller which is able to solve the synchronization problem in such a way that the pendulum reaches the desired level of energy and they move synchronously in opposite directions. In addition, the construction of Lypunov function, the local stability analysis to the proposed controller as well as the presentation of simulation results have also been stated and proved the validity of the method.

For the next step, the design of a more efficient new collision model and also the further analysis about the added simplified impulse force under new condition will be helpful to practical application of the design of the robot's gait.

9. The description and assumption on he model with knees

Next, we will extend the model to the new one with knees, and the state of the straight direction of the gait biped will be equivalent to the one of the compass-like gait biped, then the equation can be united as the unanimous form discussed formerly the model shown as Figure 25, partially resemble Compass-like waker with point masse m_H, m_s and m_t concentrated at the hip ,shanks and thighs respectively. The leg-length is L, which is divided into three parts: l_s and l_t, l_s is the distance from the heel to the knee of m and l_s is the distance from knee to the hip center m_H ,in the meanwhile,both shank and thigh are divided by their respective sub-mass center m_s and m_t into two parts,with l_s a_1 and b_1 ,with l_t a_2 and b_2. $l_s + l_t = L, a_1 + b_1 = l_s, a_2 + b_2 = l_t$ As is depicted in figure 1, three key parameters are needed to describe the configuration of the walker, q_1, q_2 and q_3. 2α is the total angle between the legs, which is defined as the "inter-leg angle", and in addition is formed during the instant when both legs are touching the ground. The slope of the ground with the horizontal is denoted by the angle γ .

The model has been made by the following assumptions: the total mass of the robot $m_C = 2m_s + 2m_t + m_H$ is constant. For the sake of analysing the model, we separate the motion into two phases, knee-free phase and knee-locked phases,whose boundaries are knee-strike and heel strike,that is the period between knee-strike and heel-strike is knee-locked stage and the period between heel-strike and knee-strike is knee-free stage.

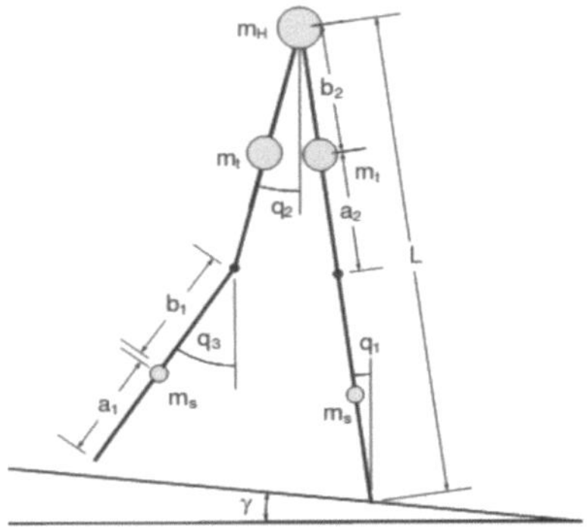

Fig. 25. Model of a passive kneed-walker on a slope

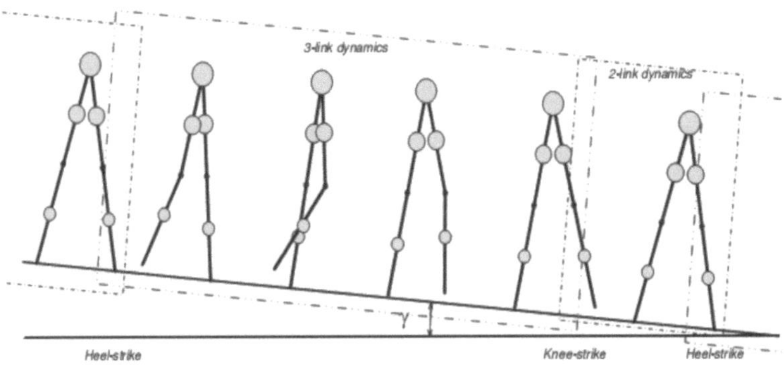

Fig. 2. Stages in a step cycle of a kneed-walker

In order to simplify the analysis and calculation of The waking model, we shoud make some assumptions at first. All masses are considered point-masses and one leg are identical with the other. The gait consists of knee-free stage and a knee-locked stage: during the knee-locked stage the robot behaves exactly like an inverted planar double pendulum with its support point being analogous to the point of suspension of the pendulum. During the knee-free stage, the stance leg remain straight while the swing-leg bends at its knee, which is different from the Compass-like walker. The robot is assumed to move on a horizontal or inclined plane surface. The impact of the swing leg with the ground is assumed to be inelastic and without sliding, so is the impact between the thigh and the shank of the swing leg, which marks the inception of the knee-locked stage. This implies that during the instantaneous transition stage the robot configuration remains un-changed, and the angular momentum of the robot about the impacting foot as well as the angular momentum of the pre-impact support leg about the hip are conserved. Thanks to angular-momentum conservation law, we can obtain some useful equations, by which some meaningful simulations will be made.

10. The applicable function on 3D model

As we have discussed before, we have finished the model of the whole integral process of the gait biped, while for the sake of the real application, a 3D model- a more anthropomorphic model should be presented necessarily.
The simulation of the 3D one is similar with the 2D model. Previously, when the roll angle and its derivative are set to zero, the equation gained from the support leg angle, non-support angle and their derivatives will be share some characteristics with the 2D robot. Moreover, reversely if the two angles and their derivatives are set to zero, the model represented by the roll angle will behave like an inverted pendulum. So the comparison between 3D model in this special condition and the 2D model is a direct way to the correct of the modeling.

The 3d dynamic walking bipedal model

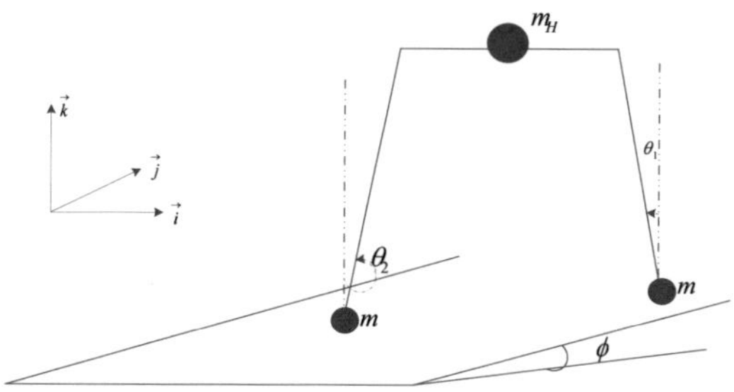

Fig. 26. 3d dynamical bipedal walker

The model consists of two legs connected by a pelvis, with pin joints at the hips[3]. The legs are point feet, shown as Figure.1. The same with 2D bipedal model, it is also equivalent to a double pendulum (more obviously in saggital plane model) with point masse m_H and m concentrated at the hip and feet respectively. The leg-length is l. The support angle θ_1, nonsupport angle θ_2 and roll angle θ_3 determine the configuration of the gait. The angle was made by the biped leg with the vertical (counterclockwise positive). 2α is the total angle between the legs, which is defined as the "inter-leg angle", and in addition is formed during the instant when both legs are touching the ground. The slope of the ground with the horizontal is denoted by the angle ϕ.
The model has been made by the following assumptions: the total mass of the robot $m_C = 2m + m_H$ is constant and equal to 20kg. For the sake of simplifying the model, all masses are considered point-masses and the legs are identical with point feet. The same as 2d bipedal model, the 3d gait also consists of swing stage and an instantaneous transition stage: during the swing stage the robot behaves exactly like an inverted planar double pendulum with its support point being analogous to the point of suspension of the pendulum. During the transition stage the support is transferred from one leg to the other. The robot is assumed to move on an inclined plane surface and the leg only swing forward and backward. The impact of the swing leg with the ground is assumed to be inelastic and without sliding [4]. This implies that during the instantaneous transition stage the robot configuration remains un-changed, and the angular momentum of the robot about the

impacting foot as well as the angular momentum of the pre-impact support leg about the hip is conserved. These conservation laws lead to a discontinuous change in robot velocity.

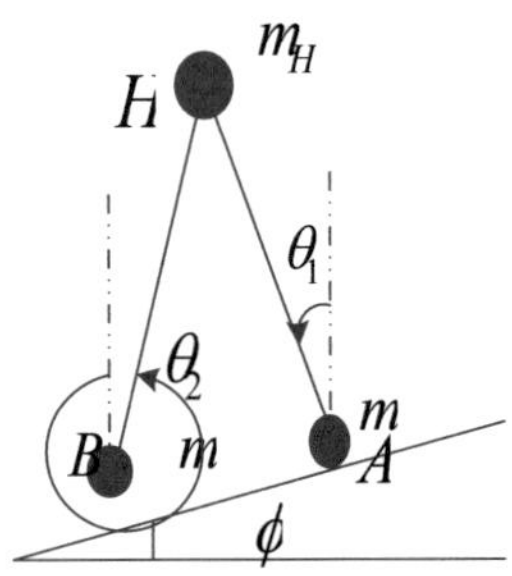

Fig. 27. Sagittal planar 3d model

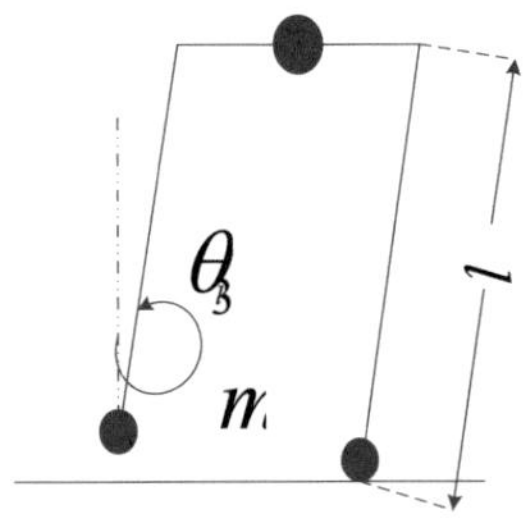

Fig. 3. Frontal 3d model

The new dynamic equations of the swing stage are similar to the well-known double pendulum equations. Since the legs of the robot are assumed identical, the equations are similar regardless of the support leg considered with the variation of the following.

$$M(\theta)=\begin{pmatrix} (m_H+m)L^2 & -mL^2\cos(\theta_1-\theta_2) & 0 \\ -mL^2\cos(\theta_1-\theta_2) & mL^2 & 0 \\ 0 & 0 & m_H L^2\cos^2\theta_1+mL^2(\cos\theta_1-\cos\theta_2)^2 \end{pmatrix}$$

$$N(\theta,\dot{\theta})=\begin{pmatrix} 0 & -mL^2\dot{\theta}_2\sin(\theta_1-\theta_2) & mL^2\dot{\theta}_3\sin\theta_1(\cos\theta_1-\cos\theta_2)-m_H L^2\dot{\theta}_3\cos\theta_1\sin\theta_1 \\ mL^2\dot{\theta}_1\sin(\theta_1-\theta_2) & 0 & -mL^2\dot{\theta}_3\sin\theta_2(\cos\theta_1-\cos\theta_2) \\ L^2\dot{\theta}_1[-(m_H+m)\sin2\theta_1+2m\cos\theta_2\sin\theta_1] & 2mL^2\dot{\theta}_2(-\cos\theta_1\sin\theta_2+\sin\theta_2\cos\theta_2) & 0 \end{pmatrix}$$

$$g(\theta)=\begin{pmatrix} -m_H gL\sin\theta_1\cos\theta_3 - mgL\sin\theta_1\cos\theta_3 \\ mgL\sin\theta_2\cos\theta_3 \\ -m_H gL\cos\theta_1\sin\theta_3 - mgL\cos\theta_1\sin\theta_3 + mgL\cos\theta_1\sin\theta_3 \end{pmatrix}$$

The parameters used for our simulations are $l=1\,m$, $m_H=2\,m=10$kg. Since no dissipation takes place during swing stage, thus the total mechanical energy E of the robot is conserved during this stage.

$$\frac{d}{dt}\frac{\partial L(\theta,\dot{\theta})}{\partial\dot{\theta}}-\frac{\partial L(\theta,\dot{\theta})}{\partial\theta}=0$$

Where the Lagrangian $L(\theta,\dot{\theta})$ is the difference between the kinetic energy and the potential energy of the robot: $L(\theta,\dot{\theta})=K(\theta,\dot{\theta})-P(\theta)$, the right-hand side term of (3) is 0, since the robot is completely passive. The new equation of the gait biped are given as the following and the results presented before can be applied perfectly.

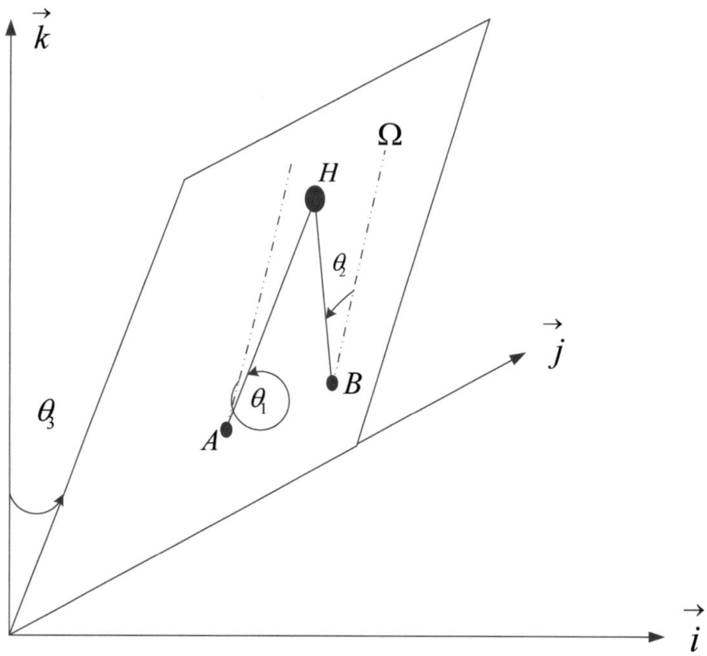

Fig. 19. Model in the 3d space

In order to calculate the energy of the robot, we simply consider the dynamics of the three distinct masses:

$$K(\theta,\dot{\theta})=\frac{1}{2}m_H\left\|\vec{V_H}\right\|^2+\frac{1}{2}m\left\|\vec{V_B}\right\|^2+\frac{1}{2}m\left\|\vec{V_B}\right\|^2 \quad (34)$$

$$P(\theta) = m_H gL\cos\theta_1\cos\theta_3 + mgL(\cos\theta_1\cos\theta_3 - \cos\theta_2\cos\theta_3) + P_0 \tag{35}$$

Where $\vec{V}_H$, $\vec{V}_A$ and $\vec{V}_B$ are the velocities of the point masses. In the frame $[\vec{i},\vec{j},\vec{k}]$ depicted on Fig.3, these vectors are given by:

$$\overrightarrow{V_H} = (-L\dot{\theta}_1\sin\theta_1\sin\theta_3 + L\dot{\theta}_3\cos\theta_1\cos\theta_3)\vec{i} + L\dot{\theta}_1\cos\theta_1\vec{j} + (-L\dot{\theta}_1\sin\theta_1\cos\theta_3 - L\dot{\theta}_3\cos\theta_1\sin\theta_3)\vec{k}$$

$$\begin{aligned}\overrightarrow{V_B} = &(-L\dot{\theta}_1\sin\theta_1\sin\theta_3 + L\dot{\theta}_3\cos\theta_1\cos\theta_3 + L\dot{\theta}_2\sin\theta_2\sin\theta_3 - L\dot{\theta}_3\cos\theta_2\cos\theta_3)\vec{i}\\ &+(L\dot{\theta}_1\cos\theta_1 - L\dot{\theta}_2\cos\theta_2)\vec{j} + (-L\dot{\theta}_1\sin\theta_1\cos\theta_3 - L\dot{\theta}_3\cos\theta_1\sin\theta_3\\ &+L\dot{\theta}_2\sin\theta_2\cos\theta_3 + L\dot{\theta}_3\cos\theta_2\sin\theta_3)\vec{k}\end{aligned}$$

$$\vec{V}_A = 0$$

The Transition equation

Since our robot is constituted of only two links, the condition of conservation of angular momentum leads to only two equations:

$$m_H\overrightarrow{V_{H^-}} \times \overrightarrow{A^-H^-} + m\overrightarrow{V_{B^-}} \times \overrightarrow{A^-B^-} = m_H\overrightarrow{V_{H^+}} \times \overrightarrow{A^+H^+} + m\overrightarrow{V_{B^+}} \times \overrightarrow{A^+B^+} \tag{36}$$

$$m\overrightarrow{V_{A^=}} \times \overrightarrow{H^-A^-} + m\overrightarrow{V_{B^-}} \times \overrightarrow{H^-B^-} = m\overrightarrow{V_{A^+}} \times \overrightarrow{H^+A^+} + m\overrightarrow{V_{B^+}} \times \overrightarrow{H^+B^+} \tag{37}$$

Where points H, A and B are respectively the hip, the mass center of the support leg, the mass center of the non-support leg. $\vec{V}_H$, $\vec{V}_A$ and $\vec{V}_B$ are respectively the velocity vectors at H, A and B. The superscripts - and + indicate respectively pre-impact and post-impact variables.All the vectors appearing are given by:

$$\begin{aligned}\overrightarrow{V_{H^-}} = &(L\dot{\theta}_1\sin\theta_1\sin\theta_3 - L\dot{\theta}_3\cos\theta_1\cos\theta_3)\vec{i} + (-L\dot{\theta}_1\cos\theta_1)\vec{j} + (-L\dot{\theta}_1\sin\theta_1\cos\theta_3\\ &-L\dot{\theta}_3\cos\theta_1\sin\theta_3)\vec{k}\end{aligned}$$

$$\begin{aligned}\overrightarrow{V_{A^-}} = &(L\dot{\theta}_1\sin\theta_1\sin\theta_3 - L\dot{\theta}_3\cos\theta_1\cos\theta_3 - L\dot{\theta}_2\sin\theta_2\sin\theta_3 + L\dot{\theta}_3\cos\theta_2\cos\theta_3)\vec{i} +\\ &(-L\dot{\theta}_1\cos\theta_1 + L\dot{\theta}_2\cos\theta_2)\vec{j} + (-L\dot{\theta}_1\sin\theta_1\cos\theta_3 - L\dot{\theta}_3\cos\theta_1\sin\theta_3\\ &+L\dot{\theta}_2\sin\theta_2\cos\theta_3 + L\dot{\theta}_3\cos\theta_2\sin\theta_3)\vec{k}\end{aligned}$$

$$\overrightarrow{V_{B^-}} = 0$$

$$\overrightarrow{V_{H^+}} = (L\dot{\theta}_1^+ \sin\theta_1^+ \sin\theta_3^+ - L\dot{\theta}_3^+ \cos\theta_1^+ \cos\theta_3^+)\vec{i} + L\dot{\theta}_1^+ \cos\theta_1^+ \vec{j}$$
$$+(-L\dot{\theta}_1^+ \sin\theta_1^+ \cos\theta_3^+ - L\dot{\theta}_3^+ \cos\theta_1^+ \sin\theta_3^+)\vec{k}$$

$$\overrightarrow{V_{B^+}} = (L\dot{\theta}_1^+ \sin\theta_1^+ \sin\theta_3^+ - L\dot{\theta}_3^+ \cos\theta_1^+ \cos\theta_3^+ - L\dot{\theta}_2^+ \sin\theta_2^+ \sin\theta_3^+ + L\dot{\theta}_3^+ \cos\theta_2^+ \cos\theta_3^+)\vec{i}$$
$$+(L\dot{\theta}_1^+ \cos\theta_1^+ - L\dot{\theta}_2^+ \cos\theta_2^+)\vec{j} + (-L\dot{\theta}_1^+ \sin\theta_1^+ \cos\theta_3^+ - L\dot{\theta}_3^+ \cos\theta_1^+ \sin\theta_3^+$$
$$+L\dot{\theta}_2^+ \sin\theta_2^+ \cos\theta_3^+ + L\dot{\theta}_3^+ \cos\theta_2^+ \sin\theta_3^+)\vec{k}$$

$$\overrightarrow{V_{A^+}} = 0$$

$$\overrightarrow{A^-H^-} = -L\cos\theta_2^- \sin\theta_3^- \vec{i} - L\sin\theta_2^- \vec{j} + L\cos\theta_2^- \cos\theta_3^- \vec{k}$$

$$\overrightarrow{H^-A^-} = L\cos\theta_2^- \sin\theta_3^- \vec{i} + L\sin\theta_2^- \vec{j} - L\cos\theta_2^- \cos\theta_3^- \vec{k}$$

$$\overrightarrow{A^+H^+} = -L\cos\theta_1^+ \sin\theta_3^+ \vec{i} + L\sin\theta_1^+ \vec{j} + L\cos\theta_1^+ \cos\theta_3^+ \vec{k}$$

$$\overrightarrow{H^+B^+} = L\cos\theta_2^+ \sin\theta_3^+ \vec{i} - L\sin\theta_2^+ \vec{j} - L\cos\theta_2^+ \cos\theta_3^+ \vec{k})$$

$$\overrightarrow{A^+B^+} = (L\cos\theta_1^+ \sin\theta_3^+ + L\cos\theta_2^+ \sin\theta_3^+)\vec{i} + (L\sin\theta_1^+ - L\sin\theta_2^+)\vec{j}$$
$$+(L\cos\theta_1^+ \cos\theta_3^+ - L\cos\theta_2^+ \cos\theta_3^+)\vec{k}$$

$$m_H[-\dot{\theta}_3^- \cos\theta_1^- \sin\theta_2^- \sin\theta_3^- - \dot{\theta}_1^- \cos\theta_3^- \cos(\theta_1^- - \theta_2^-)] = m_H(\dot{\theta}_1^+ \cos\theta_3^+ + \dot{\theta}_3^+ \cos\theta_1^+ \sin\theta_1^+ \sin\theta_3^+)$$
$$+m[\cos\theta_3^+(\theta_1^+ + \dot{\theta}_2^+) + (\dot{\theta}_1^+ - \dot{\theta}_2^+)\cos\theta_3^+ \cos(\theta_1^+ - \theta_2^+) - \dot{\theta}_3^+ \sin\theta_3^+ \sin(\theta_1^+ + \theta_2^+)$$
$$+\dot{\theta}_3^+ \sin\theta_3^+(\cos\theta_1^+ \sin\theta_1^+ + \cos\theta_2^+ \sin\theta_2^+)]$$

$$-m_H \dot{\theta}_3^- \cos\theta_1^- \cos\theta_2^- = -m_H \dot{\theta}_3^+ \cos^2\theta_1^+ - m\dot{\theta}_3^+ (\cos\theta_1^+ - \cos\theta_2^+)^2$$

$$\dot{\theta}_1^- \cos\theta_3^- \cos(\theta_1^- - \theta_2^-) - \dot{\theta}_2^- \cos\theta_3^- + \dot{\theta}_3^- \sin\theta_3^- \sin\theta_2^- (\cos\theta_1^- - \cos\theta_2^-) = \ldots$$
$$-\dot{\theta}_1^+ \cos\theta_3^+ \cos(\theta_1^+ - \theta_2^+) + \dot{\theta}_2^+ \cos\theta_3^+ + \dot{\theta}_3^+ \sin\theta_3^+ \sin\theta_2^+ (\cos\theta_2^+ - \cos\theta_1^+)$$

$$\dot{\theta}_3^- \cos\theta_1^- \cos\theta_2^- - \dot{\theta}_3^- \cos^2\theta_2^- = \dot{\theta}_3^+ \cos\theta_1^+ \cos\theta_2^+ - \dot{\theta}_3^+ \cos^2\theta_2^+$$

And for the assumption of the angles between hip and legs are all constant with 90° , there is extra torque at H.
We get the following compact equation between the pre-impact and post-impact angular velocities:

$$Q_{-}(\theta)\dot{\theta}^{-} = Q_{+}(\theta)\dot{\theta}^{+}$$

With matrices $Q_{-}(\theta)$ and $Q_{+}(\theta)$ given by:

$$Q_{-}(\theta) = \begin{pmatrix} -m_H \cos\theta_3^- \cos(\theta_1^- - \theta_2^-) & 0 & -m_H \cos\theta_1^- \sin\theta_2^- \sin\theta_3^- \\ m\cos\theta_3^- \cos(\theta_1^- - \theta_2^-) & -m\cos\theta_3^- & m\sin\theta_3^- \sin\theta_2^- (\cos\theta_1^- - \cos\theta_2^-) \\ 0 & 0 & -m_H \cos\theta_1^- \cos\theta_2^- \end{pmatrix}$$

$$Q_{+}(\theta) = \begin{pmatrix} m_H \cos\theta_3^+ + m\cos\theta_3^+[1-\cos(\theta_1^+ - \theta_2^+)] & m\cos\theta_3^+[1-\cos(\theta_1^+ - \theta_2^+)] \\ -m\cos\theta_3^+ \cos(\theta_1^+ - \theta_2^+) & m\cos\theta_3^+ \\ 0 & 0 \end{pmatrix.$$
$$\begin{matrix} m_H \cos\theta_1^+ \sin\theta_1^+ \sin\theta_3^+ + m\sin\theta_3^+[-\sin(\theta_1^+ + \theta_2^+) + \cos\theta_1^+ \sin\theta_1^+ + \cos\theta_2^+ \sin\theta_2^+] \\ m\sin\theta_3^+ \sin\theta_2^+ (\cos\theta_2^+ - \cos\theta_1^+) \\ -m_H \cos^2\theta_1^+ - m(\cos\theta_1^+ - \cos\theta_2^+)^2 \end{matrix}\Bigg)$$

By doing the transition of the equation, we can avail the consequences of the model to the real robot model.

11. Conclusions and future work

The focus of the work is a relative further study of the passive gait of a compass-like, planar, biped robot with knees on inclined slopes. An analysis about the distribution of the energy and also the conversion law between the swing leg and the stance leg during the process of the steady robot gaits, have been discussed in the paper. Phase-position property corresponds to the limit cycle, the coupling properties between two legs, the existence of the culmination points which produced in the course of the conversion of KE and PE are also the topic of the research. To a certain slope angle ϕ, one and only one stable limit cycle exists.
The research of the paper will have positive significance in getting better aware of the law and global property to biped gaits of the robot. The model we adopt here is quite applicable, how to enlarge the initial value attraction region of the limit cycle as well as how to apply the efficient control on the robot combined with its own property with the least energy possible will guide our further research direction.

12. Acknowledgement

This work is supported by the Research Fund for the Doctoral Program of Higher Education of China under grant 20090061120050 as well as the Forward and Crossing Research Fund for the Basic Theory in Jilin University under Grant 421033205410.

13. References

[1] T. McGeer, "Passive dynamic walking", International Journal of Robotics Research, 9(2), pp. 62-82, 1990.

[2] S.H. Collins, M. Wisse and A. Ruina, "A 3-d passive dynamic walking robot with two legs and knees", International Journal of Robotics Research, 20(7), pp. 607-615, 2001.

[3] A. Goswami, B. Espiau and A. Keramane, "Limit cycles and their stability in a passive bipedal gait", Proc. IEEE Conference on Robotics and Automation, pp. 246-251, 1996.

[4] M. Garcia, A. Chatterjee, A. Ruina and M. Coleman, The simplest walking model: stability, complexity, and scaling", ASME Journal of Biomechanical Engineering, 120(2), pp. 281-288, 1998.

[5] H. Ohta, M.. Yamakita and K. Furuta, "From passive to active dynamic walking", IEEE Conference on Decision and Control, pp. 3883-3885, 1999.

[6] A. Goswami, "Kinetic quantification of gait symmetry based on bilateral cyclograms", Proc. XIXth Congress of the International Society of Biomechanics, pp. 56-62, 2003.

[7] A. Goswami, B. Thuilot and B. Espiau, "A study of the passive gait of a compass-like biped robot: symmetry and chaos", International Journal of Robotics Research, 17(12), pp. 1282-1301, 1998.

[8] J.W. Grizzle, G. Abba and F. Plestan, "Asymptotically stable walking for biped robots: analysis via systems with impulsive effects", IEEE Transactions on Robotics and Automation Control, 46(1), pp. 51-64, 2001., pp. 51-64, 2001.

[9] A. Goswami, B. Tuilot and B. Espiau, "Compass like bipedal robot part I: stability and bifurcation of passive gaits", INRIA Research Report, 1996.

[10] G. Bhatia, M.W. Spong, "Hybrid control for smooth walking of a biped with knees and torso", Proc. 2004 IEEE Conference on Control Applications, pp. 88-96, 2004.

[11] Y.F. Zheng, J. Shen, F.R. Sias, "A motion control scheme for a biped robot to climb sloping surfaces", Proc. IEEE Conference on Robotics and Automation, 2(4), pp814-816, 1988.

[12] I. A. Hiskens, "Stability of hybrid system limit cycle: application to the compass gait biped robot", Proceedings of 40th IEEE Conference on Decision and Control, pp. 774-779, 2001.

[13] A. Goswami, B. Tuilot and B. Espiau, "Compass like bipedal robot part I: stability and bifurcation of passive gaits", INTRIA Research Report, 1996.

[14] M.W. Spong, G. Bhatia, "Further Results on Control of the Compass Gait Biped", IEEE International Conference on Intelligent Robots and Systems Proceedings, pp. 1933-1938, 2003.

[15] M. Garcia, A. Chatterjee, A. Ruina and M. Coleman, "The simplest walking model: stability, complexity, and scaling", ASME Journal of Biomechanical Engineering, 120(2), pp. 281-288, 1998.

11

Dynamic Joint Passivization for Bipedal Locomotion

Shohei Kato and Minoru Ishida
Nagoya Institute of Technology
Japan

1. Introduction

A lot of research on humanoid robots or biped robots has been conducted. This research focused on enabling robots to walk very smoothly, similar to the way humans walk, which is highly energy efficient. Motion control using a central pattern generator (CPG) has attracted much attention as an effective control mechanism for biped robots to achieve human-like walking (e.g., Kato & Itoh (2005); Kotosaka & Schaal (2000); Miyakoshi et al. (1998); Taga (1995a;b); Taga et al. (1991)). The CPG is modeled mathematically to a neural rhythm generator that exists at a relatively low level of the central nervous system, such as the spinal cord of animals. This motion control using the CPG has generated various motions: walking by Ishida et al. (2009a;b); Itoh et al. (2004); Nakamura et al. (2005); Taki et al. (2004), step by Miyakoshi et al. (1998), and drum motions by Kotosaka & Schaal (2000). Taga (1995a;b) proposed a neuro-musculo-skeletal model based on the CPG, and it enabled a biped robot to have a human gait in two dimensions. Among the researchers of highly energy-efficient walking, McGeer (1990) was the first to study passive dynamic walking (PDW). A PDW robot walks forward by placing the foot on the ground and riding on the supporting leg, which rolls forward as an inverted pendulum mounted on the supporting foot. At the same time, it places the swing foot forward by moving the swing leg in a pendular arc, so that it makes the foot strike a ground when the mechanism is in a configuration identical to that at the beginning of a step. If the dynamic characteristic of the robot and the environment (e.g., the slope and velocity when the walking begins) agree, then the PDW robot achieves highly energy-efficient walking without any actuator control. Sugimoto & Osuka (2004) proposed a control method for quasi passive dynamic walking (Quasi-PDW). Quasi-PDW means that the robot usually does PDW without any input torque, and the actuators of the robot are used for ensuring walking stability only when the walking begins or when a disturbance occurs. Haruna et al. (2001) researched a PDW robot with a torso. CPG-based motion control inputs some torque to all the joints of the robot's lower limbs regardless of gait. When we think about human walking, we take into account the joints of the swing leg without any input torque. Active walking needs to be mixed with PDW for robot walking. Quasi-PDW is an example of a mixture of active walking and PDW. Quasi-PDW is applicable to the robot whose dynamic characteristic suits to PDW. However, there is a lot of robots that can not satisfy a dynamic characteristic for PDW. To achieve this mixture with the robots, we added the mechanism of PDW to an active walking robot. In this chapter, we describe a motion control method based on the mixture of CPG and PDW (Ishida et al. (2009a;b)), that is, the

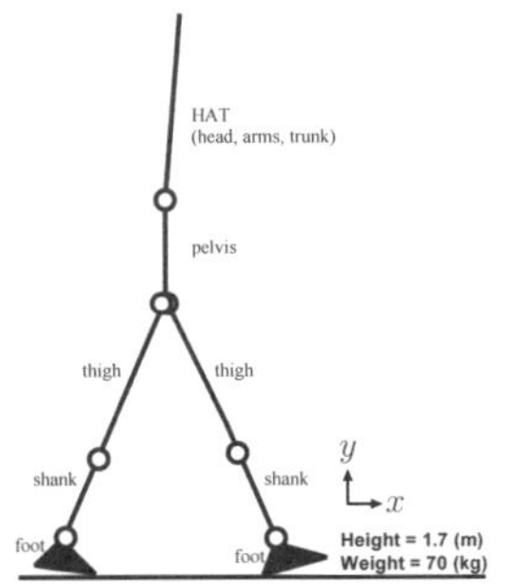

Fig. 1. Link Structure of Robot (2D).

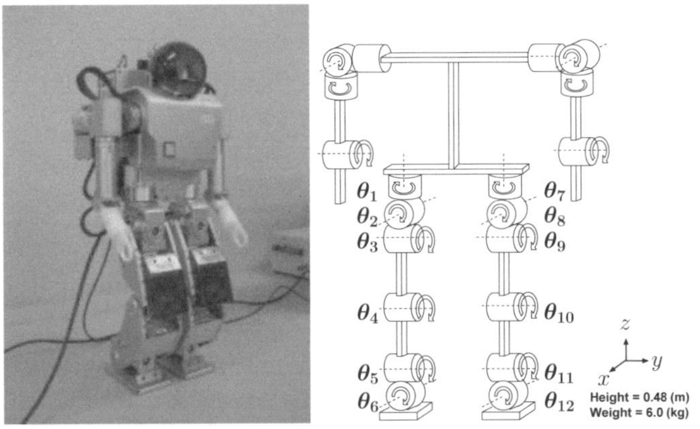

Fig. 2. HOAP-1 (left) and Its Link Structure (right).

dynamic passivization of joint control, achieving robust and energy-efficient walking. We focused on robot walking on a downhill slope.

2. The Robot model and its motion control primitive

2.1 The link structure of humanoid robot

The section describes two link models of humanoid robot: two dimensional link model in sagittal plane and three dimensional link model of entire body.

2.1.1 Two dimensional model

The model of the human body is composed of the HAT (head, arms, and trunk), pelvis, thighs, shanks, and feet (shown in Figure 1). There are seven joints, two each at the hips, knees, and ankles, and one at the trunk.

2.1.2 Three dimensional model

We also consider the motion control of a humanoid robot, HOAP-1 (Murase et al. (2001)), shown in Figure 2. HOAP-1 has 6-DOFs in each leg. The coxa joint has three degrees of freedom; pitch, yaw, and roll, the knee joint has one degree of freedom; pitch, and the ankle joint has two degrees of freedom; pitch, and roll. The height and weight of the robot and 48[cm] and 6[kg].

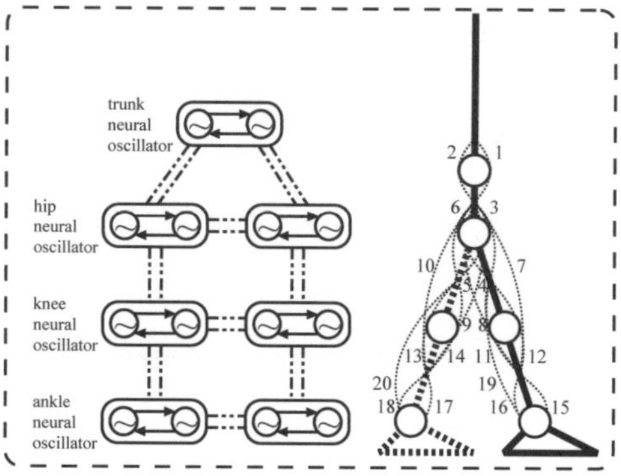

Fig. 3. The neural system (left) and the musculo-skeletal system (right) (two dimensional version).

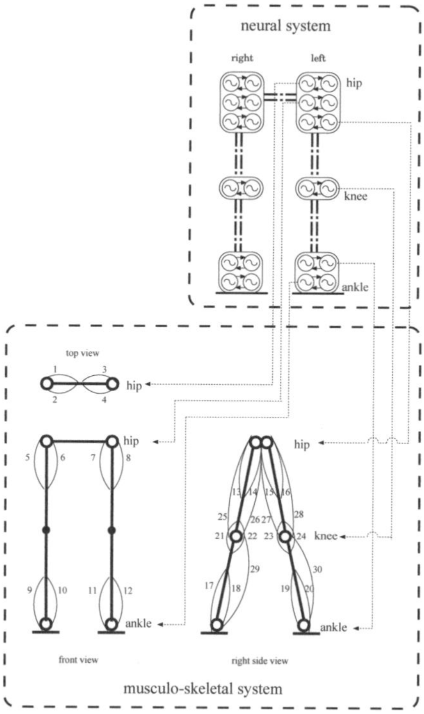

Fig. 4. The neural system and the musculo-skeletal system (three dimensional version).

2.2 Neural rhythm generator

Walking movement is periodical. In this research, we control walking movement using CPG, which is often used in generating periodical movement. CPG is modeled mathematically to the neural rhythm generator which exists at a relatively low level of the central nervous system such as the spinal cord of animals. Standout feature of CPG, it is synchronized its inner state with rhythmic input from outside in term of phase. Using this feature, therefore walking movement having robustness for changes of environment is able to be generated. CPG is composed of multi-neurons which inhibit each other. The mathematical model of a neuron is represented as following system of differential equation.

$$\tau_i \dot{u}_i = -u_i - b \cdot f(v_i) + \sum_{j=1}^{n} w_{ij} f(u_i) + u_0 + S_i, \quad (1)$$

$$\tau_i' \dot{v}_i = -v_i + f(u_i), \quad (2)$$

$$f(x) = max(x, 0), \quad (3)$$

where u_i is the inner state of i-th neuron; v_i is a variable which represents the degree of the adaptation or self-inhibition effect of the i-th neuron; τ_i and τ_i' are time constants of the inner state and the adaptation effect of the i-th neuron, respectively; b is a coefficient of the adaptation effect; w_{ij} is a connecting weight from the j-th neuron to the i-th neuron; u_0 is an external input with a constant rate; and S_i is the local and global sensory information that is sent to the i-th neuron. A neuron excited by u_0 is oscillated by self-inhibition and cross-inhibition, and $f(u)$ is output of neuron. For more precise, please refer to Matsuoka (1985) and Matsuoka (1987).

2.3 Neuro-musculo-skeletal system

In this research, we adopted the neuro-musculo-skeletal system proposed by Taga (1995a) for a motion control method based on CPG in the robot. The neuro-musculo-skeletal system is composed of two dynamical systems: a neural system and a musculo-skeletal system. The neural system is composed of CPG network, and the musculo-skeletal system is composed of skeletons considered muscles surrounding to them. The system can generate flexible and adaptable walking movement through the mutual interaction among the neural system, the musculo-skeletal system and environment.

In this chapter, we propose CPG-based walking motion generation considering two styles in neuro-musculo-skeletal system: walking in sagittal plane (in two dimensions) and walking with real lower body (in three dimensions).

2.3.1 Two dimensional model

In the neural system, the neural rhythm generator consists of seven neural oscillators in accordance with the robot's link structure shown in Figure 1. The neural oscillators are allocated to seven joints: the trunk and the pairs of the hips, knees, and ankles, shown in Figure 3 (left). Two neurons at a neural oscillator each have a flexion and extension effect on muscles corresponding to the CPG. In the musculo-skeletal system, the skeletons match the robot's link structure. There are six single-joint muscles and three double-joint muscles for each of the limbs and two for the upper body. Figure 3 (right) shows the configuration of the muscles. Two neurons in the neural oscillators alternately activate the antagonist muscles.

2.3.2 Three dimensional model

For the three dimensional link model (see Figure 2 (left)), neural oscillators are allocated to twelve joints: the pairs of the coxas (pitch, yaw, and roll), knees, and ankles (pitch and roll). Two neurons at a neural oscillator each have a flexion and extension effect on muscles corresponding to the CPG. In the musculo-skeletal system, the skeletons match the robot's link structure. There are twelve single-joint muscles and three double-joint muscles for each of the limbs. Figure 4 shows the configuration of the muscles (30 muscles in total).

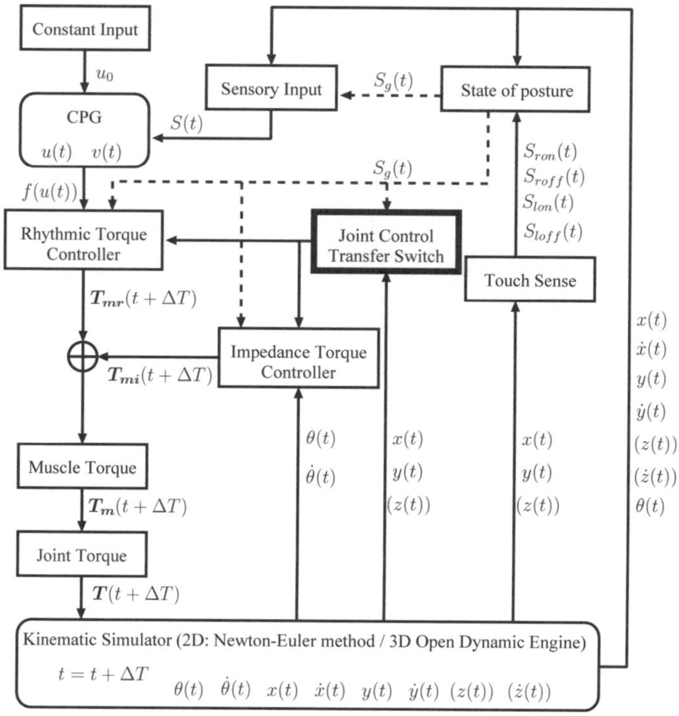

Fig. 5. Block diagram of motion control system based on dynamic passivization of joint control

2.4 Motion control based on neuro-musculo-skeletal system

Figure 5 shows a block diagram of the motion control system based on the dynamic passivization of joint control. In this research, we added a joint control transfer switch to the neuro-musculo-skeletal system proposed by Taga for a motion control method. Joint control transfer switch is described in Section 3.1. If the transfer switch is passive, then it nullifies the input torque in the swing leg for the rhythmic torque controller and for the impedance torque controller. If it is active, then it enables the input torque in the swing leg for the rhythmic torque controller and for the impedance torque controller. The system performs the motion control based on the iteration of the following processes:

1. First, output $f(u(t))$ of the CPG in time t is excited by constant input u_0 to the neuron. The rhythmic torque controller generates rhythmic torque $T_{mr}(t+\Delta T)$ from $f(u(t))$, sensory input $S(t)$ of the robot at time t, and the output of the joint control transfer switch in time t.

2. The impedance torque controller generates impedance torque $T_{mi}(t+\Delta T)$ to maintain a standing position from joint angle $\theta(t)$, joint angular velocity $\dot{\theta}(t)$, and the sensory input of the robot at time t.

3. The muscle torque $T_m(t+\Delta T)$ is generated from $T_{mr}(t+\Delta T)$ and $T_{mi}(t+\Delta T)$.

4. The joint torque $T(t+\Delta T)$ is calculated from $T_m(t+\Delta T)$.

5. The kinematics simulator generates the motion of the robot when joint torque $T(t+\Delta T)$ is given to the robot. Then, the simulator calculates joint angle $\theta(t+\Delta T)$, joint angular

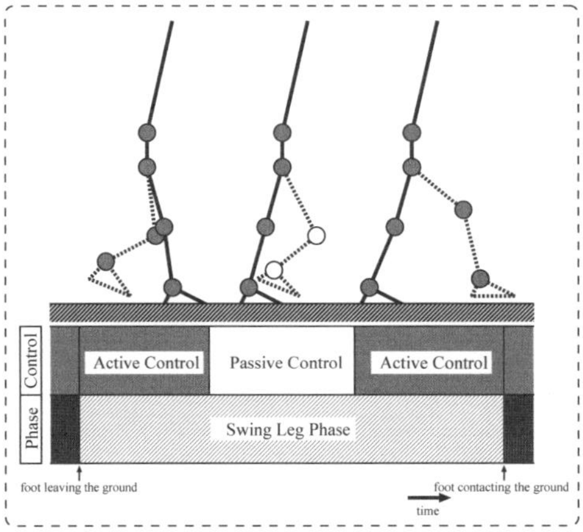

Fig. 6. Dynamic passivization of joint control

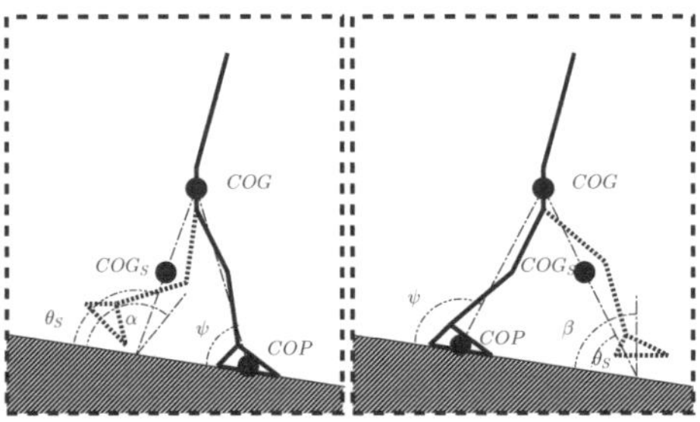

Fig. 7. Postures while walking

velocity $\dot{\theta}(t+\Delta T)$, the coordinates $p(t+\Delta T) = (x(t+\Delta T), y(t+\Delta T))^T$ ($p(t+\Delta T) = (x(t+\Delta T), y(t+\Delta T), z(t+\Delta T))^T$ for 3D simulation), and velocity $\dot{x}(t+\Delta T)$, $\dot{y}(t+\Delta T)$ ($\dot{z}(t+\Delta T)$ for 3D simulation) of each link after motion. The simulator sets the time forward for ΔT.

6. The flags of the foot contacting the ground S_{ron}, S_{roff}, S_{lon}, S_{loff} are obtained by the touch sense. The state of posture $S_g(t)$ is updated by them and by the output of the kinematics simulator.

3. Dynamic passivization of joint control

In this paper, we describe a "Joint Control Transfer Switch" that is switched to "ACTIVE" or "PASSIVE" according to the environment and the posture information for adding the mechanism of PDW to the motion control method based on CPG. Our intention was to make the joint control of the swing leg temporarily passive in the swing leg phase. Figure 6 shows a concept chart of the dynamic passivization of the joint control. The important part is the passive phase time and the switch timing of the joint control.

3.1 Joint control transfer switch

The slope and posture information is used as information that decides the switch timing of the joint control (see Figure 7). The body's center of gravity (COG) and the swing leg's center of gravity (COG_S) are obtained from the posture information while walking. Angle θ_S is

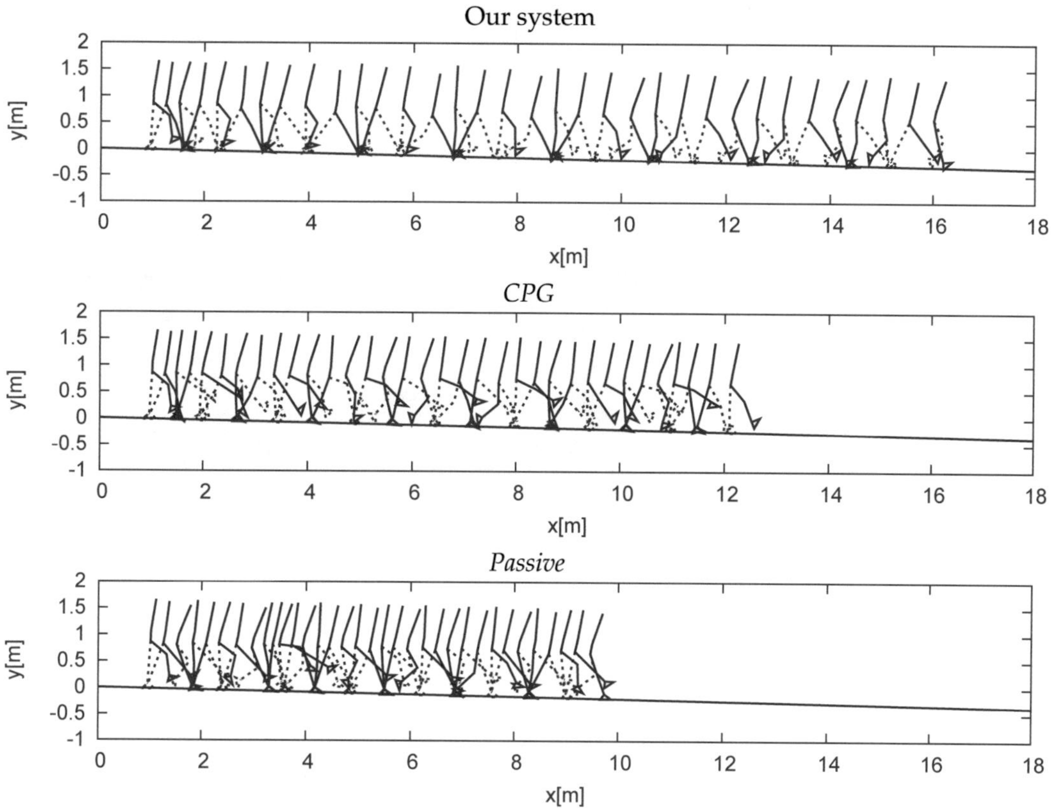

Fig. 8. Snapshots of gaits on 2 percent downhill slope with our system (top), *CPG* (middle), and *Passive* (bottom)

calculated from the *COG*, COG_S, and the slope. In addition, angle ψ is calculated from the *COG*, center of pressure (*COP*), and the slope.
The joint control transfer switch changes joint control to active or passive according to the following conditions:

- $\psi \leq \frac{\pi}{2}$
  - $\theta_S \geq \alpha$: ACTIVE
  - $\theta_S < \alpha$: PASSIVE
- $\psi > \frac{\pi}{2}$
  - $\theta_S > \beta$: PASSIVE
  - $\theta_S \leq \beta$: ACTIVE,

where α and β are set to an appropriate value according to a dynamic characteristic of the robot and a slope ($0 \leq \alpha, \beta \leq \pi$).

4. Walking in sagittal plane

We conducted a walking control experiment to test the effectiveness of our method. Firstly, this section reports walking control performances in sagittal plane using 2D link model shown in Figure 1. In the experiments, we used the neuro-musculo-skeletal system proposed by Taga

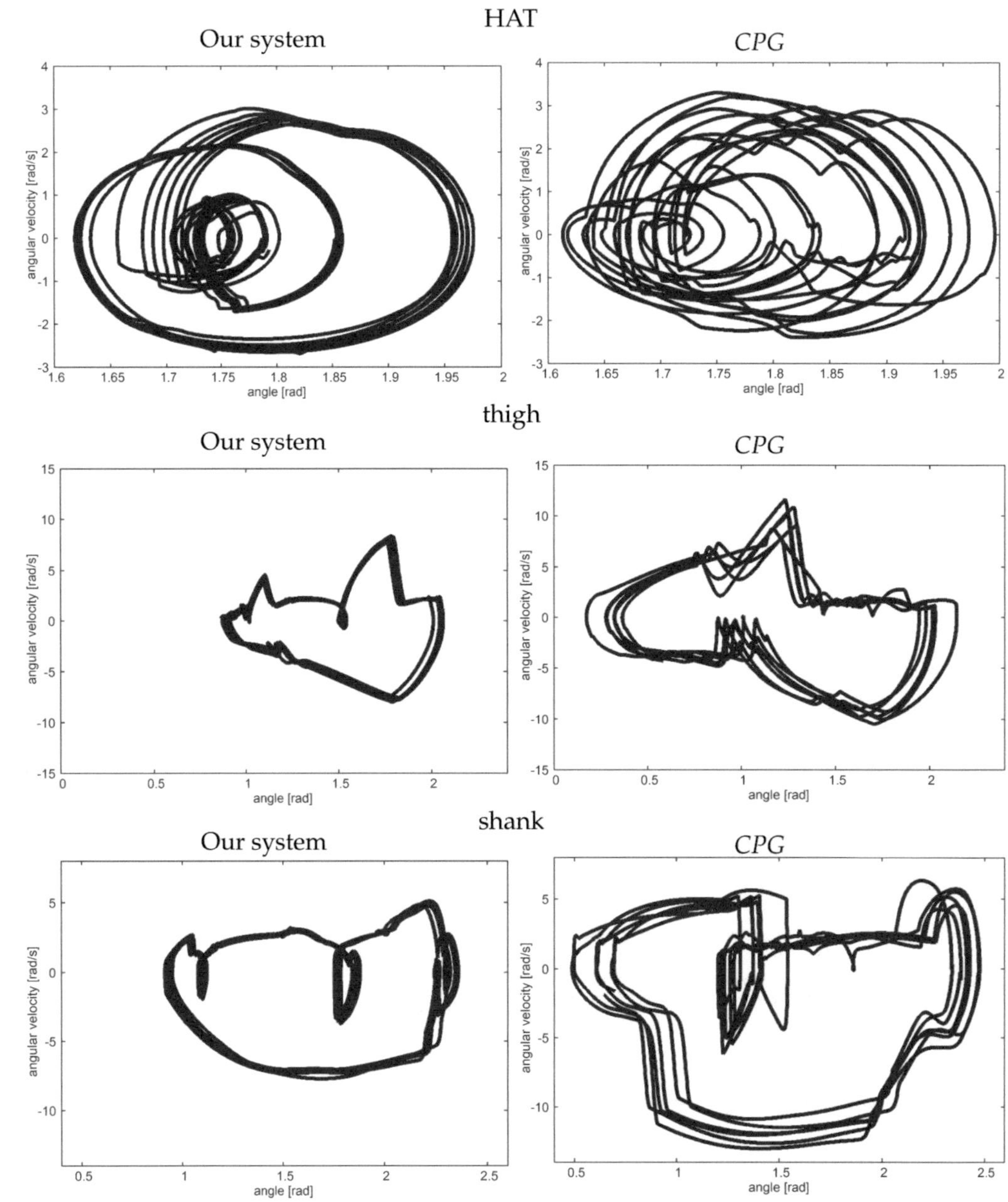

Fig. 9. Phase plots with our system (left) and *CPG* (right)

	Our system	*CPG*	*Passive*
Travel distance [m]	**15.4**	11.7	9.6
Sum total of input torque [Nms]	2.1E+03	2.2E+03	**1.6E+03**
Locomotion cost [Ns]	**1.4E+02**	1.9E+02	1.6E+02
α [rad]	1.79	-	-
β [rad]	0.24	-	-

Table 1. Results of first experiment

	Our system	*CPG*
$\lambda_{1\theta_1}$ (HAT)	**7.5E-02**	1.1E-01
$\lambda_{1\theta_2}$ (pelvis)	**9.3E-03**	1.3E-02
$\lambda_{1\theta_3}$ (right thigh)	**1.0E-01**	2.3E-01
$\lambda_{1\theta_4}$ (left thigh)	**1.6E-01**	2.3E-01
$\lambda_{1\theta_5}$ (right shank)	**1.2E-01**	2.7E-01
$\lambda_{1\theta_6}$ (left shank)	**2.2E-01**	2.4E-01
$\lambda_{1\theta_7}$ (right foot)	**1.0E-01**	1.5E-01
$\lambda_{1\theta_8}$ (left foot)	1.7E-01	**1.4E-01**
average	**1.2E-01**	1.7E-01

Table 2. Maximum lyapunov exponent

and a control method where the joint control was set to passive during the swing leg phase for comparison with our method. The former method is labeled "*CPG*" and the latter method is labeled "*Passive*". Because this robot could not do PDW in this environment, PDW was excluded from the objects of comparison.

4.1 Optimizing parameters

We optimized the common parameters of all the methods and parameters (α and β) of our method with simulated annealing with advanced adaptive neighborhood (SA/AAN) by Miki et al. (2002) prior to conducting the walking experiments (e.g., Itoh et al. (2004); Nakamura et al. (2005); Taki et al. (2004)). We optimized the rhythmic torque parameter as common parameters of all the methods. The rhythmic torque Tmr_j that acts on j-th muscle is defined by the following equation:

$$Tmr_j = (r_{part} \cdot S_{on} + (1 - r_{part}) \cdot S_{off}) \cdot p_{part} \cdot f(u_i), \tag{4}$$

where r and p are rhythmic torque parameter, *part* is a type of muscle, $S_{on}(S_{off})$ is flag of the contacting (leaving) the ground. In the experiments, the walking time and the locomotion cost were used for optimizing the performance. The value of the locomotion cost is defined by the following equation:

$$Cost = \frac{1}{L} \sum_{i=1}^{M} \int_0^{Time} |T_i(t)| dt, \tag{5}$$

where T_i is the input torque of the i-th joint, L is the travel distance, M is the number of joints, and *Time* is the simulation time. We perturbed parameters 10,000 times simultaneously. Each simulation took 10 seconds. We set the value of reference in Taga (1995a) for the common parameters of all the methods other than rhythm torque parameter.

4.2 Generating locomotion based on the dynamic passivization of joint control

First of all, we determined the effects of the dynamic passivization of the joint control. In this experiment, we used a 2 percent downhill slope.

Figure 8 shows the gaits over 10 seconds on a 2 percent downhill slope. In this figure, the snapshots of the gait were traced every 0.3 seconds. Table 1 shows the travel distances, the sum total of the input torques, the locomotion costs, and the parameters of our method (α and β). Our system's walks were longer than the other's. Two methods, our system and the *Passive* system, having the mechanism of passivization of joint control generated walking that was more energy efficient than that of *CPG*. When our system is compared with the *Passive* system, we found that our system consumed more torque because its passive phase time was shorter. However, the travel distance increased more than the increment of the consumption torque, and our system reduced the locomotion cost.

4.3 Gait stability analysis

Next, we determined the gait stability. In this section, the two motion control methods of our system and *CPG* were used. In the following experiments, the *Passive* system was excluded from the comparison because it is a special example of our system and our system reduced the locomotion cost more than the *Passive* system in section 4.2. Figure 9 shows the phase plots of parts of the body in the frontal plane for 9.0 seconds from 1.0 second after the walking begins: the HAT (head, arms, and trunk) (top), the thigh of right leg (middle), and the shank of right leg (bottom). The horizontal axes represent the absolute angle in the radian, and the vertical axes represent the angular velocities. Our system generates steady periodic motion because its phase plots are more periodic than those of *CPG*.

We analyzed the gait stability using the maximum lyapunov exponent. Table 2 shows the maximum lyapunov exponent (Alligood et al. (1996)) of parts of the body. In this table, the maximum lyapunov exponents of our system's gait are smaller than those of *CPG*. The walking using our system is steadier than the *CPG*.

4.4 Adaptive walking on various slopes

The performance of the systems on various downhill slopes was then examined at a 2 percent interval with a 0 to 18 percent downhill slope. In this section, the parameters were optimized beforehand in each environment.

The experimental results demonstrated that our system and the *CPG* can walk for 10 or more seconds on downhill slopes with a 0 to 16 percent downhill. Our system and the *CPG* could not walk for 10 or more seconds on a 18 percent downhill slope. Figure 10 shows the gaits over 10 seconds on a 16 percent downhill slope. In this figure, the snapshots of the gait were traced every 0.3 seconds. Our system walks farther than *CPG* on a 16 percent downhill slope, as the results in the preceding section also indicate. Figure 11 shows the locomotion cost of each method on each downhill slope. Our system generates the energy-efficient walking on each downhill slope if it can walk for 10 or more seconds. Figure 12 shows the parameters (α, β, the maximum value of θ_S (θ_S^{max}), and the minimum value of θ_S (θ_S^{min})) on each downhill slope. If α exceeds θ_S^{max} (14 and 16 percent downhill slopes), the joint control transfer switch changes the joint control to passive at the same time the foot leaves the ground. If β falls below θ_S^{min} (2, 10, 14, 16 percent downhill slope), the joint control transfer switch changes the joint control to active at the same time as the foot touches the ground. Our system generates energy-efficient walking on various (from 0 to 12 percent) downhill slopes because it uses the

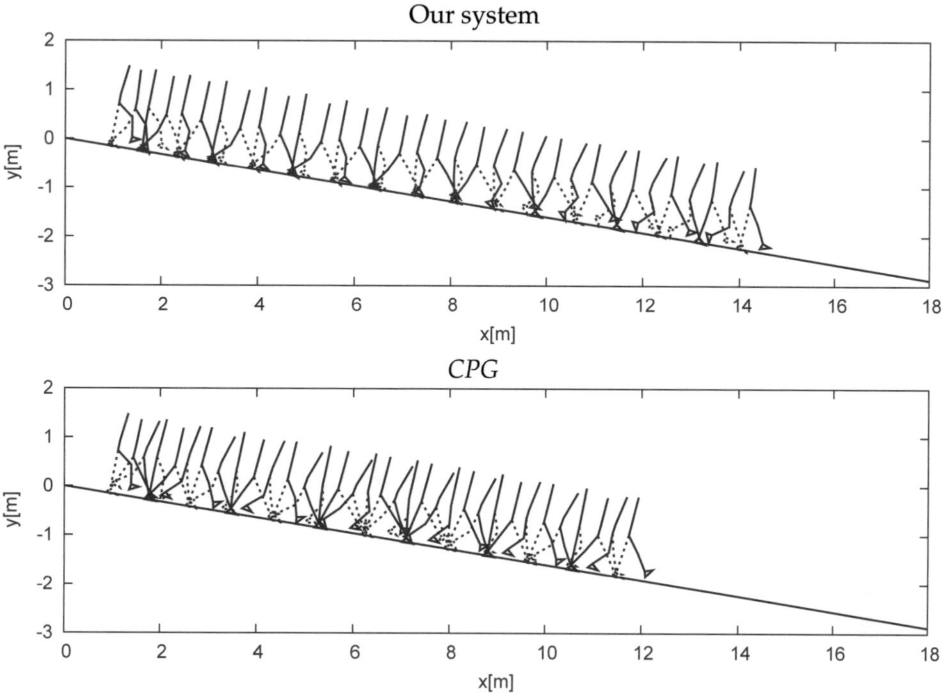

Fig. 10. Snapshots of gaits on 16 percent downhill slope with our system (top) and *CPG* (bottom)

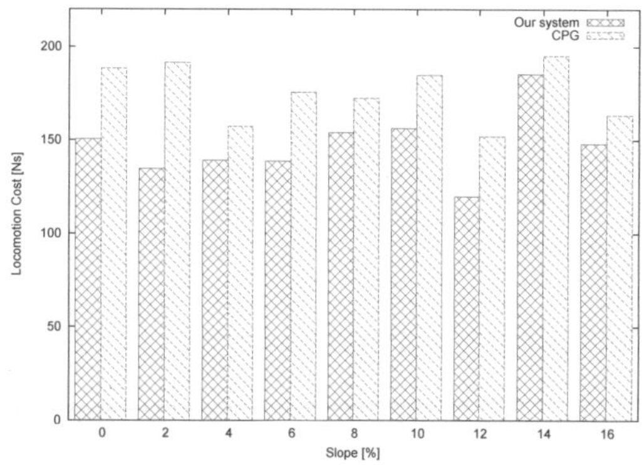

Fig. 11. Locomotion costs

passive phase appropriately. Our system generates walking that has been made passive on a 14 or 16 percent downhill slope for all periods of the swing leg phase.

4.5 Adaptive walking on uneven terrain

Next, we conducted a walking control experiment on uneven terrain. The profile of the downhill slope $y_g(x)$ is described by

$$y_g(x) = \begin{cases} -0.02x & (x < x_0) \\ -a(x - x_0) - 0.02x_0 & (x \geq x_0), \end{cases} \quad (6)$$

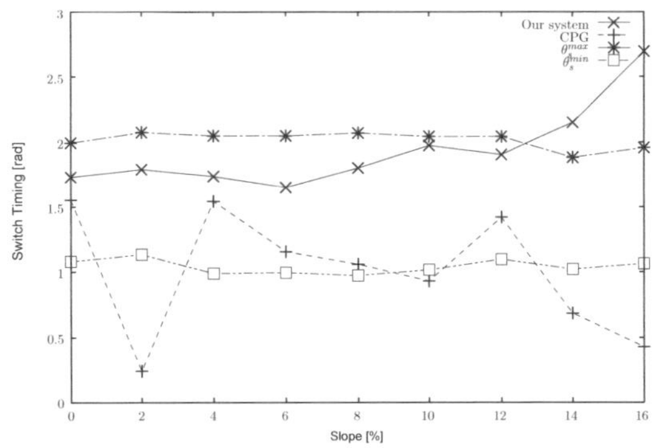

Fig. 12. Parameters of our method

a	variation	Our system	*CPG*
0	-0.0200	✓	✓
0.0025	-0.0175	✓	✓
0.0050	-0.0150	✓	✓
0.0075	-0.0125	✓	✓
0.0100	-0.0100	✓	×
0.0125	-0.0075	✓	✓
0.0150	-0.0050	✓	✓
0.0175	-0.0025	✓	×
0.0200	0	✓	×
0.0225	0.0025	✓	×
0.0250	0.0050	✓	×
0.0275	0.0075	✓	✓
0.0300	0.0100	✓	✓
0.0325	0.0125	✓	×
0.0350	0.0150	×	×
0.0375	0.0175	×	×
0.0400	0.0200	×	×

Table 3. Results of walking experiment on uneven terrain

where a ($0 \leq a \leq 4$) is the slope of the terrain at a 0.0025 interval, and x_0 is the position at which the slope of the terrain changes. In this experiment, we set $x_0 = 5$. We used each parameter that was obtained in the preceding section. Table 3 shows the experimental results. In this table, "✓" indicates that the robot could walk for 10 or more seconds. "×" means that the robot could not walk for 10 or more seconds; it fell down. Our system has robustness that is as good as *CPG*'s on uneven terrain. We found that our system did not detract from the robustness of CPG. Figure 13 shows the gaits over 10 seconds; $a = 0.0325$. In this figure, the snapshots of the gaits are traced every 0.3 seconds. Our system performs gaits that are more stable than those of *CPG*.

4.6 Comparing robot gait with human gait

Finally, we compared the robot and human gaits. The data on the human gait were measured using motion capture.

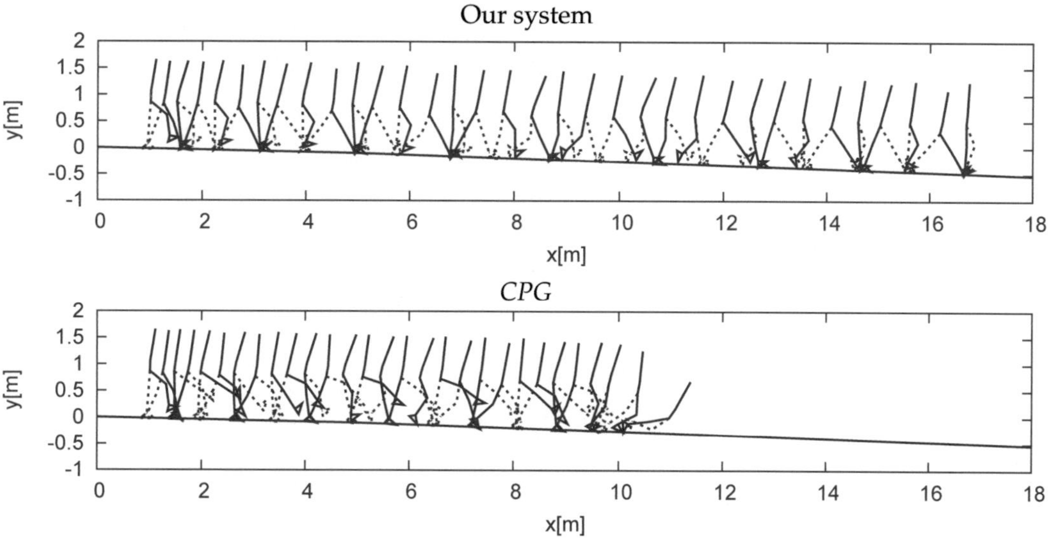

Fig. 13. Snapshots of gaits on uneven terrain with our proposed (top) and *CPG* (bottom)

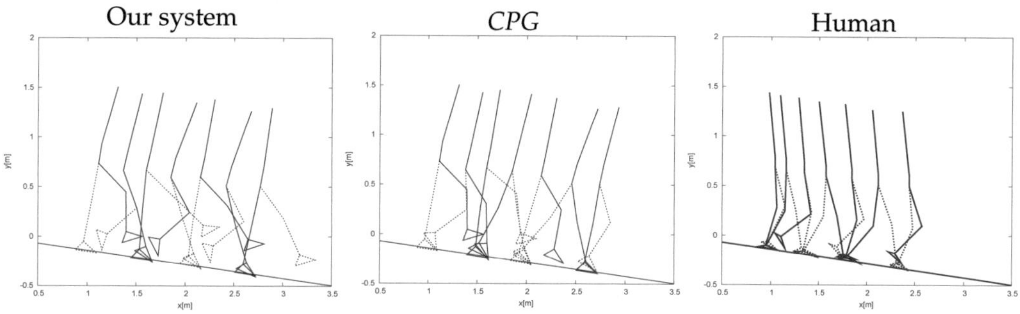

Fig. 14. Snapshots of gaits on 14 percent downhill slope with our system (left), *CPG* (center), and Human (right)

Figure 14 shows the gait over 2 seconds on a 14 percent downhill slope. In this figure, the snapshots of the gait were traced every 0.3 seconds. The our system's and *CPG*'s gaits are vorlage. In contrast, human's gait is backward tilting.

The error of the vertical component of the COG's trajectory for the robot gait and that for the human gait were calculated. Figure 15 shows the trajectories of the COG. In this figure, the trajectory of our system's gait is closer to the trajectory of a human gait than the *CPG*'s, and it is more periodically steady than *CPG*'s as well. The mean absolute error of our system is 0.0122[m], and that of *CPG* is 0.0285[m]. Our system's gait is closer to a human gait than *CPG*'s.

5. Three dimensional bipedal walking

In this section, for the expansion of the sophisticated CPG-PDW-mixture based motion control mechanism, we describe a motion control method for three dimensional biped robots shown in Figure 2 using dynamic passivization of the joint control. The motion control primitive and

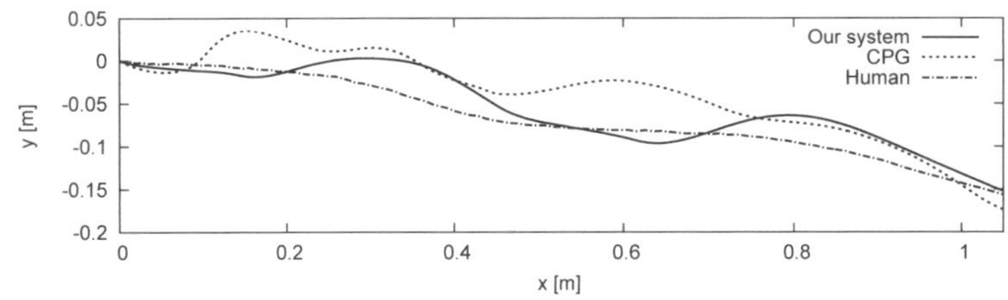

Fig. 15. Trajectories of *COG*

	Our system	*CPG*
Travel distance[m]	**0.507**	0.411
Sum total of input torque[Nms]	102.2	**93.34**
Locomotion cost[Ns]	**201.8**	227.2

Table 4. Results of First Experiment (3D walk)

the dynamic passivization mechanism and their parameters optimization are in an analogous manner of two dimensional way (described in Section 2 and Section 3).

5.1 Generating locomotion based on the dynamic passivization of joint control

We determined the effects of the dynamic passivization of the joint control. In this experiment, we used a level ground.
Figure 16 shows the gaits over 10 seconds on a level ground. In this figure, the snapshots of the gait were traced every 1.0 seconds. Table 4 shows the travel distances, the sum total of the input torques, the locomotion costs. Our system that has the mechanism of passivization of joint control generated walking that was longer than that of *CPG*. The sum total of the input torques of *CPG*'s walking is less than that of Out system's walking. However, the travel distance increased more than the increment of the consumption torque, and our system reduced the locomotion cost. We confirmed that our system generated energy efficient walking.
Figure 17 shows the trajectories of center of gravity of the robot (horizontal plane component) by the two methods. In this figure, the solid and broken lines show the trajectory of our system and *CPG*, respectively. The Stride of our system's gait is longer than that of *CPG*'s gait, because our system appropriately nullifies the input torque in the swing leg. Therefore, our system generated walking that was longer than that of *CPG*.

5.2 Gait stability analysis

We determined the gait stability. We analyzed the gait stability using the lyapunov exponent (Alligood et al. (1996)). The lyapunov exponent is a method that measures a trajectory instability of reconstructed attractor. If a maximum lyapunov exponent λ_1 that was calculated by this analysis is positive and a smaller value, the result indicates that the system acquires a stable gait. In this section, the attractors were reconstructed with the longitudinal data of the body's center of gravity while walking. The time delays for attractor reconstruction were selected as the first zero of the autocorrelation function (Albano et al. (1988)). We set the embedding dimension $m = 3$ (Takens (1981)). Table 5 shows the maximum lyapunov exponent of the body's center of gravity. In this table, the maximum lyapunov exponents of our system's gait are smaller than those of *CPG*. The walking using our system is steadier than the *CPG*.

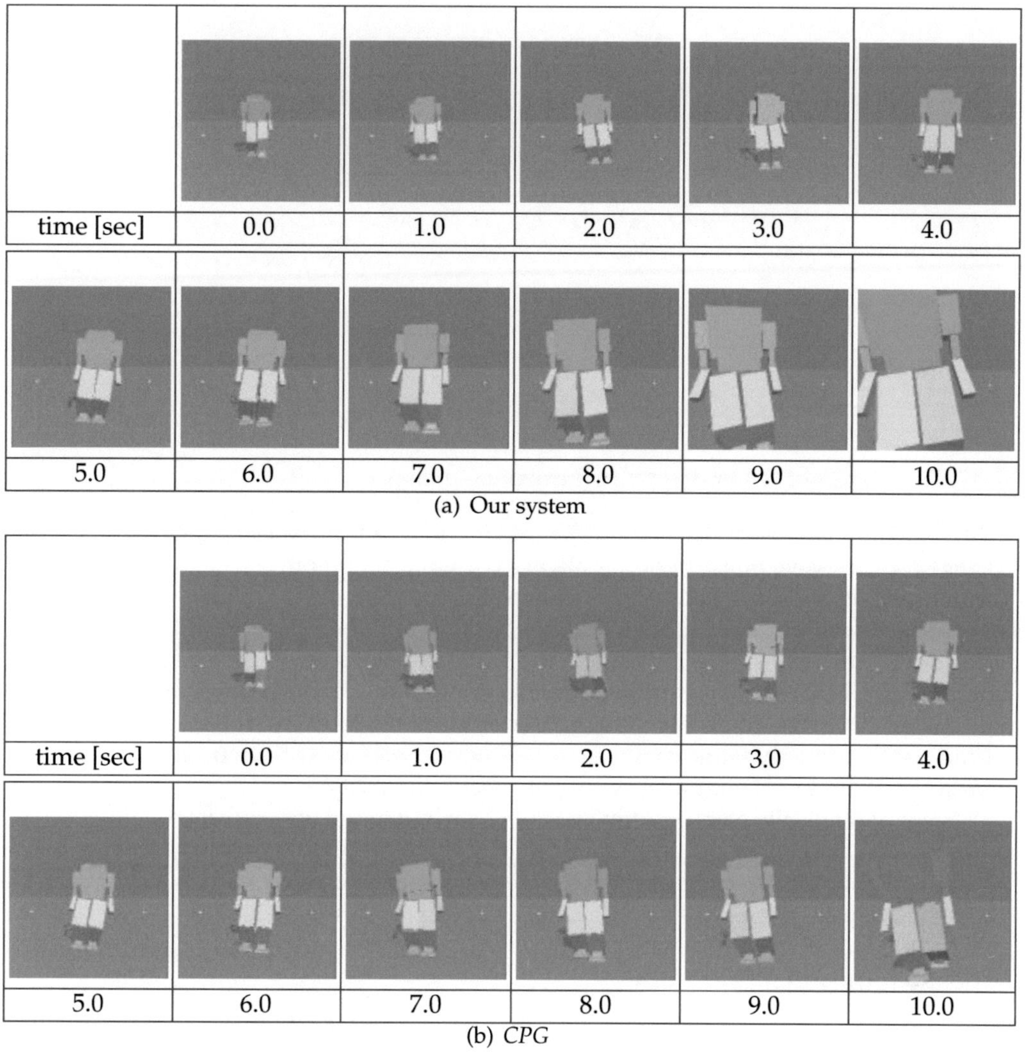

Fig. 16. Snapshots of 3D gaits

	Our system	*CPG*
$\lambda 1_{COG_y}$	**0.0027**	0.0062
$\lambda 1_{COG_z}$	**0.0676**	0.1085

Table 5. Maximum Lyapunov Exponent (3D walk)

6. Related work

In one of the mixture of active walking and PDW, there is walking that is called "ballistic walking". Ballistic walking is supposed to be a human walking model suggested by Mochon & McMahon (1980). They got the idea from the observation of human walking data, in which the muscles of the swing leg are activated only at the beginning and the end of the swing phase. Ogino et al. (2003) proposed a motion control method for energy efficient walking with

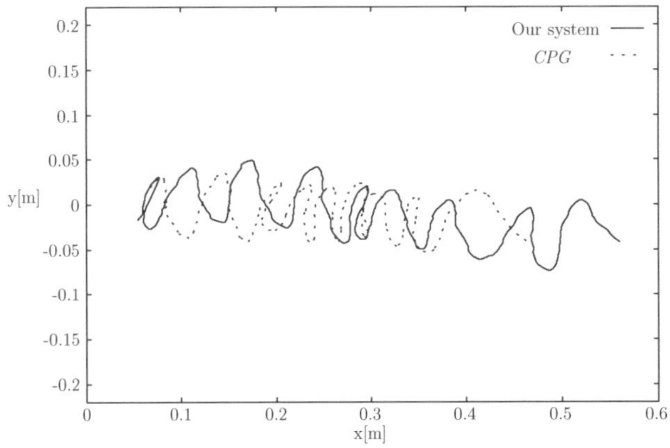

Fig. 17. Trajectories of center of gravity (horizontal plane)

ballistic walking. However, the method changes motion control statically. Even if the step time is changed, the passive phase period is always constant. Sugimoto & Osuka (2004) proposed a control method for quasi passive dynamic walking (Quasi-PDW). Quasi-PDW means that the robot usually does PDW without any input torque, and the actuators of the robot are used for ensuring walking stability only when the walking begins or when a disturbance occurs. Therefore, Quasi-PDW robots can generate energy efficient walking. However, Quasi-PDW is applicable to the robot whose dynamic characteristic suits to PDW. There are a lot of robots that can not satisfy a dynamic characteristic. Quasi-PDW robots have trouble to change the actions (e.g., changing of course, changing of speed, stop motion).

Our system dynamically changes joint control according to the pose information of robot and environment. Therefore, if the step time is changed, the robot can appropriately change joint control to passive. Our system need not to satisfy a dynamic characteristic to PDW, can easily change the actions.

7. Conclusions and future work

We described a motion control method for 2D and 3D biped robots based on a mixture of CPG and PDW, that is, dynamic passivization of joint control. We conducted walking control experiments to test the effectiveness of our method, and it demonstrated superior gaits. In gait stability analysis, we conducted that our system generated more stable gait than *CPG*'s. We conducted walking control experiments on various downhill slopes, and our method was superior here as well. In experiments on uneven terrain, our method generated robust walking that was better than *CPG*'s. We compared the robot and human gait, and our system had a trajectory that more closely modeled human walking than *CPG*.

In future work, we will create a motion control method that accounts for dynamic passivization of joint control other than in the swing leg. We will analyze the factor that the motion control using our system improved gait stability and robustness.

8. Acknowledgments

This work was supported in part by the Ministry of Education, Science, Sports and Culture, Grant–in–Aid for Scientific Research under grant #20700199.

9. References

Albano, A. M., Muench, J., Schwarts, C., Mees, A. I. & Rapp, P. E. (1988). Singular-value decomposition and the grassberger-procaccia algorithm, *Physical Review A* 38(6): 3017–3026.

Alligood, K. T., Sauer, T. D. & Yorke, J. A. (1996). *Chaos: an introduction to dynamic systems*, Springer.

Haruna, M., Ogino, M., Hosoda, K. & Asada, M. (2001). Yet another humanoid walking. -passive dynamic walking with torso under simple control-, *Proceedings of the 2001 IEEE/RSJ International Conference on Intelligent Robots and Systems (IROS-2001)*.

Ishida, M., Kato, S., Kanoh, M. & Itoh, H. (2009a). Generating locomotion for biped robots based on the dynamic passivization of joint control, *Proceedings of the 2009 IEEE International Conference on Systems, Man, and Cybernetics (SMC 2009)*, pp. 3251–3256.

Ishida, M., Kato, S., Kanoh, M. & Itoh, H. (2009b). Three dimensional bipedal walking locomotion using dynamic passivization of joint control, *Proceedings of the 2009 IEEE International Symposium on Micro-NanoMechatronics and Human Science (MHS 2009)*, pp. 580–585.

Itoh, Y., Taki, K., Kato, S. & Itoh, H. (2004). A stochastic optimization method of cpg-based motion control for humanoid locomotion, *Proceedings of the IEEE Conference on Robotics, Automation and Mechatronics (RAM 2004)*, pp. 347–351.

Kato, S. & Itoh, H. (2005). Soft computing approaches to motion control for humanoid robots, *International Symposium on Micro-Nano Mechatronics and Human Science*, pp. 47–52. (as an invited paper).

Kotosaka, S. & Schaal, S. (2000). Synchronized robot drumming by neural oscillators, *Proceedings of the International Symposium on Adaptive Motion of Animals and Machines*.

Matsuoka, K. (1985). Sustained oscillations generated by mutually inhibiting neurons with adaption, *Biological Cybernetics* 52: 367–376.

Matsuoka, K. (1987). Mechanisms of frequency and pattern control in the neural rhythm generators, *Biological Cybernetics* 56: 345–353.

McGeer, T. (1990). Passive dynamic walking, *The International Journal of Robotics Research* 9(2): 62–81.

Miki, M., Hiroyasu, T. & Ono, K. (2002). Simulated annealing with advanced adaptive neighborhood, *Proceedings of the 2nd International Workshop on Intelligent Systems Design and Applications (ISDA 2002)*, pp. 113–118.

Miyakoshi, S., Taga, G., Kuniyoshi, Y. & Nagakubo, A. (1998). Three dimensional bipedal stepping motion using neural oscillators - towards humanoid motion in the real world -, *Proceedings of the 1998 IEEE/RSJ International Conference on Intelligent Robots and Systems (IROS'98)*, pp. 84–89.

Mochon, S. & McMahon, T. A. (1980). Ballistic walking, *Journal of Biomechanics* 13: 49–57.

Murase, Y., Yasukawa, Y., Sakai, K. & et al. (2001). Design of a compact humanoid robot as a platform, *Proc. of the 19-th conf. of Robotics Society of Japan*, pp. 789–790. (in Japanese), http://pr.fujitsu.com/en/news/2001/09/10.html.

Nakamura, T., Kato, S. & Itoh, H. (2005). A motion learning method condsider joint stiffness for biped robots, *The 3rd International Conference on Computational Intelligence, Robotics and Autonomous Systems (CIRAS'2005)*, pp. CD–ROM (RH3–1).

Ogino, M., Hosoda, K. & Asada, M. (2003). Learning energy efficient walking with ballistic walking, *Proceedings of the 2nd International Symposium on Adaptive Motion of Animal and Machines (AMAM'2003)*, pp. CD–ROM (ThP–I–5).

Sugimoto, Y. & Osuka, K. (2004). Walking control of quasi passive dynamic walking robot "quartet iii" based on continuous delayed feedback control, *Proceedings of IEEE International Conference on Robotics and Biomimetics (Robio 2004)*.

Taga, G. (1995a). A model of the neuro-musculo-skeletal system for human locomotion i. emergence of basic gait, *Biological Cybernetics* 73: 97–111.

Taga, G. (1995b). A model of the neuro-musculo-skeletal system for human locomotion ii. real-time adaptability under various constraints, *Biological Cybernetics* 73: 113–121.

Taga, G., Yamaguchi, Y. & Shimizu, H. (1991). Self-organized control of bipedal locomotion by neural oscillators in unpredictable environment, *Biological Cybernetics* 65: 147–159.

Takens, F. (1981). Detecting strange attractors in turbulence, *Lecture Notes in Mathematics* 898: 366–381.

Taki, K., Itoh, Y., Kato, S. & Itoh, H. (2004). Motion generation for bipedal robot using neuro-musculo-skeletal model and simulated annealing, *Proceedings of the IEEE Conference on Robotics, Automation and Mechatronics (RAM 2004)*, pp. 699–703.

12

A Bio-Robotic Toe & Foot & Heel Model of a Biped Robot for More Natural Walking: Foot Mechanism & Gait Pattern

Jungwon Yoon[1], Gabsoon Kim[2], Nandha Handharu[1] and Abdullah Özer[1]
[1]*School of Mechanical and Aerospace Engineering and ReCAPT*
Gyeongsang National University,
[2]*Control Instrumentation Engineering Department*
Gyeongsang National University,
Republic of Korea

1. Introduction

Humans possess a complex physical structure and can perform difficult movement tasks. Over the past few decades, many researchers around the world have concentrated on achieving human-like artificial mobility or dexterity either on humanoid robots or during the implementation of robotic assistive devices. In particular, humanoid-type robots mainly focused on hands to understand the mechanical and dynamical functions of ourselves. On the other hand, there have been few researches to achieve human like foot. Until now, human-like skillful mobility has not been achieved on humanoid robots, since the robotic feet are far from adaptation to keep stable contact on the ground and the current kinematic structures of a humanoid foot is different from that of a real human foot. Stability related issues have been the main goal for humanoid robots in relevant researches. Initially, humanoid robots were built so that they can walk stably with flat foot (Sakagani et al., 2002; Okada et al., 2004). These initial walking patterns were optimized for the highest stability, and the resulting walking pattern had knee bending and flat-feet walking. A more advanced strategy was developed for generating biped walking pattern involving heel strike and toe off motion in (Huang et al., 2001). However, because of the mechanism's limitation the knee bending walking patterns were always chosen for the benefit of stability, thus making it less natural. Today, more advanced control approaches, faster and more powerful actuators, and more sophisticated walking pattern generation strategies have helped the research goal to be shifted to pursue more natural walking patterns for biped robots, with the expectation that someday humanoid robot can coexist with human.
To improve walking capacity of humanoid-type robots, toe mechanisms with 1-dof was suggested earlier, (Ahn et al., 2003; Takahashi et al., 2004). For walking in a straight direction, 1-dof toe mechanism can supply faster walking for a robot. In addition, relative toe motion can increase the naturalness of robot walking and help to reduce the load on the knee joints, where high force and speed are required to achieve robot locomotion (Nishiwaki & Kagami, 2002). However, the foot device with 1-dof toe mechanism cannot

adapt turning during walking and may not satisfy safe and natural walking under uneven terrain conditions, (Takao et al., 2003). Takao et. al concluded through gait experiments that the robotic foot mechanisms with multi degree-of-motions can contribute smooth and effective body motions in the stance phase. A foot mechanism with such dexterous motions at toe, foot, and heel will certainly enable the humanoid robots to perform more efficient and skillful movements on various terrains. Therefore, it will be advantageous to develop foot mechanisms with multi-dof motions for humanoid-type robots to satisfy natural walking actions on various terrain situations. Even though serial actuations of rotary motors at foot joints can generate multi-dof motions, it is difficult to implement the active joints due to the high torque by the heavy humanoid weight. Thus, a serial-parallel mechanism can be a good solution to satisfy the desired performance of foot devices in humanoid robots since it can produce high rigidity, compactness, and precise resolution, as compared to serial mechanisms. In addition, the foot mechanism for humanoid-type robots should have multi-platforms to allow relative rotations between the toe and the foot, and to generate heel motions.

This chapter deals with a toe & foot& heel model that can allow a humanoid robot to walk more naturally, closer to a normal human. A foot device with a 4-DOF parallel mechanism is suggested to generate human-like foot motions. The mechanism for the toe & foot& heel motions for the humanoid consists of several toe platforms using a serial-parallel hybrid mechanism; a foot platform, a heel platform, corresponding limbs to the platforms, and the base, which is located at the humanoid shin. The suggested foot platform can generate pitch & roll motions at ankle position using 2-dof-driving parallel mechanism with two linear actuators fixed to the base. The toe and heel joint motions can be implemented by attaching 6-dof serial joints between the platform and the base, and by connecting the corresponding platform and the foot platform with a revolute joint. As a result, the suggested foot mechanism with more dexterous functions can adjust the biped robot's walking movements during the stance phase of gait. Together with its design advantage, the suggested toe, foot, and heel model can also facilitate more natural walking patterns for biped robots. A new alternative methodology to generate gait pattern online with knee stretched motion utilizing toe and heel joints will also be presented within this chapter.

2. The mechanism description

2.1 Human foot

The set of foot and ankle is mechanically very complex since it has 26 bones, 33 joints, more than 100 muscles, tendons, and ligaments (Guihard &Gorce, 2004). All these mechanical elements collaborate to offer the body support, balance, and mobility. Structurally, the foot has three main parts: The hindfoot, the midfoot, and the forefoot. The hindfoot is composed of three joints and connects the midfoot to the ankle. It lets the foot to move up and down (-50°- 20°) and provides slight rotations. The midfoot has five irregular shaped tarsal bones, which acts as a shock absorber. The forefoot has five phalanges of toes and their connecting long bones called metatarsals. Their maximum rotation range is about 90° when the toes are totally raised.

2.2 Mechanism description for toe, foot, and heel models

The mechanism for the toe & foot & heel motions of the humanoid robot consists of several toe platforms, a foot platform, and a heel platform and corresponding limbs to the platforms, revolute joints between the toes and the foot platform, and a revolute joint between the heel and the foot platform as shown in Figure 1. Base of each platform is located at a humanoid shin.

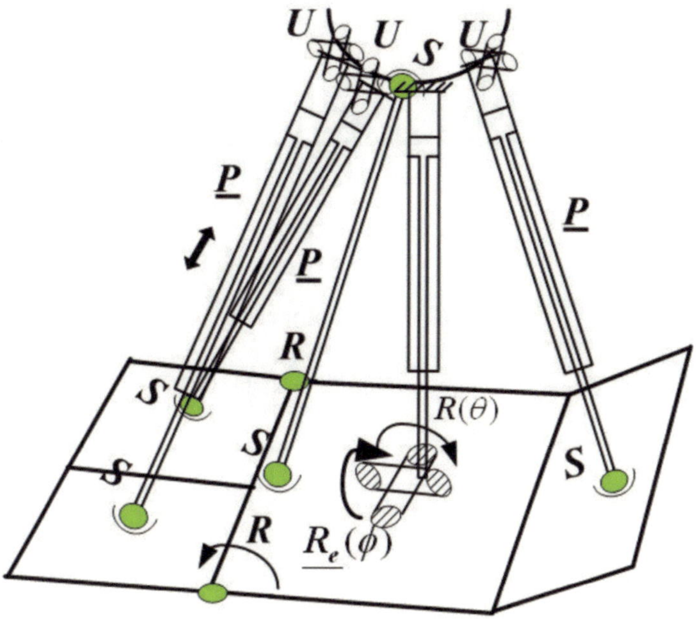

Fig. 1. A novel foot mechanism with toe, foot, and heel motions

In the figure, the letters *P*, *R*, *U*, and, *S* represent prismatic, revolute, universal, and spherical joints, respectively. An underlined letter represents an actuated joint. The numbers of toe and heel joint are selected as two and one, respectively. One limb with 6-dof serial joints (S-P-U) is attached to toe and heel platforms, respectively, while middle limb ($\underline{P}_e$-$\underline{R}_e$-R) and four-bar limb (S-S) are attached to the foot platform. The middle limb with equivalent 3-dof serial joint ($\underline{P}_e$-$\underline{R}_e$-R) is driven by the 2-dof driving mechanism that is equivalent active serial prismatic and revolute joints ($\underline{P}_e$-$\underline{R}_e$) with two base-fixed prismatic actuators. The four-bar limb will allow the foot platform to generate a pitch motion (θ) according to active prismatic joint motions ($\underline{P}_e$) of the 2-dof driving mechanism. In result, the foot platform can generate two rotations with equivalent 2-dof serial joint ($\underline{R}_e$-$\underline{R}_e$). A toe joint motion can be implemented by attaching 6-dof serial joints (S-$\underline{P}$-U) between the toe platform and the base, and by connecting the toe and the foot platform with a revolute joint. Similarly, a heel joint motion can be implemented by attaching a 6-dof serial joints (S-$\underline{P}$-U) limb and a revolute joint to the rear part of the foot platform

Since toe and heel platforms have one limb with one 6-dof serial joint (S-$\underline{P}$-U) and the foot platform has 2-dof serial joint ($\underline{R}_e$-$\underline{R}_e$), the final output displacement of each platform is dependent only on that of the foot platform with its 2-dof serial joint ($\underline{R}_e$-$\underline{R}_e$), which is the intersection of the special Plücker of two limbs. Therefore, the suggested mechanisms have five degrees of freedom in total when the toe joints are two and the heel joint is one. The suggested foot model can generate pitch & roll motions at the ankle position of a humanoid robot, toe joints motions, and a heel joint motion. Toe joint motions can be extended by attaching another toe platform with an additional 6-dof limb to the foot platform.

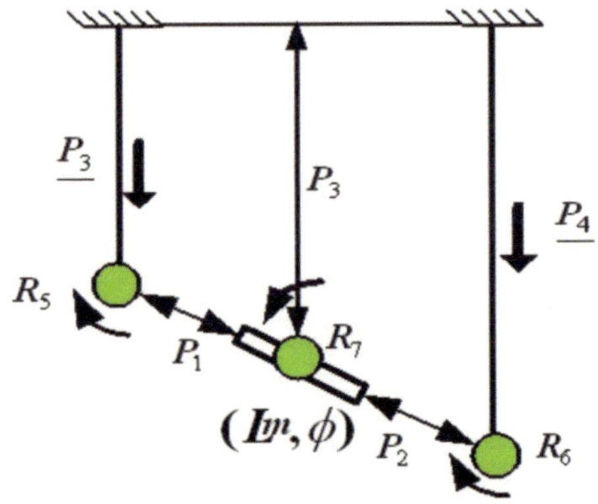

(a) T-R type joint array

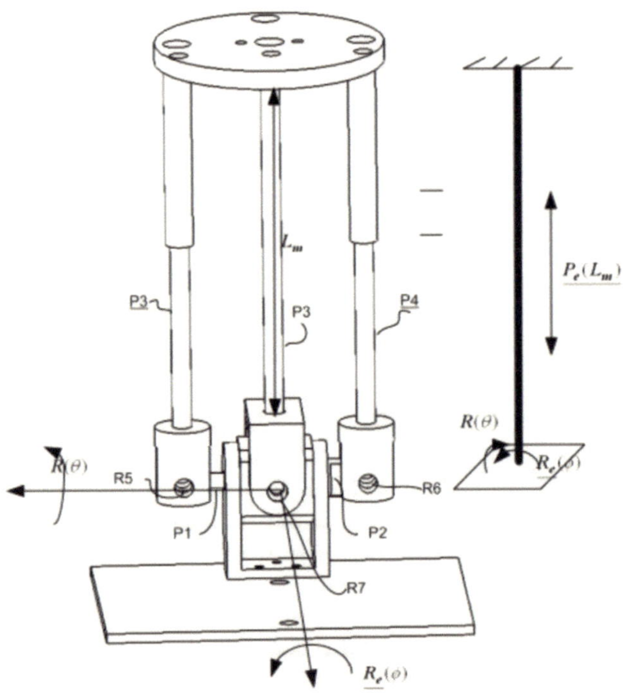

(b) T-R type Implementation

Fig. 2. T-R type 2-dof-driving mechanism

Figures 2 shows a T-R type 2-dof driving mechanism that can generate 2-dof translational and rotational motions of the platform using two base-fixed prismatic actuators. Figure 2(a) shows the joint array of the T-R type 2-dof-driving mechanism. The T-R-type driving mechanism consists of passive prismatic joints (P_1 and P_2) between revolute joints (R_5 and R_6) at the upper ends of the active prismatic joints ($\underline{P_3}$ and $\underline{P_4}$) fixed to the base and end-effector of the driving mechanism, and includes a passive prismatic joint (P_3) between the revolute joint R_7 of the end-effector and base plate. P_3 allows the end-effector to move in the z-direction only and R_7 allows the end-effector to rotate about the y-axis only. A CAD model of the T-R-type driving mechanism is shown in Figure 2(b). Consequently, the T-R-type driving mechanism is conceptually equivalent to active serial prismatic and revolute joints ($\underline{P_e}$-$\underline{R_e}$). Also, R-R-type mechanism will be generated by attaching an additional rotation bar, which will change a translation motion to a rotation as shown in Figure 3.

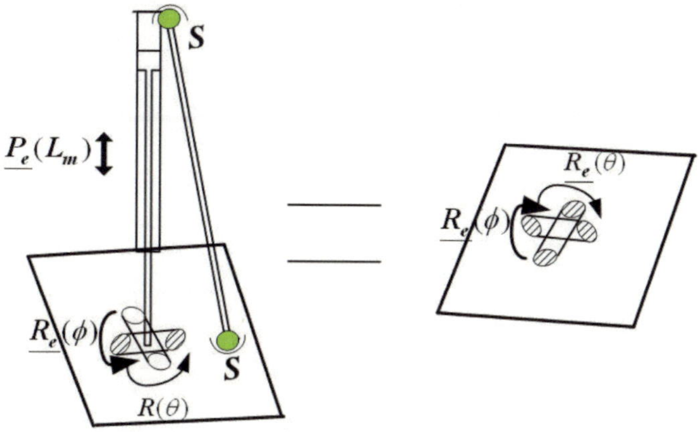

Fig. 3. R-R type 2-dof driving mechanisms

Figure 4 shows the link-pair relationship diagram for the 2-dof-driving mechanism. The white boxes represent passive joints and the hatched boxes represent active joints. Lines between letters represent links. It is possible to consider each driving mechanism as equivalent to the two actuated joints ($\underline{P_e}$-$\underline{R_e}$ *or* $\underline{R_e}$-R_e) in terms of the number and type of degrees of motion. The motions of the given mechanism can be verified using Grübler's formula as:

$$M = d(n - g - 1) + \sum_{i=1}^{g} f_i \tag{1}$$

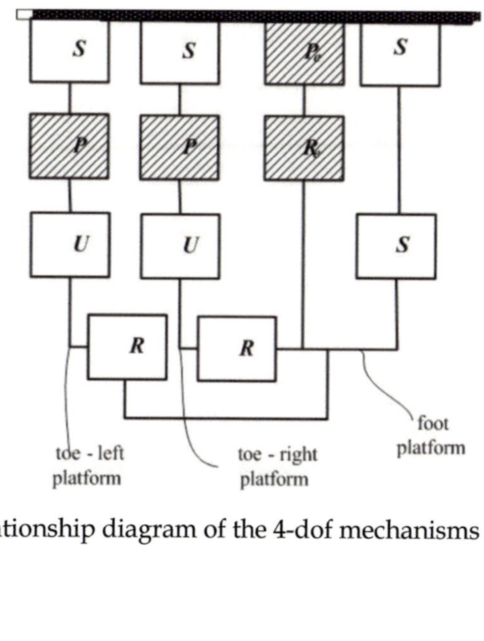

Fig. 4. The link-pair relationship diagram of the 4-dof mechanisms with three platforms

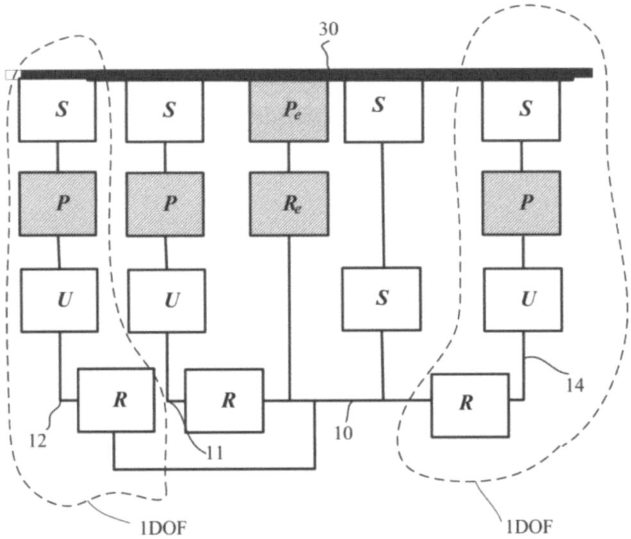

Fig. 5. The joint structure of the mechanism with the N+2 DoF

If the driving mechanisms are considered equivalent to serial ($\underline{P_e}$-$\underline{R_e}$) joints and the number of the toe platform and the heel platform is two and zero, the mobility of the 4-dof mechanisms with three platforms as shown in Figure 4 is:

$$M_{4dof} = 6(10 - 12 - 1) + 1\times6 + 2\times2 + 3\times4 = 4$$

The Figure 5 shows that the numbers of the toe platform and the heel platform with the foot platform determine the mobility of the mechanism. If the platform numbers is N except the foot platform, then, the mobility of the mechanism can be computed as;

$$\begin{aligned} M_{(2+N)dof} &= 6((4+3N)-(4+4N)-1)+1\times 2N+ \\ &2\times N+3\times(2+N)+3\times(2+N)=2+N \end{aligned} \quad (2)$$

Therefore, the mobility of the mechanism can be generalized as follows:

$$M = 2 + N$$

where M is the mobility of the suggested mechanism and N is the number of platforms with a 6-joint (S-$\underline{P}$-U) limb and a revolute joint. For example, If N=3, the mobility becomes five as follows;

When N=3,

$$M_{(2+3)dof} = 6(13 - 16 - 1) + 1\times 8 + 2\times 3 + 3\times 5 = 5$$

Figure 6 shows that when N=3, the mechanism can generate two toe joints motions and foot motions with two rotations and one heel joint motion.

Figure 7 shows the developed humanoid foot which can generate pitch motions of the two toe platforms and the heel platform with relative rotations between the corresponding platform and the foot platform. The mechanism can generate two rotations at an ankle with a $\underline{R}$-$\underline{R}$ type 2-dof-driving mechanism motion.

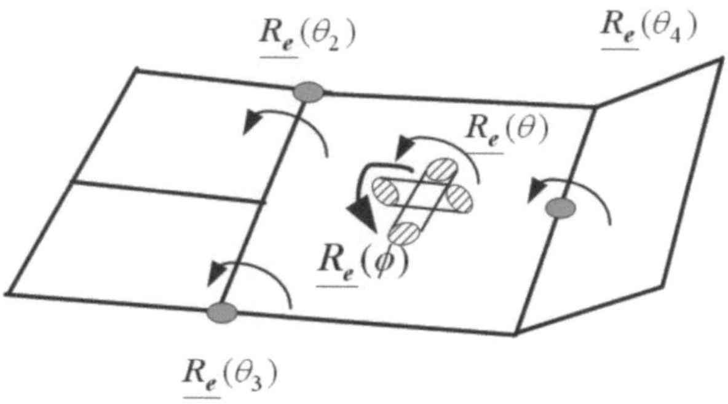

Fig. 6. Platform motions with N=3

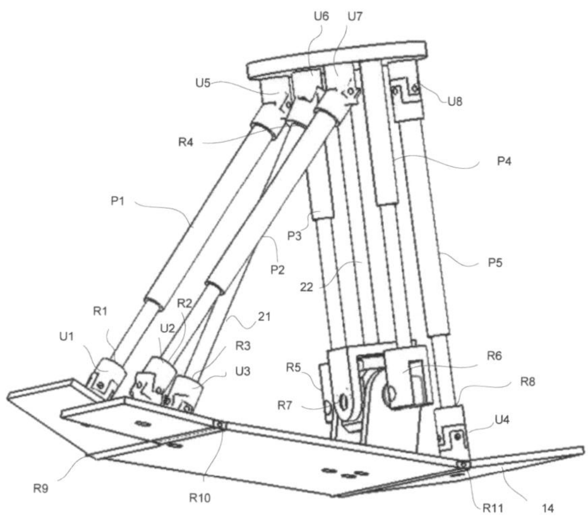

Fig. 7. The developed humanoid foot

3. Walking simulations with a biped robot

3.1 Kinematics of a biped robot with the foot models

The kinematic relationships of a biped robot with the foot models are derived to generate walking trajectory. Forward and inverse kinematic equations are used to calculate the posture of robot and angles of each joint. Coordinate system of the biped robot with foot models with two toe joints and one heel joint is shown in Figure 8. The base coordinates $\{O_b\}$ is located on the ground in the middle point of the feet, and the truck coordinates $\{O_M\}$ is located on the middle point of waist. The left $\{O_{f,L}\}$ and right $\{O_{f,R}\}$ foot coordinates are located on the foot platform. The waist coordinate $\{O_w\}$ is located on the upper position of each limb.

The inverse kinematics computes angles $\phi_{1,i}, \theta_{2,i}, \theta_{3,i}, \theta_{4,i}, \phi_{5,i}, \nu_{6,i}$ of each joint of the biped robot given the position and orientation of the waist center, and the toe joints $\theta_{tr,i}, \theta_{tl,i}$ and the heel joint $\theta_{h,i}$, where $\phi_{1,i}, \theta_{2,i}, \theta_{3,i}$ are pitch and roll angles of the ankle joint, knee joint angle, respectively. $\theta_{4,i}, \phi_{5,i}, \nu_{6,i}$ are the joint angles of the pelvis, and i is the sub-suffix for left L and right R. For inverse kinematics of the biped robot, the equation can be derived as follows;

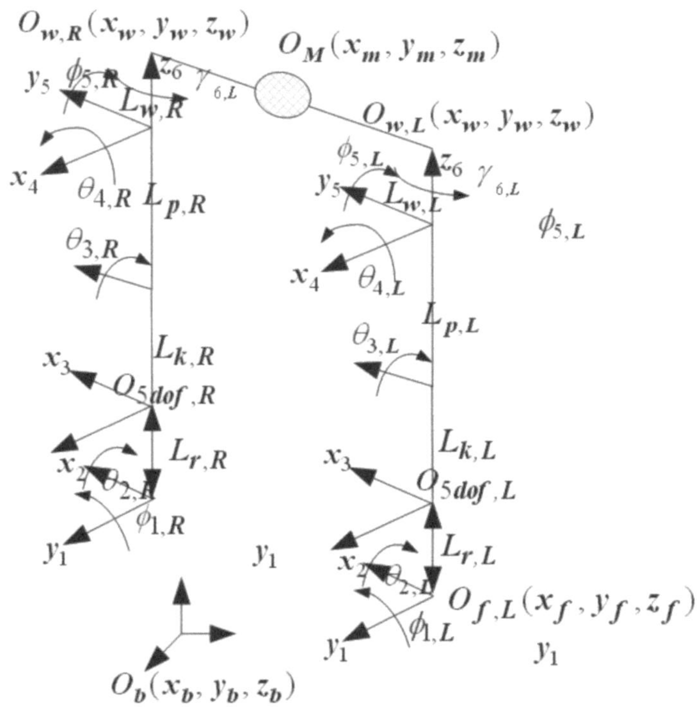

Fig. 8. Coordinate system of the biped robot

$$O_{t,i} = R_y(\phi_{1,i})R_x(\theta_{2,i})T_f^3(0,0,L_{k,i}+L_{r,i})$$
$$R_x(\theta_{3,i})T_3^4(0,0,L_{p,i})R_x(\theta_{4,i})R_y(\phi_{5,i})T_5^w(0,0,L_w)$$
$$(\boldsymbol{i} = \boldsymbol{L}, \boldsymbol{R}) \tag{3}$$

where $O_t(x_t, y_t, z_t)$ is the vector from the foot coordinate $O_{f,i}$ to waist coordinate $O_{w,i}$, and L_r, L_k, L_p, and L_w are the average length of the 2dof driving mechanism, the length of calf, thigh, and waist, respectively. $\boldsymbol{R}$ is the rotation matrix, $T_f^3(x,y,z)$ is the translation matrix from local reference frame $\boldsymbol{O}_f$ (x_f, y_f, z_f) of the foot platform to the 3-axis mobile reference frame $\boldsymbol{O}_3$ (x_3, y_3, z_3). If $L_{w,i}$ is equal to zero, then the equation (3) can be simplified into equation (4)

$$x_t = (-\sin(\phi_1)\sin(\theta_2)\sin(\theta_3) + \sin(\phi_1) \cos(\theta_2)\cos(\theta_3))L_p + \sin(\phi_1)\cos(\theta_2)L_{kr} \tag{4-a}$$

$$y_t = (-\cos(\theta_2)\sin(\theta_3) - \sin(\theta_2)\cos(\theta_3))L_p - \sin(\theta_2)L_{kr} \tag{4-b}$$

$$z_t = (-\cos(\phi_1)\sin(\theta_2)\sin(\theta_3) + \cos(\phi_1)\cos(\theta_2) \cos(\theta_3))L_p + \cos(\phi_1)\cos(\theta_2)L_{kr} \tag{4-c}$$

From the equations (4-a) and (4-c), the roll angle ϕ_1 of ankle joints can be computed as;

$$\phi_1 = \tan^{-1}(x_t / z_t)$$

From the equation (4-b) and equation (4-c), the equation (5) can be derived by deleting the parameter θ_3 and utilizing the MATLAB symbolic toolbox (Mathworks).

$$\begin{aligned} & f_1 = (y_t \cos(\phi_{1,i}) \sin(\theta_2) - \cos(\theta_2) z_t - (L_k + L_r) \\ & \cos(\phi_1))^2 - (-\cos(\phi_1)^2 y_t^2 \cos(\theta_2)^2 - 2\cos(\phi_1)\cos(\theta_2) \\ & y_t z_t \sin(\theta_2) - z_t^2 + z_t^2 \cos(\theta_2)^2 + \cos(\phi_1)^2 L_p^2) \end{aligned} \quad (5)$$

Then, the kinematic relationships of the 5-dof foot mechanism are shown in Figure 9. A local reference frame for the foot platform, $O_f(x_f, y_f, z_f)$, is located at the center of the foot platform. A local reference frames for the toe-left O_f (x_f, y_f, z_f) and the toe-right O_{tr} (x_{tr}, y_{tr}, z_{tr}) are centered at each toe platform, respectively. The base of the 5-dof foot mechanism, O_{5dof} $(x_{5dof}, y_{5dof}, z_{5dof})$, mobile reference frames is located on the z-axes of reference frame O_f of its platform. b is the distance of y-axis from center position O_f of the foot platform to the toe-revolute joint, a is the distance of y-axis from the revolute joint to center position O_{tl} of the toe platform, and c is the distance of x-axis from the foot platform joint to center position O_{tl} of the toe platform .

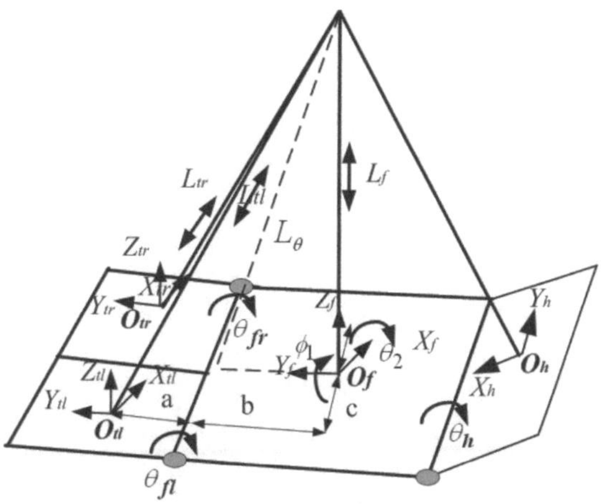

Fig. 9. The kinematic model of the foot mechanism

The height L_f from the foot reference frame to O_{5dof} reference frame have following relation

$$L_\theta^2 = b^2 + L_f^2 - 2bL_f \cos(\theta_2 + \pi / 2) \quad (6)$$

where L_θ is a constant distance value from center revolute joint to O_{5dof} reference frame. From equation (6), the height L_f can be computed as

$$L_f = -b\sin(\theta_2) + \sqrt{b^2 \sin(\theta_2)^2 - b + L_\theta^2} \quad (7)$$

Then, the equation (7) can be rearranged into equation (8).

$$f_2=b^2+(L_f)^2-2bL_f\cos(\theta_2+\pi/2)-L_\theta^2 \tag{8}$$

Since the equations (5) and (8) have only unknown parameters of θ_2 and L_f, these two nonlinear equations can be solved by Newton-Rapson's numeric method. If the parameter θ_2 and L_f are solved, then θ_3 can be computed by utilizing the equation (4-c). Next, the lengths of the actuators of the 2-dof driving mechanism are then obtained as:

$$L_{f1}=L_f+\frac{L_{base}\tan\phi}{2} \quad L_{f2}=L_f-\frac{L_{base}\tan\phi}{2} \tag{9}$$

where L_{f1}, L_{f2}, and L_{base} are the active lengths of the left and right actuators, and the distance between the two active prismatic joints of the 2-dof mechanism, respectively.
If the toe joint angles θ_{tr}, θ_{tl} and the heel joint angle θ_h are given, the actuator's length of each platform can be computed as follows;

The mobile reference frame $\boldsymbol{O_{5dof}}$ can be represented in the base reference frame (X_b, Y_b, Z_b) as:

$$O_{5dof}=O_f+R_y(\phi_1)R_x(\theta_2)T_f^{5dof}(0,0,L_f) \tag{10}$$

The coordinate position $\boldsymbol{O_{tl}}$ of the toe-left mobile reference frame can be represented in the base reference frame (X_b, Y_b, Z_b) as:

$$\begin{aligned}O_{tl}&=R_y(\phi_1)R_x(\theta_2)T_r^m(-c,b,0)\\&+R_y(\phi_{1,i})R_x(\theta_{2,i})T_m^{tl}(0,a,0)\end{aligned} \tag{11}$$

Similarly, the coordinate position $\boldsymbol{O_{tr}}$ of the toe-right mobile reference frame can be represented in the base reference frame (X_b, Y_b, Z_b) as:

$$\begin{aligned}O_{tr}&=R_y(\phi_1)R_x(\theta_2)T_r^m(c,b,0)\\&+R_y(\phi_{1,i})R_x(\theta_{2,i})T_m^{tl}(0,a,0)\end{aligned} \tag{12}$$

Subsequently, the toe-left actuator length can be easily computed as follows;

$$L_{tl}=\sqrt{|O_{tl}-O_a|} \tag{13}$$

Also, the toe-right actuator length can be easily computed as follows;

$$L_{tr}=\sqrt{|O_{tr}-O_a|} \tag{14}$$

Using the above method, the heel actuator length can be computed as follows;

$$L_h=\sqrt{|O_{heel}-O_a|} \tag{15}$$

3.2 The models of normal gait cycle

Normal gait is defined as series of rhythmic alternating movements of the limbs and trunk (Perry, 1992). The gait cycle is the activity that occurs between heel strike by one extremity and the subsequent heel strike on the same side. The gait cycle consists of stance and swing phases. The entire period during which the foot is on the ground is the stance phase. Conversely, during the swing phase, the foot that is stepping forward is not in contact with any object. Stance phase accounts for approximately 60% of a single gait cycle, while swing phase accounts for approximately 40%. During the stance phase, the human foot performs a rolling motion on the ground. Throughout the motion trajectory, one foot lands on its heel at some heel-strike angle a (loading phase). Then, the foot stays flat on the ground during mid-stance (mid-stance phase). After the sole of the foot has made contact with the ground, the heel begins to rise from the ground with relative rotation at the metatarsal joint, and the contact moves to the toe with a toe-off angle λ (terminal stance). Finally, the sole of the foot flattens before the foot is lifted from the ground (pre-swing phase). To simulate these foot trajectories on a planar surface during stance phase, the platform variables θ_f and θ_r of the footpad mechanism can be defined for each subphase during the stance phase, as shown in Figure 10. Note that the left and right toe joints are identical and during loading phase, the toe joint angle θ_t and the foot pitch angle θ_f are identical, while the foot pitch angle θ_f and the heel joint angle θ_h are identical during terminal stance

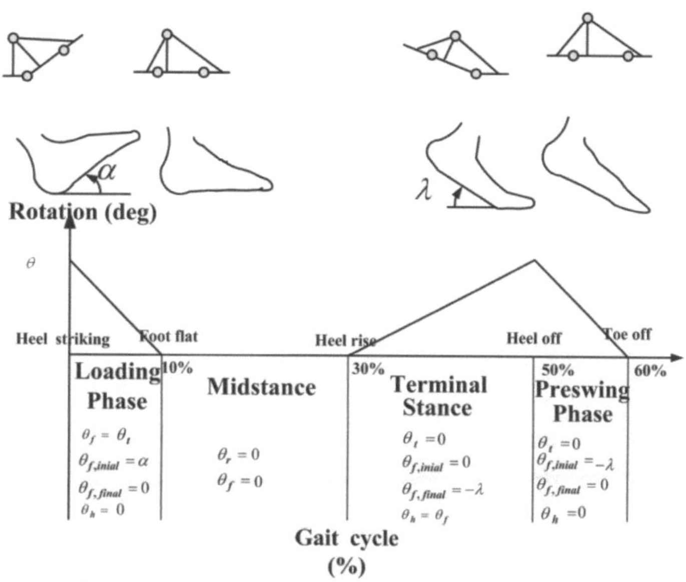

Fig. 10. Foot trajectory during stance phase

3.3 Trajectory generation of the humanoid robot with the suggested toe, foot, and heel models

If heel-strike angle α, toe-off angle λ, step length are 20°, and 30cm, the foot trajectory configuration at a planar surface with respect to the gait cycle is obtained as shown in Figure 11. These simulations based on gait analysis showed that the suggested mechanism with toe, foot, and heel models can generate natural foot motions, including relative rotations at the toe and the heel during the stance phases.

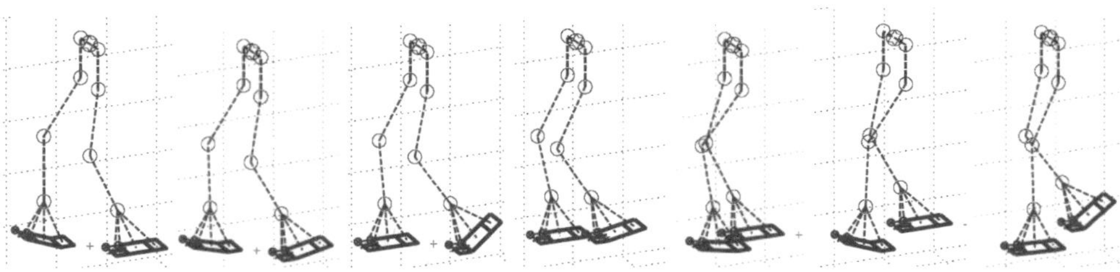

(a) (b) (c) (d) (e) (f) (g)

Fig. 11. The walking simulations of a biped robot with toe, foot, and heel models during the stance phase

4. Online gait pattern generation

Traditionally, the ZMP based trajectory method was used by researchers to generate gait patterns for humanoid robots. Approaches using simplified biped robot model such as rolling mass model or inverted pendulum model were usually being utilized to solve the ZMP equations, (Nishiwaki et al., 2002). However, there could be difficulties in some cases to achieve the necessary high hip accelerations to realize the desired ZMP trajectories. In such cases, since the trunk has the biggest mass value, energy consumption increases, and control for task execution of the upper limbs becomes difficult, (Huang et al., 2001). Moreover, no matter how well the algorithm can make the biped robot follow the desired ZMP trajectory, the motion result of the hip itself cannot be assured. This has motivated several researchers to find alternative ways of generating gait patterns for biped robots. Huang et al. proposed a gait pattern algortihm that control the hip motion with adjustable hip parameters that achieved highest stability margin, instead of designing the ZMP trajectory first. This way, the motion result can be directly controlled with those parameters. Though Huang's method has some advantages compared to the ZMP trajectory method, it required an optimization scheme to select best hip parameters that can obtain gait pattern with the highest stability margin which made the algorithm unable to generate the gait pattern online. Advancement has been made with this method, (Peng et al., 2004). However, since typical walking pattern has to be defined first, there will be limitation on the variety of the gait patterns. Hence, it is very important to improve effective gait pattern algorithms.

In terms of foot mechanism more important developments were made as discussed in (Ramzi et al., 2003; Nishiwaki et al., 2002; Yoon et al., 2007; Ki Ahn et al., 2003). These robots had modified feet using extra joints in their toe positions. There are many researches concerning toe joints in the biped robot research. Faster walking, longer steps and more degrees of freedom were obtained in these research studies. Knee stretch motion has often been related to the naturalness of biped walking. Recently some researchers tried to achieve stable walking patterns involving knee stretch motion (Kurazume et al., 2005; Ogura et al., 2006). There is even a commercial product for humanoid robot that is able to walk with straight knee (Garage, 2008). The research utilized a conventional biped robot with flat foot to generate straight walking pattern (Kurazume et al., 2005). One of the main reasons why knee stretch motion for biped walking is hard to achieve is that during the knee stretch, the inverse kinematics solution for the leg becomes singular. This research has not been able to avoid this singularity and utilized "if-then" commands on its algorithm. In (Ogura et al., 2006), the researchers tried to achieve knee stretch walking by adding extra joint in the humanoid robot. In this way the singularity can be overcome by the extra joint in the waist owing to losing some DOF in the knees. To achieve knee stretch walking pattern, we will attempt to add extra joint in the heel here. It is true that human doesn't have heel joint, but gait analysis research shows that human walking sequence has heel strike motion and the knee stretch occurs during this phase of walking (Perry, 1992). During this motion, human heel acts as an extra support region. Based on these facts, we decided to add extra heel joint to produce some support region. Figure 12 shows the sequence of foot support areas during stance phase. With the extra heel joint, we propose a walking pattern strategy that enables the knee stretch motion which can avoid singularity. The loss of degree of freedom in the knee can be overcome through the existence of the heel joint. Our algorithm can generate stable knee stretch walking patterns without singularity. With knee-stretched motion, biped robot walking pattern will not only become more similar to human, but also will require less torque and thus making the pattern more energy efficient as human walking itself is optimized to energy efficiency (Kurazume et al., 2005; Ogura et al., 2006). Moreover, with our proposed mechanism, utilizing extra addition of heel joint is comparably less sophisticated than those biped robots, which have waist joints. We studied our algorithm with a biped robot in computer simulation. The walking trajectory generated through this method has successfully generated knee stretch walking patterns. Because of the addition of the heel joints, not only the support area during double support phase was increased, but singularity during the knee stretch motion has been also avoided. As the computer simulation results, the proposed gait pattern has showed better performance compared to the common walking gait in terms of joint torque requirements and energy consumptions.

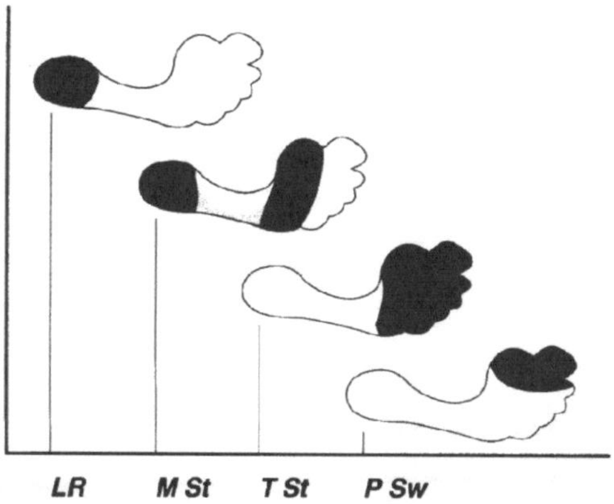

Fig. 12. Human sequence of foot support areas during stance phase (Perry, 1992).

4.1 Biped robot model and walking cycle

The suggested biped robot model has 5 degrees of freedom in each leg, with the extra 2 DOF coming from the toe and heel joints. The center of mass in each link is considered to be right in the center of each link. The biped configuration is shown in Figure 13. Biped walking is a periodic phenomenon. A complete walking cycle is composed of double-support phase and single support phase. The cycle starts from the beginning of the double support phase, where the heel strike motion occurs, and ends after the swing leg finishes its swing phase. Figure 13 also shows a complete walking cycle along with the biped configuration. The walking pattern that we discuss in this chapter covers only the motion in the sagittal plane.

Given the foot and hip trajectories, toe-heel and knee trajectories were obtained from the kinematic constraints of the mechanism.

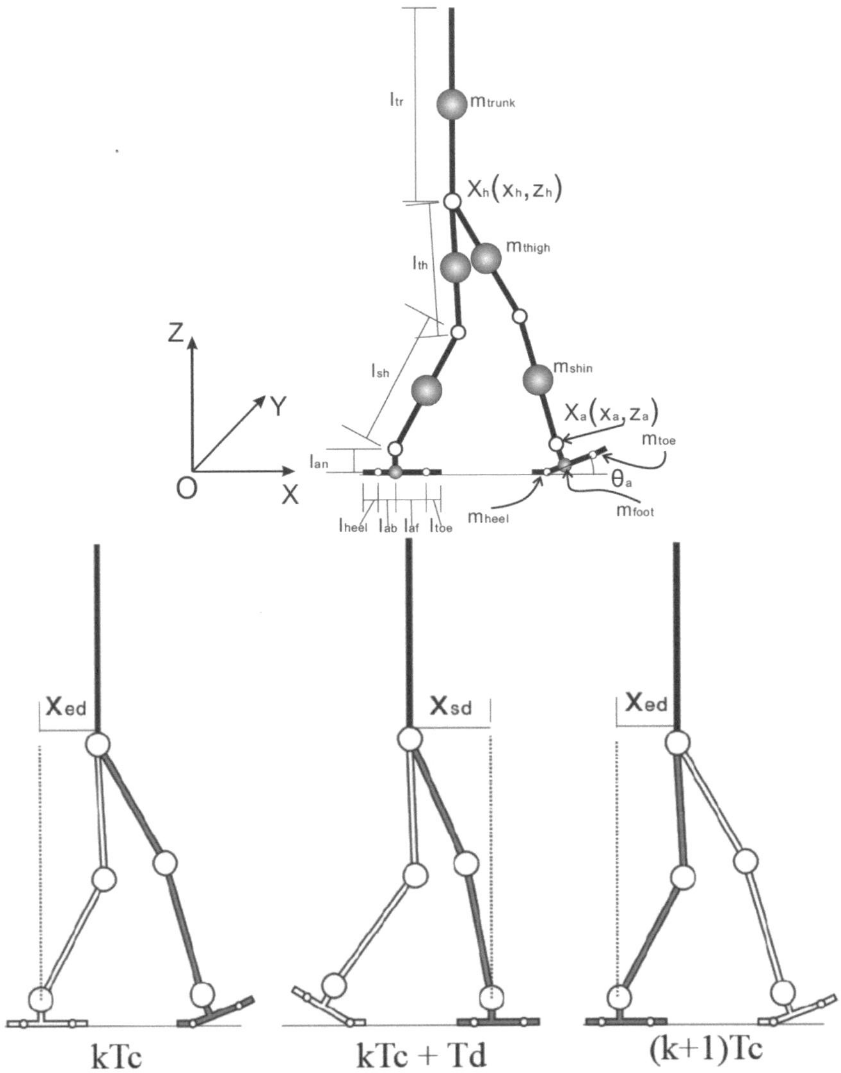

Fig. 13. Model of the biped robot and a complete walking cycle.

4.1.1 Initial foot trajectory

We planned for the initial foot trajectory as mentioned in (Huang et al., 2001). In sagittal plane each foot trajectory can be represented by vector $X_a = [x_a(t), z_a(t), \theta_a(t)]^T$, where $(x_a(t), z_a(t))$ is the coordinate of the ankle, and $\theta_a(t)$ denotes the angle between the foot and the x-axis. A similar vector can also denote the toe and heel trajectories. For toe $X_{toe} = [x_{toe}(t), z_{toe}(t), \theta_{toe}(t)]$, where $(x_{toe}(t), z_{toe}(t))$ is the coordinate of the toe, and $\theta_{toe}(t)$ denotes the angle between the toe and the foot. And, for heel $X_{heel} = [x_{heel}(t), z_{heel}(t), \theta_{heel}(t)]$, where $(x_{heel}(t), z_{heel}(t))$ is the coordinate of the heel, and

$\theta_{heel}(t)$ denotes the angle between the heel and the foot. Figure 14 shows the detailed configuration of each vector of the links in the foot.

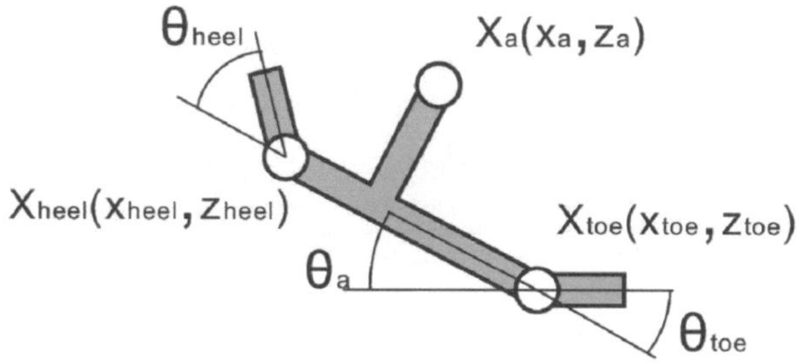

Fig. 14. Link vector configuration of the biped foot

The foot trajectory can be expressed by the function with respect to time, with a walking period of T_c , as

$$x_a(t) = \begin{cases} kD_s, & t = kT_c \\ kD_s + l_{an}\sin q_b + \cdots \\ \quad l_{af}(1 - \cos q_b), & t = kT_c + T_d \\ kD_s + L_{ao}, & t = kT_c + T_m \\ (k+2)D_s + l_{an}sinq_f - \cdots \\ \quad l_{ab}(1 - cosq_f), & t = (k+1)T_c \\ (k+2)D_s, & t = (k+1)T_c + T_d \end{cases}$$

$$z_a(t) = \begin{cases} l_{an}, & t = kT_c \\ l_{af}\sin q_b + l_{an}\cos q_b, & t = kT_c + T_d \\ H_{ao}, & t = kT_c + T_m \\ l_{ab}\sin q_f + l_{an}\cos q_f, & t = (k+1)T_c \\ l_{an}, & t = (k+1)T_c + T_d \end{cases} \tag{16}$$

$$\theta_a(t) = \begin{cases} 0, & t = kT_c \\ q_b, & t = kT_c + T_d \\ q_f, & t = (k+1)T_c \\ 0, & t = (k+1)T_c + T_d \end{cases}$$

where T_c, T_d, T_m are the period of one walking step, double support phase, and when the foot at the maximum height respectively. L_{ao} and H_{ao} are the x and z positions of the foot when it reaches the highest position , q_b and q_f are the initial toe off and heel strike angles respectively. While the vector position of the toe and heel will always be at the tip of the foot, the angle trajectories of the toe and heel are not the same as the foot. The toe and heel trajectories were designed so that it will land parallel with the ground, they can be expressed as

$$\begin{bmatrix} x_{toe}(t) \\ z_{toe}(t) \end{bmatrix} = \begin{bmatrix} x_a(t) \\ z_a(t) \end{bmatrix} + R_y(\theta_a(t)) \begin{bmatrix} l_{af} \\ -l_{an} \end{bmatrix}$$

$$\begin{bmatrix} x_{heel}(t) \\ z_{heel}(t) \end{bmatrix} = \begin{bmatrix} x_a(t) \\ z_a(t) \end{bmatrix} + R_y(\theta_a(t)) \begin{bmatrix} -l_{ab} \\ -l_{an} \end{bmatrix}$$

$$\theta_{toe}(t) = \begin{cases} 0, & t = kT_c \\ -q_b, & t = kT_c + T_d \\ 0, & t = (k+1)T_c \end{cases} \tag{17}$$

$$\theta_{heel}(t) = \begin{cases} 0, & t = kT_c \\ -q_f, & t = kT_c + T_d \\ 0, & t = (k+1)T_c \end{cases}$$

where R_y is the rotation matrix about y axis. To ensure the continuity of the trajectories, a third order cubic spline interpolation is implemented for each trajectory for generating continuous trajectory function.

After the initial foot, toe and heel trajectories are obtained, the support area of the stable margin througout the whole walking cycle can also be determined. This has been used later on as the base of the ZMP trajectory definition. Using this strategy, the increment of the support area during double support phase compared to the walking pattern without utilizing toe and heel joints can be shown. Figure 15 shows the difference between both the cases.

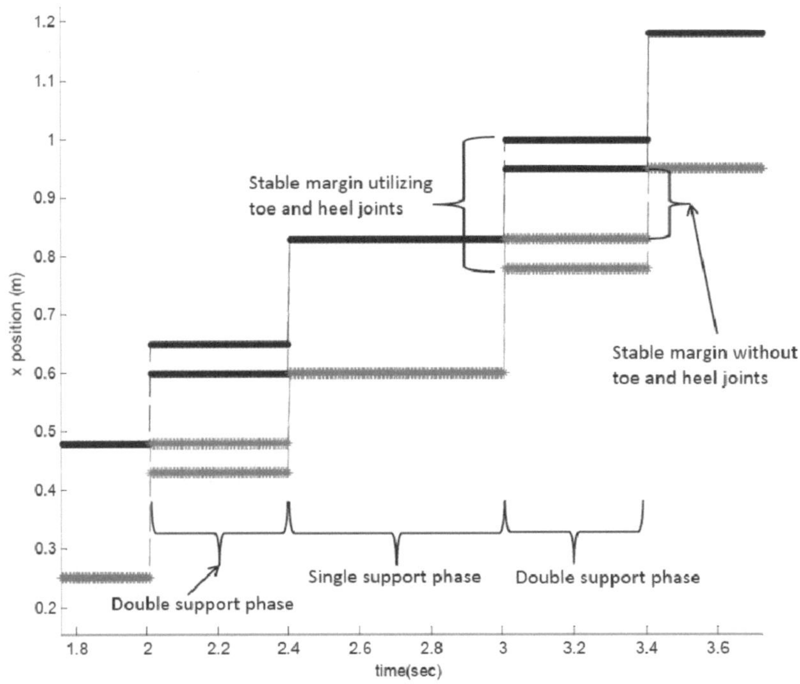

Fig. 15. Comparison of support area

4.2 Online trajectory generation

4.2.1 Desired ZMP trajectory and CoM trajectory

Zero moment point (ZMP) is the main criterion in biped robots for stable walk. We defined the ZMP trajectory after obtaining the support area. We chose moving ZMP trajectory instead of a fix one. The ZMP slides from one stable point to another stable point in each of the phases of the walking sequence. The ZMP trajectory can be expressed as

$$
\begin{aligned}
&p_x(t) = \begin{cases} m_1(t - t_1) + x_1, & kT_c < t < kT_c + T_d \\ m_2(t - t_3) + x_3, & kT_c + T_d t < t < (k+1)T_c \end{cases} \\
&m_1 = \frac{x_2 - x_1}{t_2 - t_1}, \qquad m_2 = \frac{x_4 - x_3}{t_4 - t_3}, \\
&t_1 = (k+1)T_c, \qquad t_2 = (k+1)T_c + T_d, \\
&x_1 = kD_s + l_{af} + (l_{toe}/2), \\
&x_2 = (k+1)D_s - l_{ab} - (l_{heel}/2), \\
&t_3 = (k+1)T_c + T_d, \qquad t_4 = (k+2)T_c, \\
&x_3 = (k+1)D_s - l_{ab} - (l_{heel}/2), \\
&x_4 = (k+1)D_s + l_{af} + (l_{toe}/2),
\end{aligned} \tag{18}
$$

where p_x is the ZMP position in x direction.
The center of mass (CoM) trajectory was obtained by solving a differential equation from the simplified model of the biped robot as mentioned in (Choi et al., 2007). The ZMP equation was expressed as

$$p_x = c_x - (1/\omega_n^2)\ddot{c_x} \tag{19}$$

where $\omega_n \cong \sqrt{g/c_z}$ is the natural frequency of the simplified biped walking robot system, (c_x, c_z) is the vector position of CoM and g is the value of gravity. By solving (19) with respect to (18), the x trajectory of CoM (c_x) will be obtained. In our method we keep the center of mass (CoM) trajectory at a fix height, thus the z trajectory of CoM (c_z) is fixed. The hip trajectory will be determined from the CoM trajectory as

$$
\begin{aligned}
x_h(t) &= c_x(t), \\
z_h(t) &= c_z(t) - (l_{tr}/2)
\end{aligned} \tag{20}
$$

where $(x_h(t), z_h(t))$ is the vector position of the hip.

4.2.2 Hip trajectory

The hip trajectory obtained by the above method can indeed satisfy the desired ZMP trajectory mentioned in (18) for simplified model of the biped robot. But once we apply the trajectory for the complete system of biped robot, the limited hip motion has caused the non capability of the trajectory to achieve all of the desired ZMP trajectory.
The parameterization of the hip trajectory mentioned in (Huang et al., 2001) is applied to make the adjustments. The only difference from (Huang et al., 2001) is that we utilize the parameter value x_{ed} and x_{sd} from the previous hip trajectory obtained above. In this way, no optimization scheme needs to be used. Those parameters are used to generate new hip trajectory by the same method of interpolation. And, since those parameters obtained from an already stable trajectory, we can also guarantee the stability of the new hip trajectory. Figure 16 shows the difference between adjusted CoM trajectory and the initial one.
After the hip trajectory is decided, the knee trajectory is decided by the inverse kinematics formula. Thus, all links trajectories are determined and the initial walking pattern is produced.

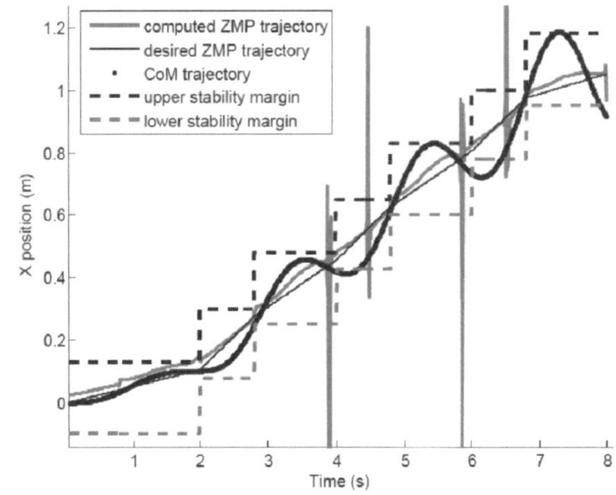

(a) Initial CoM trajectory

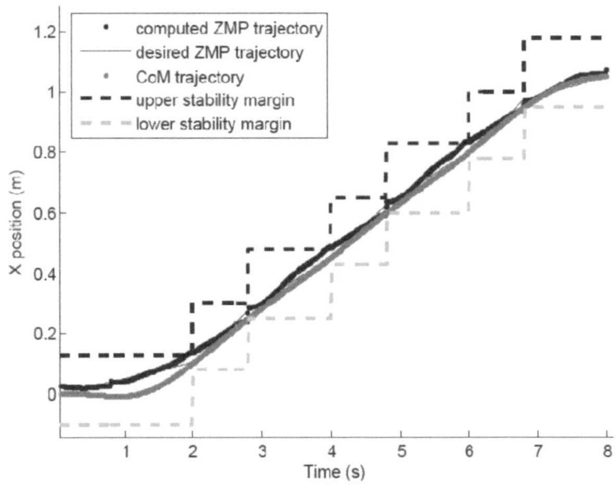

(b) Adjusted CoM trajectory

Fig. 16. Comparison between initial and adjusted CoM trajectories

4.3 Knee stretch motion

In human walking cycle, the knee angle reaches its minimum value during the heel strike motion (Perry, 1992). Here we introduce time parameters T_k, T_{ka} and T_{kb}, which mark the time when knee angle reach its minimum value, the beginning and ending time of the knee-stretch motion period. In this section we will explain how the knee-stretch motion is performed and how it can prevent the singularity.

4.3.1 Knee trajectory

After deciding the initial trajectory, the initial knee angle trajectory $(\theta_k(t))$ is modified during this knee-stretch period. Let (θ_{k1} and θ_{k2}) be the values of knee angle at $kT_k - T_{ka}$ and $kT_k - T_{kb}$, so that the new knee angle trajectory can be expressed as

$$\theta_k(t) = \begin{cases} \theta_{k1}, & t = kT_k - T_{ka} \\ 0, & t = kT_k \\ \theta_{k2}, & t = kT_k - T_{kb} \end{cases} \tag{21}$$

and the new continuous knee angle trajectory is obtained by cubic spline interpolation.

4.3.2 Inverse kinematic solution and singularity avoidance

The initial trajectory is obtained by specifying ankle trajectory and hip trajectory, and the knee trajectory is obtained by inverse kinematics solution. In sagittal plane, this scheme is exactly the same as 2 link planar mechanism, and if the knee angle is zero, the inverse kinematics solution will produce imaginary value known as singularity. In our algorithm, we will keep the hip trajectory as it is, and the ankle trajectory is modified as follows. During the time period of ($kT_k - T_{ka} < t < kT_k - T_{kb}$), instead of determining the knee position from the ankle and hip trajectories, we utilize the heel position and the modified knee angle trajectories to find the solution of knee and ankle position through the inverse kinematics problem. In this way, the system will become as three-link planar mechanism. As a result, the loss of the degree of freedom at the knee position can still be overcome by extra DOF at the heel joint. This can be explained as shown in Figure 17.

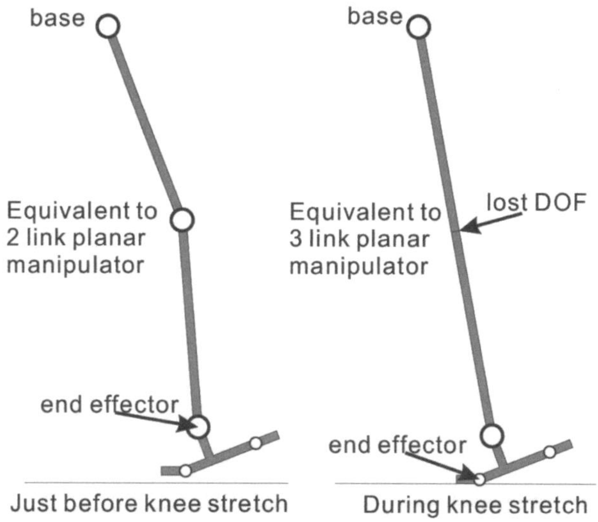

Fig. 17. Extra DOF at heel joint that avoid singularity.

After modifying the knee angle trajectory and obtaining modified parameters from previous steps, new knee and foot trajectories are obtained. The modification also changes the initial heel strike angle q_f. A complete algorithm description is shown in Figure 18.

4.4 Simulation result

The simulations have been performed in ADAMS. This simulation was intended to verify the effectiveness of the algorithm in terms of stability, torque requirements and energy consumptions. Comparisons have been done between the proposed gait pattern and a common gait pattern of the biped robot. The parameters of the biped robot (Figure 13) were set according to Table I. The walking speed was 2 s/step with the step length of 0.35 m/step.

Figure 19 is for the stick figures and knee angle trajectories obtained from the algorithm. Both results were compared with those in human which indicates a similarity compared to common gait pattern. Although they do not perfectly match, it can be observed that the knee, foot and foot angle trajectories show quite similar trends to those in human. Figure 20 shows the comparison between the knee angle of biped robot and human in the same 100% walking cycle, though not all parts of the trajectory are the same, similarity can be seen in some period of time during the knee stretch motion and our proposed pattern can also generate 0° knee angle. During around 40% of the walking cycle, which is at the toeoff motion, our proposed pattern is different from human. This was due to the fact that humans also generate an almost knee stretch motion during toe-off.

The knee stretch walking not only gives natural and humanlike walking pattern but also better performance in the knee torque requirement (Kurazume et al., 2005; Ogura et al., 2006). This gives an opportunity to employ smaller actuators for the knee. Figure 9 shows knee joint torque comparison between the common gait and the proposed gait, it is clearly shows that the proposed gait has less torque requirement. Figure 21 shows knee joint energy consumption comparison between the common gait and the proposed gait. The proposed gait shows better performance than the common gait. Table II summarize the torque requirement and energy consumption comparisons between the common gait and the proposed gait. All of the joint torque requirements of the proposed gait pattern is less than those in the common gait pattern. The addition of the heel and toe joints did not show high torque requirements, so it will make it possible for real implementation. In terms of energy consumptions, the overall result shows that the proposed gait pattern shows better performance compared to the common gait.

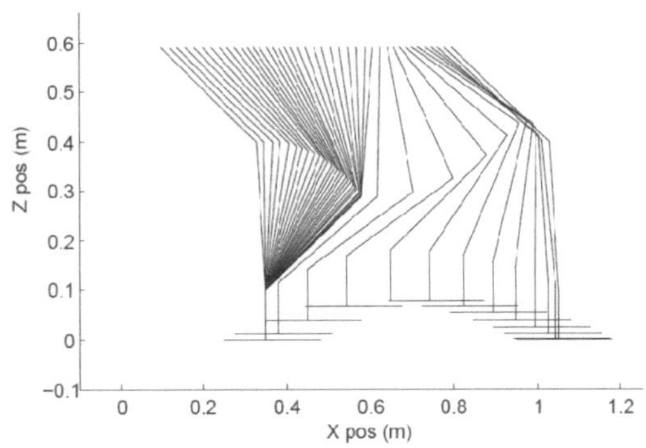

(a) Common gait pattern stick figure

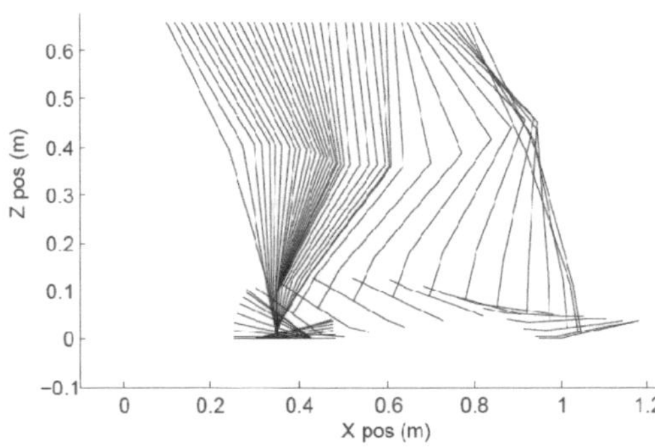

(b) Proposed gait pattern stick figure

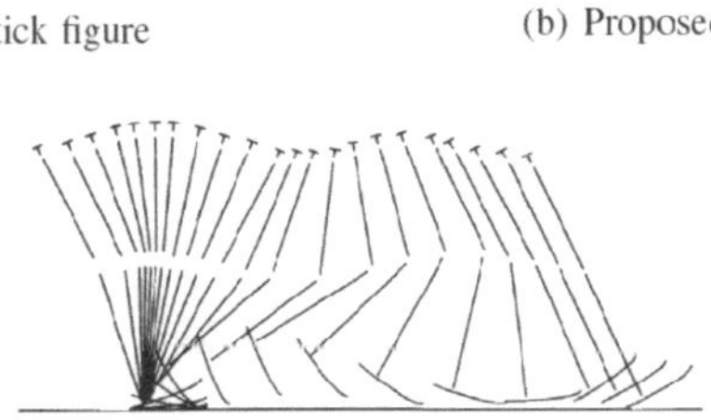

(c) Human gait pattern stick figure

Fig. 19. Stick figure comparison with human.

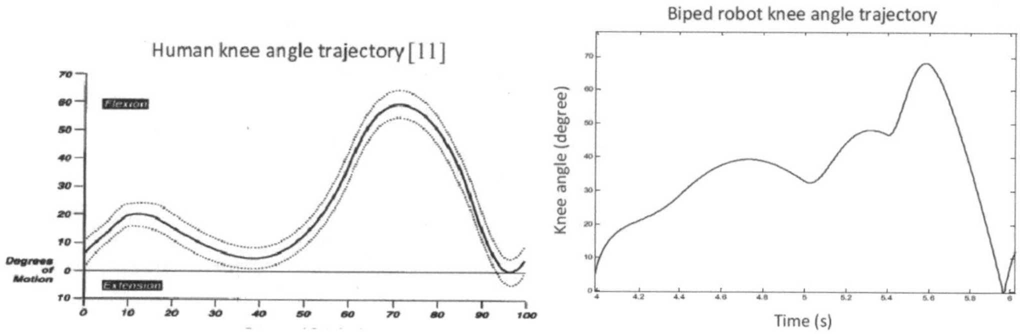

Fig. 20. Knee angle comparison with human.

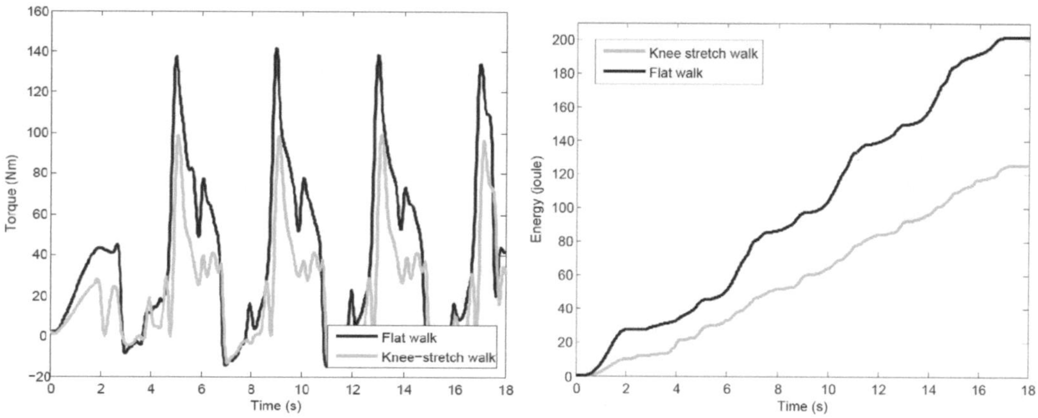

Fig. 21. Knee torque and energy consumption comparison between the common gait and the proposed gait.

Length (cm)	l_{tr}	l_{th}	l_{sh}	l_{an}	l_{ab}	l_{af}	l_{toe}	l_{heel}
	50	30	30	10	5	8	5	5

Weight(kg)	m_{tr}	m_{th}	m_{sh}	m_{foot}	m_{toe}	m_{heel}
	43	10	5.7	2.5	0.3	0.3

inertia(kgm^2)	I_{tr}	I_{th}	I_{sh}	I_{foot}	I_{toe}	I_{heel}
	1.69	0.02	0.08	0.008	0.002	0.002

Table I Parameters of The Biped Robot Model

	Joints	Common gait	Proposed gait
Torque Requirements (Nm)	hip	52.24	49.36
	knee	141.34	99.46
	ankle	75.19	69.13
	toe	-	14.42
	heel	-	19.62

	Joints	Common gait	Proposed gait
Energy Consumption (Joule)	hip	161.19	106.26
	knee	201.71	125.56
	ankle	72.39	133.24
	toe	-	21.29
	heel	-	6.19
	Total	435.29	392.54

Table II Summary of The Simulation Results

Figure 22 shows a frame by frame animation of the resulting algorithm. It can be noticed that the knee stretch occurs during the heel strike motion. The animation showed that the biped robot can walk stably without falling.

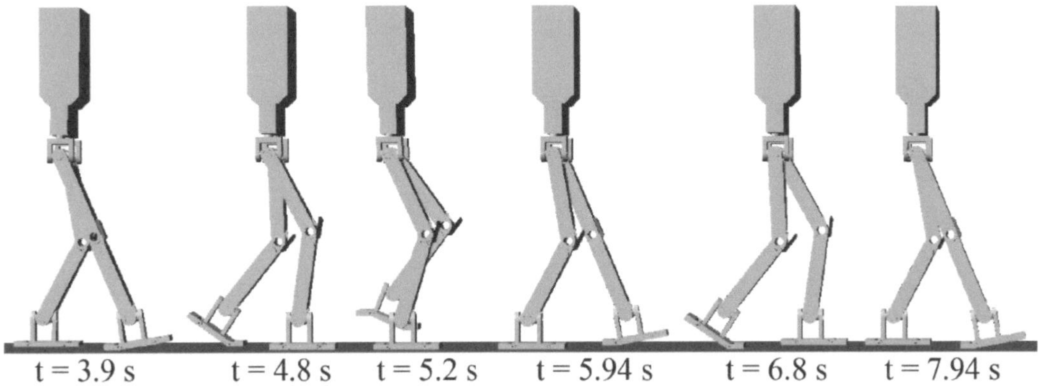

Fig. 22. Sequence of successive frame showing knee stretch walking pattern.

5. Conclusions

During walking sequence, human heels act as passive joints that create some support area, which helps the stability of human walking. This research attempts to replace human-heel like mechanism with heel joint in the biped robot foot. The existence of heel joints in the biped robot feet has two main advantages. The first one is that the support area during double support phase will be increased. Secondly, singularity during knee stretch motion can be avoided.

This chapter presents a new mechanism for toe&foot&heel motions with multi-platforms using a serial-parallel hybrid mechanism. The suggested mechanism can generate pitch and roll motions at ankle position, during toe and heel joint motions. These motions are adequate for natural foot and ankle movements of a humanoid robot. The developed foot device with the suggested mechanism will allow humanoid robots to walk more stably and in ways that are more natural. A new alternative method for generating knee stretch walking pattern for biped robot utilizing toe and heel joints has been also presented with this chapter, which has several advantages over previous algorithms. The proposed algorithm was verified using computer simulation, and better performances have been obtained as compared to common gait pattern. The improvements for the effectiveness of the proposed algorithm in terms of joint torque requirements as well as energy consumptions have been presented.

It is observed that the utilization of toe and heel joints increases the stability margin during the double support phase, thus giving more freedom in designing the walking patterns. Furthermore, the addition of heel joint has the advantage of avoiding singularity during the knee stretch motion, because the loss of degree of freedom in the knee can still be compensated by the existence of extra dof in the heel joint. While the previous researchers have also suggested the addition of waist joint to avoid singularity, compared to waist joint, two heel joints are comparably smaller and less complicated. For existing biped robots, it will be a lot easier to modify the foot than to modify the waist. As a final point, the pattern obtained from this method shows similarities with those of human walking pattern, thus can provide more natural walking for the robot.

6. Acknowledgment

This work was supported in part by the Korea Research Foundation funded by the Korean Government (Ministry of Education and Human Resources Development) under Grant KRF-2008- 331-D00027 and in part by the Priority Research Centers Program through the National Research Foundation of Korea funded by the Ministry of Education, Science and Technology under Grant 2009-0094016.

7. References

Ahn C.; Lee, M.; Go, S. (2003). Development of a biped robot with toes to improve gait pattern, *Proc. of the IEEE/ASME Int. Conf. on Advanced Intelligent Mechatronics (AIM2003)*, pp729-734.

Choi,Y; Kim, D; Oh, Y. & You, B.J. (2007). Posture/ Walking Control for Humanoid Robot Based on Kinematic Resolution of CoM Jacobian With Embedded Motion, IEEE Trans. on Robotics, vol. 23, no. 6, pp. 1285-1293.

Guihard, M. & Gorce, P. (2004). Biorobotic foot model applied to BIPMAN robot, IEEE International Conference on Systems, Man and Cybernetics, pp6491-6496.

Huang, Q.; Yokoi,K.; Kajita, S.; Kaneko, K.; Arai, H.; Koyachi, N.; Tanie, K. (2001). Planning Walking Patterns for a Biped Robot, IEEE Trans. on Robotics and Automation, vol.17, no.3, pp. 280-289.

Ki Ahn, C.; Cheol Lee, M.; Jo Go, S. (2003). Development of a Biped Robot with Toes to Improve Gait Pattern, Proceedings of the IEE/ASME International Conference on Advanced Intelligent Mechatronics.

Kurazume, R.; Tanaka, S; Yamasi, M; Hasegawa, T.; Yoneda, K. (2005) Straight Legged Walking of a Biped Robot, in Proc. of the IEEE/RSJ International Conference on Intelligent Robots and Systems, Alberta, Canada, pp. 3095-3101 .

Nishiwaki, K. & Kagami, S. (2002). Toe joint that enhances bipedal and full body motion, IEEE Int. Conf. on Rob. and Auto.(ICRA 2002), pp3105-3110.

Nishiwaki, K.; Kagami, S.; Kuniyoshi, Y.; Inaba, M.; Inoue, H. (2002). Online Generation of Humanoid Walking Motion Based on a Fast Generation Method of Motion Patterns that Follows Desired ZMP, Proc. of the IEEE/RSJ International Conference on Intelligent Robots and Systems, Lausanne, Switzerland, October 2002, pp. 2684.

Nishiwaki, K.; Kagami, S.; Kuniyoshi, Y.; Inaba, M.; Inoue, H. (2002). Toe Joints that Enhance Bipedal and Fullbody Motion of Humanoid Robots, Proc. of the IEEE International Conference on Robotics & Automation, Washington, DC.

Ogura, Y.; Lim, H & Takanishi, A. (2003) Stretch Walking Pattern Generation for a Biped Humanoid Robot, in Proc. of the IEEE/RSJ International Conference on Intelligent Robots and Systems, Las Vegas, USA, pp. 352-357.

Ogura, Y.; Shimomura, K.; Kondo, H; Morishima, A.; Okubo, T; Momoki S.; Lim, H.; Takanishi, A. (2006). Humanlike Walking with Knee Stretched, Heel-contact and

Toe-off Motion by a Humanoid Robot, in Proc. of the IEEE/RSJ International Conference on Intelligent Robots and Systems, Beijing, China, pp. 3976-3981.

Okada, K.; Ogura, T.; Haneda, A.; Kousaka, D.; Nakai, H.; Inaba, M.; Inoue, H. (2004). Integrated System Software for HRP2 Humanoid, Proceedings of the 2004 IEEE International Conference on Robotics & Automation, New Orleans, LA.

Peng Z.; Huang, Q.; Zhao, X.; Xiao, T.; Li, K. (2004). Online Trajectory Generation Based on off-line Trajectory for Biped Humanoid, Proceedings of the 2004 IEEE International Conference on Robotics and Biomimetics, Shenyang, China, August 2004, pp. 752.

Perry, J. (1992). Gait Analysis:Normal and Pathological Function , Slack Inc., Thorofare, N. J., 1992.

R. Garage, (2008) http://www.robogarage. com/english/robo/chroino.html.

Ramzi, S.; Olivier, S.; Shuuji, Kajita; Kazuhito, Y. & Abderrahmane, K. (2006). Faster and Smoother Walking of Humanoid HRP-2 with Passive Toe Joints, Proc. of the IEEE/RSJ International Conference on Intelligent Robots and Systems, Beijing, China, pp. 4909- 4914.

Sakagami, Y.; Watanabe R.; Aoyama C.; Matsunaga S.; Higaki N.; Fujimura K. (2002). The intelligent ASIMO : System Overview and Integration, Proceedings of the 2002 IEEE/RSJ Intl. Conference on Intelligent Robots and Systems EPFL, Lausanne, Switzerland.

Takahashi, T. & Kawamura, A. (2204) Posture control using foot toe and sole for biped walking robot Ken, AMC, Maribor, Slovenica, pp.437-442.

Takao, S; Ohta, H.; Yokokohji, Y.; Yoshikawa, T. (2004). Function analysis of human-like mechanical foot, using mechanically constrained shoes, IEEE/RSJ Int. Conf. on Intelligent Robots and System (IROS2004), pp3847-3852.

www.mathworks.com

Yoon, J. W.; Handharu, N.; Kim, G.S. (2007). A Bio-Robotic Toe and Foot and Heel Models of a Biped Robot for More Natural Walking, Proc. Of The IASTED International Conference on Modeling, Identification & Control, Innsbruck, Austria.

13

Optimal Gait Generation in Biped Locomotion of Humanoid Robot to Improve Walking Speed

Hanafiah Yussof[1], Mitsuhiro Yamano[2], Yasuo Nasu[2] and Masahiro Ohka[3]
[1]*Faculty of Mechanical Engineering, Universiti Teknologi MARA,*
[2]*Faculty of Engineering, Yamagata University*
[3]*Graduate School of Information Science, Nagoya University*
[1]*Malaysia*
[2,3]*Japan*

1. Introduction

Humanoid robot is a type of robot that the overall appearance is based on that of the human body. Humanoid robots include a rich diversity of projects where perception, processing and action are embodied in a recognizably anthropomorphic form in order to emulate some subset of the physical, cognitive and social dimensions of the human body and experience. The research on humanoid robots spans from stability to optimal control, gait generation, human-robot and robot-robot communication (Konno et al., 1997) (Hirai et al, 1998) (Cheng et al., 2001). In addition, humanoid robots have been also used to understand better human motion and establish working coexistence of human and humanoid robot (Althaus et al., 2004).

Humanoid robot with two legs usually have problem to stabilize its biped walk motions. In fact, one of the most sophisticated forms of legged motion is that of biped gait locomotion. Human locomotion stands out among other forms of biped locomotion chiefly in terms of the dynamic systems point of view. This is due to the fact that during a significant part of the human walking motion, the moving body is not in static equilibrium.

Biped walking robot can be classified by its gait. There are two major research areas in biped walking robot: the static gait and dynamic gait. For a biped robot, two different situations arise in sequence during the walking motion: the statically stable double-support phase in which the whole structure of the robot is supported on both feet simultaneously, and the statically unstable single-support phase when only one foot is in contact with the ground, while the other foot is being transferred from back to front. Eventually, this type of walking pattern delays the walking speed. Moreover, joint structure design in robots does not permit flexible movement like that of human being. Indeed, one motor only can rotate in one direction. Even by reducing reduction-ratio can increase the motor rotation, it will eventually reduce the torque output which is not desirable for real-time operation. Therefore, a method to control sufficient walking speed in conjunction with the biped gait trajectory is inevitably important. This is because in real-time application, the robots are likely to be required to walk faster or slower according to situation that occurred during the operation.

This chapter presents analysis results of optimal gait trajectory generation in a biped humanoid robot. The work presented in this chapter is focusing on analysis to improve biped walk quality and speed by considering reduction-ratio at joint-motor system with other physical parameters in humanoid robot's body. The analysis utilized a 21-dofs humanoid robot *Bonten-Maru II* as experimental platform.

The early sections of this chapter presents the background and motivation of current research, followed by hardware structure and design characteristics of humanoid robot *Bonten-Maru II*. The next section presents the optimization of trajectory generation using inverse kinematics for 6-dofs humanoid robot legs. A simplified approach was implemented to solving inverse kinematics problems by classifying the robot leg's joints into several groups of joint coordinate frames. To describe translation and rotational relationship between adjacent joint links, a matrix method proposed by Denavit-Hartenberg (Denavit and Hartenberg, 1955) was employed, which systematically establishes a coordinate system for each link of an articulated chain. In addition, to perform a smooth and reliable gait, it is necessary to define step-length and foot-height during transferring one leg in one step walk. The step-length is a parameter value that can be adjusted and fixed in the control system. On the other hand, the foot-height is defined by applying ellipse formulation. Interpolation by time function of the leg's start and end points using ellipse formulation provide smooth trajectory pattern at each gait.

The final section presents analysis of biped walking speed by maintaining reduction-ratio value but consider step length, hip-joint height from ground and duty-ratio as experimental parameters. Eventually, it is easy to control the walking speed by reducing or increasing the reduction-ratio at the robot joint-motor system. However, in real-time operation it is desirable to have a stable and high reduction-ratio value in order to provide high torque output to the robot's manipulator during performing tasks, such as during object manipulation, avoiding obstacle, etc. Therefore the reduction-ratio is required to remain always at fixed and high value.

2. Background and motivation

To realize human-like walking robots, many researches about the biped locomotion robot have been archived especially in prototyping biped locomotion, biped legged control and optimal gait locomotion. In these researches, dynamic and stable walking can be realized. Vukobratovic (Vukobratovic et al. 1990) have investigated the walking dynamics and proposed Zero Moment Point (ZMP) as an index of walking stability. Meanwhile, Takanishi and Hirai (Takanishi et al., 1985, Hirai et al., 1998) have proposed methods of walking pattern synthesis based on the ZMP, and demonstrate walking motion and pattern synthesis with real humanoid robots. Meanwhile, the methods to realizing dynamic walking were presented in (Hasegawa et al., 2000, Lim et al., 2000). Other achievements are in prototyping biped locomotion, biped legged control and optimal gait locomotion (Goswani et al., 1997, Capi et al., 2003).

Input energy is another important index for natural walking motion. Eventually, the energy-optimal trajectory for highly non-linear equations of a complex robot is hard to find numerically. Recently, an evolutionary optimization method, such as evolutionary programming (EP), genetic algorithm (GA) and so on have been identified to find the optimal solutions in a non-linear system. In the research with *Bonten-Maru II*, Capi (Capi et al., 2003, Nasu et al., 2007) has proposed a real time generation of humanoid robot optimal

gait by using soft computing techniques. GA was employed to minimize the energy for humanoid robot gait. The Radial Basis Function Neural Networks (RBFNN) is used for real time gait generations, which are trained based on GA data.
In order to realize optimal gait generation, several studies have been reported related with walking speed of biped robot. For example Chevallereau & Aoustin (Chevallereau & Aoustin, 2001) have studied optimal reference trajectory for walking and running of a biped robot. Furthermore, Yamaguchi (Yamaguchi et al., 1993) have been using the ZMP as a criterion to distinguish the stability of walking for a biped walking robot which has a trunk. The authors introduce a control method of dynamic biped walking for a biped walking robot to compensate for the three-axis (pitch, roll and yaw-axis) moment on an arbitrary planned ZMP by trunk motion. The authors developed a biped walking robot and performed a walking experiment with the robot using the control method. The result was a fast dynamic biped walking at the walking speed of 0.54 s/step with a 0.3 m step on a flat floor. This walking speed is about 50% faster than that with the robot which compensates for only the two-axis (pitch and roll-axis) moment by trunk motion. Meanwhile, control system that stabilizes running biped robot HRP-2LR has been proposed by Kajita (Kajita et al., 2005). The robot uses prescribed running pattern calculated by resolved momentum control, and a running controller stabilizes the system against disturbance.
Eventually, it is easy to control the walking speed by reducing or increasing the reduction-ratio at the robot joint-motor system. However, in real-time operation it is desirable to have a stable and high reduction-ratio value in order to provide high torque output to the robot's manipulator during performing tasks, such as during object manipulation, avoiding obstacle, etc.

3. Humanoid robot "*Bonten-Maru II*"

Motivated by the current state-of-the-art in humanoids research, we have previously developed a research prototype biped humanoid robot called *Bonten-Maru II*. The *Bonten-Maru II* appearance diagram and outer dimension are shown in Fig. 1. It is 1.25 [m] tall and weight 31.5 [kg], which similar to an eight or nine year old child. The *Bonten-Maru II* is a research prototype humanoid robot, and such has undergone some refinement as different research direction is considered. During the design process, some predefined degree of stiffness, accuracy, repeatability, mobility and other design factor have been taken into consideration.
The *Bonten-Maru II* was designed to mimic as much as human characteristic, especially for contribution of its joints. Figure 2 shows configuration of dofs in the robot body. The robot has total of 21 dof: 6 dof for each leg, 3 dof for each arm, 1 dof for waist and 2 dof for head. The high number of dof gives the *Bonten-Maru II* possibility to realize complex motions. Moreover, the distribution of dof which is very similar with humans gives advantages for the humanoid robot to attain human-like motion. Every joint is driven by DC servomotor with a rotary encoder and harmonic drive reduction system, and PC with Linux is utilized for control. Rotation angles of joints were recorded by the rotary encoder that installed at rear side of DC servomotor. The sampling frequency is 200 Hz. The power is supplied to each joint by timing belt and harmonic drive reduction system. Gear number at the DC servomotor side is 60; while at the harmonic drive side is 16. Therefore, it makes reduction ratio at the harmonic side to be 1:100, while overall reduction ratio is 1:333.

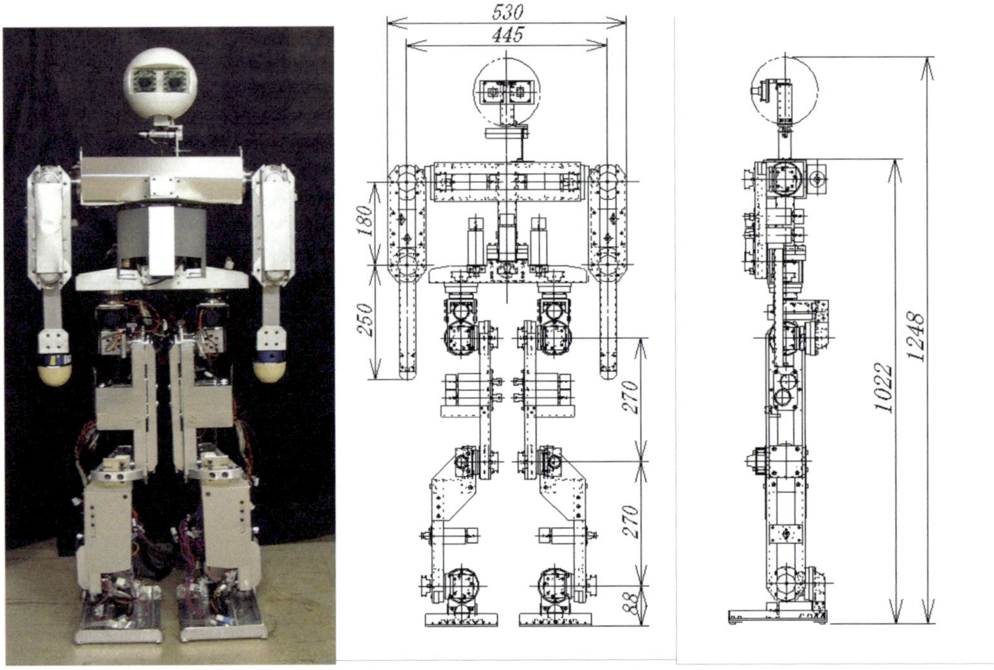

Fig. 1. *Bonten-Maru II* appearance diagram and outer dimension.

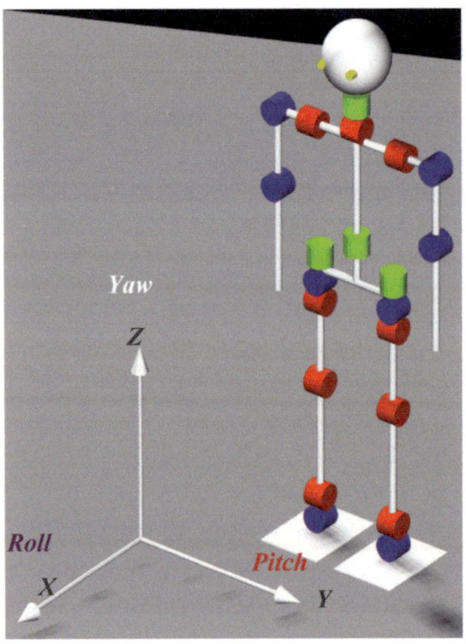

Fig. 2. Configuration of dofs in *Bonten-Maru II* humanoid robot body.

Table 1 shows range of joints rotation angle. Each joint has relatively wide range of rotation angle, especially for both leg's hip yaw which permit both legs to rotate in wide range of angle during correction of orientation and obstacle avoidance. Construction of the robot's links was also designed to mimic human's structure. The motor driver, the PC and the power supply are placed outside of the robot. At the legs side, under each foot are four pressure sensor, two at the toe and two across the heel. These provide a good indication of both contact with the ground, and the Zero Moment Point (ZMP) position. At the head part is equipped with two monochrome CCD cameras (542x492 pixels) and connected to PC by video capture board.

Axis	Range of rotation angle (deg.)
Waist (yaw)	-45 ~ 45
Hip (yaw)	-90 ~ 90
Right hip (roll)	-90 ~ 25
Left hip (roll)	-25 ~ 90
Hip (pitch)	-130 ~ 45
Knee (pitch)	-20 ~150
Ankle (pitch)	-90 ~ 60
Right ankle (roll)	-90 ~ 20
Left ankle (roll)	-20 ~ 90

Table 1. Joint rotation range at leg system in *Bonten-Maru II.*

4. Optimization of trajectory generation in humanoid's legs

Optimization of trajectory generation is nacessary to ganerate optimal biped trajectory of humanoid robot legs. It is commonly known that trajectory of robot manipulator is obtain by solving kinematic relationship between adjacent links. Robot kinematics deals with the analytical study of the geometry of a robot's motion with respect to a fixed reference coordinate system as a function of time without regarding the force/moments that cause the motion. Commonly, trajectory generation for biped locomotion robots is defined by solving forward and inverse kinematics problems (Kajita et al, 2005). In a forward kinematics problem, where the joint variable is given, it is easy to determine the end-effector's position and orientation. An inverse kinematics problem, however, in which each joint variable is determined by using end-effector position and orientation data, does not guarantee a closed-form solution.

Traditionally three methods are used to solve an inverse kinematics problem: geometric, iterative, and algebraic (Koker, 2005). However, the more complex the manipulator's joint structure, the more complicated and time-consuming these methods become. In order to optimize trajectory generation of biped walking robot, a simplified approach was proposed and implemented in 6-DOFs legs to solving inverse kinematics problems by classifying the robot's joints into several groups of joint coordinate frames at the robot's manipulator.To describe translation and rotational relationship between adjacent joint links, we employ a matrix method proposed by Denavit-Hartenberg (Denavit and Hartenberg, 1955), which systematically establishes a coordinate system for each link of an articulated chain. Optimization of kinematic solutions helps to increase calculation time and reduce memory usage of the robot control system. It also improve the stability of biped trajectory that can be utilize to ganerate and control the speed of biped walking.

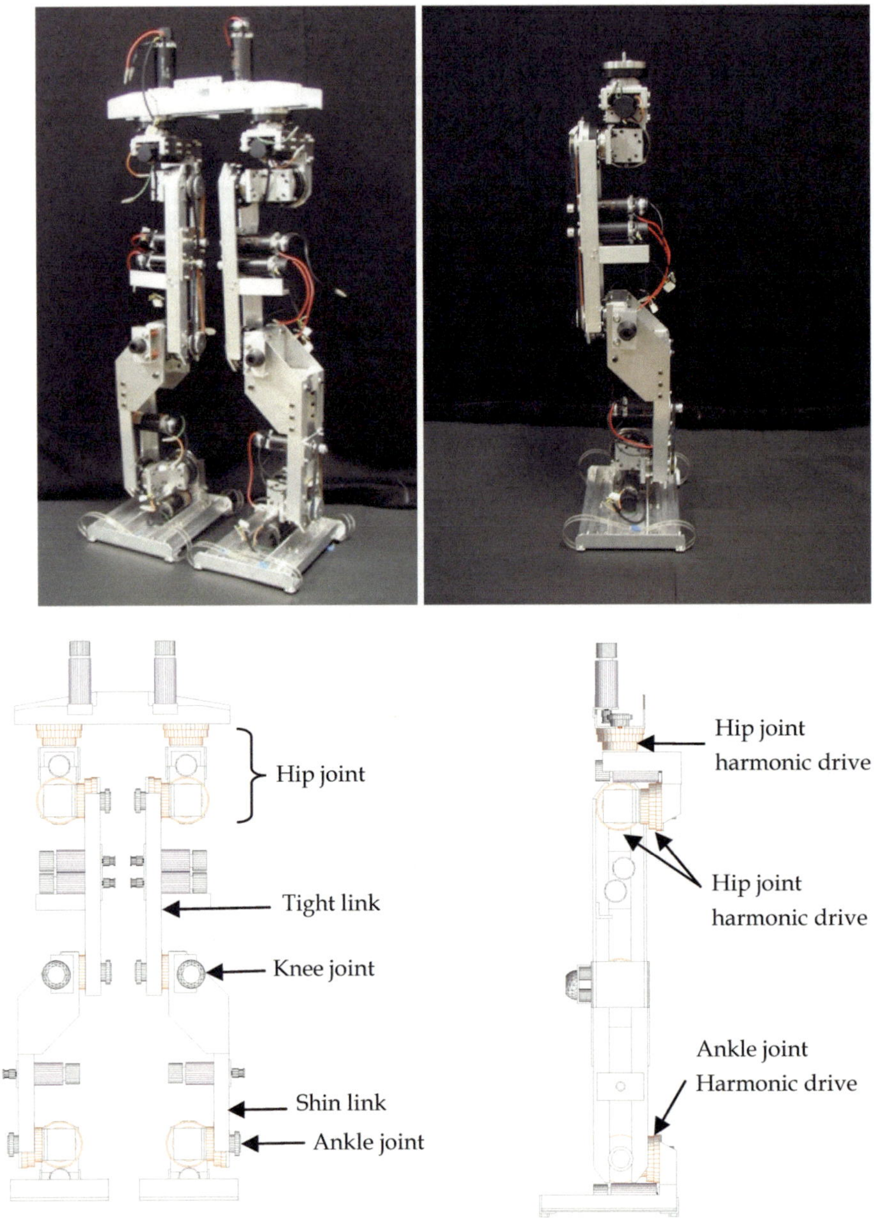

Fig. 3. Photograph of *Bonten-Maru II* lower side body and the configuration of links and joints.

4.1 Legs structure

Basic idea of legged robot is ability to perform wide and variety range of human-like gait motions. The *Bonten-Maru II* humanoid robot is designed to mimic as much as human structure, especially for its joints and links configuration to have wide range of rotation angle. Figure 3 shows photograph and diagram of the *Bonten-Maru II* lower side body.

Configuration of links, joints and harmonic drive at the *Bonten-Maru II* lower side body were shown in this figure. Each leg is consists of 6-dofs: 3 dof for hip, 1 dof for knee, and 2 dof for ankle. Hip-joint yaw is connecting each leg with the waist part.
The design of link position and joint-motor structure greatly influence the joint rotation range (Bischoff & Graefe, 2005). The link positions were configured with thigh link positioned at inner side of leg, while shin link positioned at outer side of the leg. It gives the hip joints wide rotation range to outside direction and the ankle joints also possible to rotate wider to inner side, at the same time gives better stability. Both of these links were connected with knee joint and were given specific space so that knee joint can rotate as far as 160 degree to back direction.
Configuration of the harmonic drive position at hip joints and ankle joints were installed at the rear side of roll direction so that the leg's link can swing to front direction in wide rotation angle. Moreover, both thigh links were given specific space so that when hip joint rotates to yawing direction, both links do not collide to each other. Consequently, rotation of hip joint at yaw direction can reach until 90 degree. In this research, wide rotation angle of yaw direction is required so that the robot can easily change its direction in wider angle, particularly during avoiding obstacles and operation in confined spaces.

4.2 Kinematics solutions of 6-DOFs Legs

Each of the legs in humanoid robot Bonten-Maru II has six DOFs: three DOFs (yaw, roll and pitch) at the hip joint, one DOF (pitch) at the knee joint and two DOFs (pitch and roll) at the ankle joint. In this research, only inverse kinematics calculations for the robot leg we solved. A reference coordinate is taken at the intersection point of the 3-DOF hip joint. In solving calculations of inverse kinematics for the leg, just as for arm, the joint coordinates are divided into eight separate coordinate frames as listed bellow.

Σ_0 : Reference coordinate.
Σ_1 : Hip yaw coordinate.
Σ_2 : Hip roll coordinate.
Σ_3 : Hip pitch coordinate.
Σ_4 : Knee pitch coordinate.
Σ_5 : Ankle pitch coordinate.
Σ_6 : Ankle roll coordinate.
Σ_h : Foot bottom-center coordinate.

Figure 4 shows a model of the robot leg that indicates the configurations and orientation of each set of joint coordinates. Here, link length for the thigh is l_1, while for the shin it is l_2. The Link parameters for the leg are defined in Table 2. Referring to Fig. 4, the transformation matrix at the bottom of the foot (${}^{6}_{h}\mathbf{T}$) is an independent link parameter because the coordinate direction is changeable. Here, to simplify the calculations, the ankle joint is positioned so that the bottom of the foot settles on the floor surface. The leg's orientation is fixed from the reference coordinate so that the third row of the rotation matrix at the leg's end becomes like Eq. (1).

$$P_z\text{leg} = \begin{bmatrix} 0 & 0 & 1 \end{bmatrix}^T \tag{1}$$

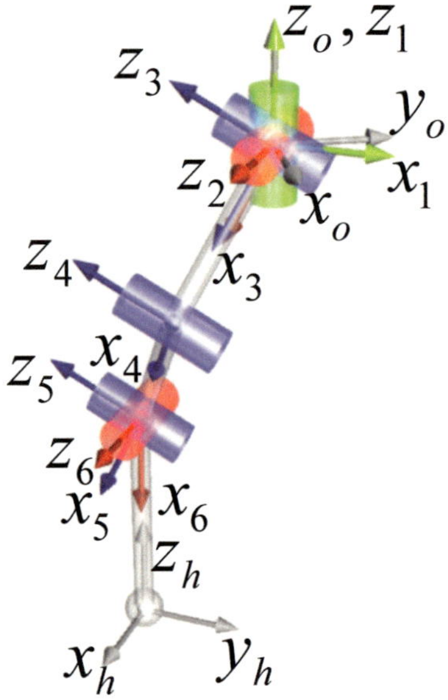

Fig. 4. Configurations of joint coordinates at the robot leg.

Link	θ_{ileg}	d	α	l
0	$\theta_{1leg}+90^o$	0	0	0
1	$\theta_{2leg}-90^o$	0	90°	0
2	θ_{3leg}	0	90°	0
3	θ_{4leg}	0	0	l_1
4	θ_{5leg}	0	0	l_2
5	θ_{6leg}	0	-90°	0
6	0	0	0	l_3

Table 2. Link parameters of the leg

Furthermore, the leg's links are classified into three groups to short-cut the calculations, where each group of links is calculated separately as follows.

i. From link 0 to link 1 (Reference coordinate to coordinate joint number 1).
ii. From link 1 to link 4 (Coordinate joint number 2 to coordinate joint number 4).
iii. From link 4 to link 6 (Coordinate joint number 5 to coordinate at the bottom of the foot).

Basically, i) is to control leg rotation at the Z-axis, ii) is to define the leg position, while iii) is to decide the leg's end-point orientation. A coordinate transformation matrix can be arranged as below.

$$ {}^{0}_{h}\mathrm{T} = {}^{0}_{1}\mathrm{T}\,{}^{1}_{4}\mathrm{T}\,{}^{4}_{h}\mathrm{T} = ({}^{0}_{h}\mathrm{T})({}^{1}_{2}\mathrm{T}\,{}^{2}_{3}\mathrm{T}\,{}^{3}_{4}\mathrm{T})({}^{4}_{5}\mathrm{T}\,{}^{5}_{6}\mathrm{T}\,{}^{6}_{h}\mathrm{T}) \tag{2} $$

Here, the coordinate transformation matrices for ${}^{1}_{4}\mathbf{T}$ and ${}^{4}_{h}\mathbf{T}$ can be defined as Eqs. (3) and (4), respectively.

$$ {}^{1}_{4}\mathrm{T} = {}^{1}_{2}\mathrm{T}\,{}^{2}_{3}\mathrm{T}\,{}^{3}_{4}\mathrm{T} = \begin{bmatrix} s_2c_{34} & -s_2s_{34} & -c_2 & l_1s_2c_3 \\ -s_{34} & -c_{34} & 0 & -l_1s_3 \\ -c_2c_{34} & c_2s_{34} & -s_2 & -l_1c_2c_3 \\ 0 & 0 & 0 & 1 \end{bmatrix} \tag{3} $$

$$ {}^{4}_{h}\mathrm{T} = {}^{4}_{5}\mathrm{T}\,{}^{5}_{6}\mathrm{T}\,{}^{6}_{h}\mathrm{T} = \begin{bmatrix} c_5c_6 & -c_5s_6 & -s_5 & l_2 + l_3c_5c_6 \\ s_5c_6 & -s_5s_6 & c_5 & l_3s_5c_6 \\ -s_6 & -c_6 & 0 & -l_3s_6 \\ 0 & 0 & 0 & 1 \end{bmatrix} \tag{4} $$

The coordinate transformation matrix for ${}^{0}_{h}\mathbf{T}$, which describes the leg's end-point position and orientation, can be shown with the following equation.

$$ {}^{0}_{h}\mathrm{T} = \begin{bmatrix} r_{11} & r_{12} & r_{13} & p_x \\ r_{21} & r_{22} & r_{23} & p_y \\ r_{31} & r_{32} & r_{33} & p_z \\ 0 & 0 & 0 & 1 \end{bmatrix} \tag{5} $$

From Eq. (1), the following conditions were satisfied.

$$ r_{13} = r_{23} = r_{31} = r_{32} = 0 \ , \ r_{33} = 1 \tag{6} $$

Hence, joint rotation angles θ_{1leg}~θ_{6leg} can be defined by applying the above conditions. First, considering i), in order to provide rotation at the Z-axis, only the hip joint needs to rotate in the yaw direction, specifically by defining θ_{1leg}. As mentioned earlier, the bottom of the foot settles on the floor surface; therefore, the rotation matrix for the leg's end-point measured from the reference coordinate can be defined by the following equation.

$$ {}^{0}_{h}\mathrm{R} = \mathrm{Rot}(z, \theta_{1\mathrm{leg}}) = \begin{bmatrix} c\theta_{1\mathrm{leg}} & -s\theta_{1\mathrm{leg}} & 0 \\ s\theta_{1\mathrm{leg}} & c\theta_{1\mathrm{leg}} & 0 \\ 0 & 0 & 1 \end{bmatrix} = \begin{bmatrix} r_{11} & r_{12} & 0 \\ r_{21} & r_{22} & 0 \\ 0 & 0 & 1 \end{bmatrix} \tag{7} $$

Here, θ_{1leg} can be defined as below.

$$ \theta_{1\mathrm{leg}} = \mathrm{atan2}\left(r_{21}, r_{22}\right) \tag{8} $$

Next, considering ii), from the obtained result of θ_{1leg}, ${}^{0}_{h}\mathbf{T}$ is defined in Eq. (9).

$$ {}^{0}_{h}\mathrm{T} = \begin{bmatrix} -s_1 & -c_1 & 0 & P_{x\mathrm{leg}} \\ c_1 & -s_1 & 0 & P_{y\mathrm{leg}} \\ 0 & 0 & 1 & P_{z\mathrm{leg}} \\ 0 & 0 & 0 & 1 \end{bmatrix} \tag{9} $$

Here, from constrain orientation of the leg's end point, the position vector of joint 5 is defined as follows in Eq. (10), and its relative connection with the matrix is defined in Eq. (11). Next, equation (12) is defined relatively.

$$ {}^{0}\mathbf{P}_5 = {}^{0}_{4}\mathbf{T}^{4}\mathbf{P}_5 = \begin{bmatrix} P_{x\text{leg}} & P_{y\text{leg}} & P_{z\text{leg}} - l_3 \end{bmatrix}^T , \tag{10} $$

$$ {}^{1}_{4}T^{4}\hat{P}_5 = {}^{0}_{1}T^{-1}{}^{0}\hat{P}_5 \tag{11} $$

$$ \begin{bmatrix} s_2c_{34} & -s_2s_{34} & -c_2 & l_1s_2c_3 \\ -s_{34} & -c_{34} & 0 & -l_1s_3 \\ -c_2c_{34} & c_2s_{34} & -s_2 & -l_1c_2c_3 \\ 0 & 0 & 0 & 1 \end{bmatrix} \begin{bmatrix} l_2 \\ 0 \\ 0 \\ 1 \end{bmatrix} = \begin{bmatrix} -s_1 & c_1 & 0 & 0 \\ -c_1 & -s_1 & 0 & 0 \\ 0 & 0 & 1 & 0 \\ 0 & 0 & 0 & 1 \end{bmatrix} \begin{bmatrix} p_x \\ p_y \\ p_z - l_3 \\ 1 \end{bmatrix} \tag{12} $$

$$ \text{Therefore,} \quad \begin{bmatrix} \hat{P}_{x_{\text{leg}}} \\ \hat{P}_{y_{\text{leg}}} \\ \hat{P}_{z_{\text{leg}}} \end{bmatrix} = \begin{bmatrix} s_2\left(l_1c_3 + l_2c_{34}\right) \\ -\left(l_1c_3 + l_2s_{34}\right) \\ -c_2\left(l_1c_3 + l_2c_{34}\right) \end{bmatrix} . \tag{13} $$

To define joint angles $\theta_{2\text{leg}}$, $\theta_{3\text{leg}}$, $\theta_{4\text{leg}}$, Eq. (13) is used . The rotation angles are defined as the following equations.

$$ \theta_{4_{\text{leg}}} = \operatorname{atan2}\left(\pm\sqrt{1-C^2}, C \right) \tag{14} $$

$$ \theta_{3_{\text{leg}}} = \operatorname{atan2}\left(\hat{p}_{xz_{\text{leg}}}, \hat{p}_{y_{\text{leg}}} \right) + \operatorname{atan2}\left(k_1, k_2\right) \tag{15} $$

$$ \theta_{2_{\text{leg}}} = \operatorname{atan2}\left(\hat{p}_{x_{\text{leg}}}, \hat{p}_{z_{\text{leg}}} \right) \tag{16} $$

Eventually, $C, \hat{p}_{xz_{\text{leg}}}, k_1, k_2$ are defined as follows.

$$ C = \frac{\hat{p}_{x_{\text{leg}}}^{\;2} + \hat{p}_{y_{\text{leg}}}^{\;2} + \hat{p}_{z_{\text{leg}}}^{\;2} - (l_1^2 + l_2^2)}{2l_1l_2} \tag{17} $$

$$ \hat{p}_{xz_{\text{leg}}} = \sqrt{\hat{p}_{x_{\text{leg}}}^{\;2} + \hat{p}_{z_{\text{leg}}}^{\;2}} \tag{18} $$

$$ k_1 = l_1 + l_2c_4, \; k_2 = -l_2s_4 \tag{19} $$

Finally, considering iii), joint angles $\theta_{5\text{leg}}$ and $\theta_{6\text{ leg}}$ are defined geometrically by the following equations.

$$ \theta_{5\text{leg}} = -\theta_{3\text{leg}} - \theta_{4\text{leg}} \tag{20} $$

$$\theta_{6leg} = -\theta_{2leg} \tag{21}$$

4.3 Interpolation and gait trajectory pattern

Application of high degree of polynomial equations will help the manipulators to perform smooth trajectory. This method called interpolation. Interpolation refers to a time history of position, velocity and acceleration for each robotic joint. A common way of making a robot's manipulator to move from start point P_0 to end point P_f in a smooth, controlled fashion is to have each joint to move as specified by a smooth function of time t. Each joint starts and ends its motion at the same time, thus the robot's motion appears to be coordinated.

In order to compute these motions, in the case of position, velocity and acceleration at start point P_0 and end point P_f are given, interpolation of end-effector's position described in time t variable function was performed using polynomial equation to generate trajectory. In this research, we employ degree-5 polynomial equation as shown in Eq. (22) to solve interpolation from start point P_0 to end point P_f. Velocity and acceleration at P_0 and P_f are defined as zero; only the position factor is considered as a coefficient for performing interpolation.

$$P(t) = a_0 + a_1 t + a_2 t^2 + a_3 t^3 + a_4 t^4 + a_5 t^5 \tag{22}$$

Time factor at P_0 and P_f are describe as $t_0 = 0$ and t_f, respectively. Here, boundary condition for each position, velocity and acceleration at P_0 and P_f are shown at following equations.

$$\left.\begin{aligned}
P(0) &= a_0 = P_o \\
\dot{P}(0) &= a_1 = \dot{P}_o \\
\dot{P}(0) &= 2a_2 = \ddot{P}_o \\
P(t_f) &= a_0 + a_1 t_f + a_2 t_f^{\,2} + a_3 t_f^{\,3} + a_4 t_f^{\,4} + a_5 t_f^{\,5} = P_f \\
\dot{P}(t_f) &= a_1 + 2a_2 t_f + 3a_3 t_f^{\,2} + 4a_4 t_f^{\,3} + 5a_5 t_f^{\,4} = \dot{P}_f \\
\ddot{P}(t_f) &= 2a_2 + 6a_3 t_f + 12a_4 t_f^{\,2} + 20a_5 t_f^{\,3} = \ddot{P}_f
\end{aligned}\right\} \tag{23}$$

Here, coefficient a_i (i = 0,1,2,3,4,5) are defined by solving deviations of above equations. Results of the deviations are shown at below Eq. (24).

$$\left.\begin{aligned}
a_0 &= y_o \\
a_1 &= \dot{y}_o \\
a_2 &= \frac{1}{2}\ddot{y}_o \\
a_3 &= \frac{1}{2t_f^{\,3}}\{20(y_f - y_0) - (8\dot{y}_f + 12\dot{y}_o)t_f + (\ddot{y}_f - 3\ddot{y}_o)t_f^{\,2}\} \\
a_4 &= \frac{1}{2t_f^{\,4}}\{-30(y_f - y_o) + (14\dot{y}_f + 16\dot{y}_o)t_f - (2\ddot{y}_f - 3\ddot{y}_o)t_f^{\,2}\} \\
a_5 &= \frac{1}{2t_f^{\,5}}\{12(y_f - y_o) - 6(\dot{y}_f + \dot{y}_o)t_f + (\ddot{y}_f - \ddot{y}_o)t_f^{\,2}\}
\end{aligned}\right\} \tag{24}$$

As mentioned before, velocity and acceleration at P_0 and P_f were considered as zero, as shown in Eq. (25).

$$\dot{P}(0) = \ddot{P}(0) = \dot{P}(t_f) = \ddot{P}(t_f) = 0\,. \tag{25}$$

Generation of motion trajectories from points P_0 to P_f only considered the position factor. Therefore, by given only positions data at P_0 and P_f, respectively described as y_0 and y_f, coefficients a_i (i = 0,1,2,3,4,5) were solved as below Eq. (26).

$$\left.\begin{aligned} a_0 &= y_o \\ a_1 &= 0 \\ a_2 &= 0 \\ a_3 &= \frac{10}{t_f^{\,3}}(y_f - y_o) \\ a_4 &= -\frac{15}{t_f^{\,4}}(y_f - y_o) \\ a_5 &= \frac{6}{t_f^{\,5}}(y_f - y_o) \end{aligned}\right\} \tag{26}$$

Finally, degree-5 polynomial function is defined as following equation.

$$y(t) = y_o + 10(y_f - y_o)u^3 - 15(y_f - y_o)u^4 + 6(y_f - y_o)u^5 \tag{27}$$

Where,

$$u = \frac{t}{t_f} = \frac{current\ time}{motion\ time}\,. \tag{28}$$

Meanwhile, to perform smooth and reliable gait trajectory in biped walk, it is necessary to define foot-height during transferring one leg in one step walk based on the acquired step-length. The foot-height is defined by applying ellipse formulation, like shown in gait trajectory pattern at Fig. 5. In walking forward and backward, the foot height at z-axis is defined in Eq. (29). Meanwhile during side-step walk, the foot height is defined in (30). Here, h is hip-joint height from the ground.

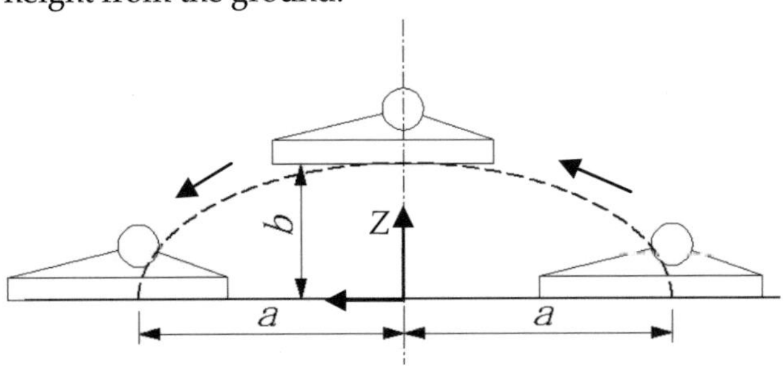

Fig. 5. Gait trajectory pattern of humanoid robot leg

$$z = b\left(1 - \frac{x^2}{S}\right)^{\frac{1}{2}} - h \tag{29}$$

$$z = b\left(1 - \frac{y^2}{S}\right)^{\frac{1}{2}} - h \tag{30}$$

In real-time operation, biped locomotion is performed by given the leg's end point position to the robot control system so that joint angle at each joint can be calculated by inverse kinematics formulations. Consequently the joint rotation speed and pattern is controlled by the above formulations of interpolation using degree-5 polynomial equations. By applying these formulations, each gait motion is performed in smooth and controlled trajectory.

4.4 Verification of biped trajectory generation by simulation

A simulation using animation that applies GnuPlot in humanoid robot *Bonten-Maru II* control system was performed to analyze and confirm of the robot joint's trajectory generation. Figure 6 presents the simulation interface of the robot's trajectory, which features a robot control process, a motion instructor process, and robot animation. In the

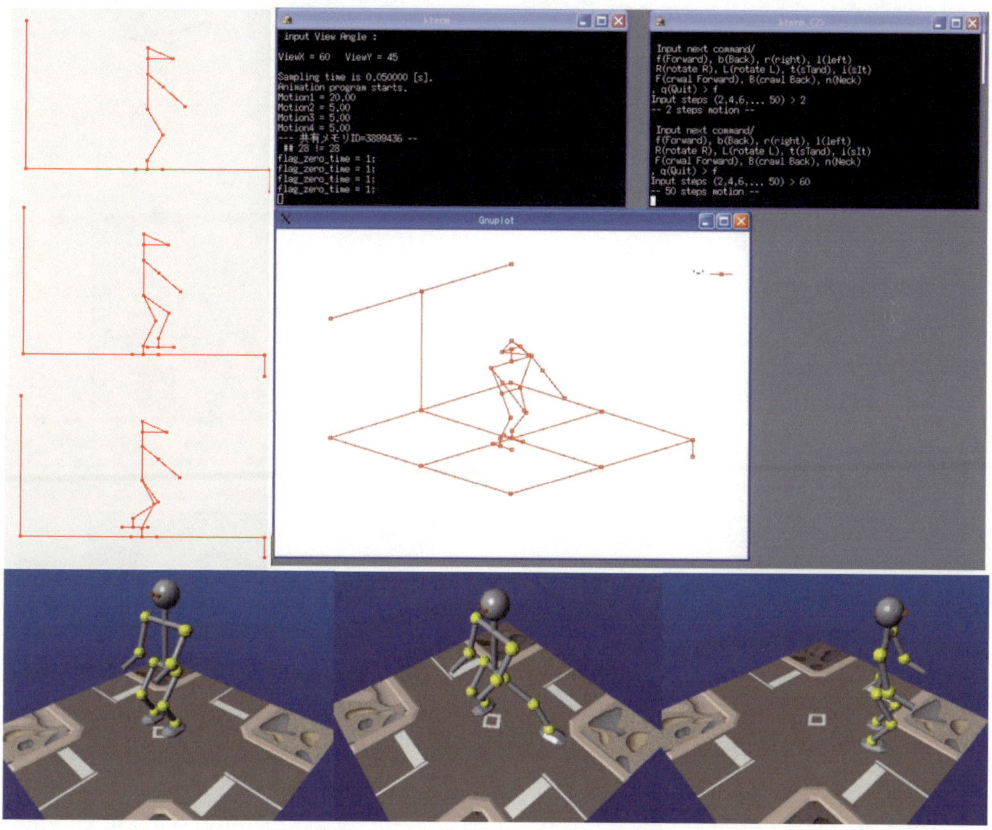

Fig. 6. Simulation interface presents robot's trajectory and motion process.

simulation, the robot performing biped locomotion of yawing motion to change its orientation by turning to back-left area.

In order to perform this motion, rotation of hip-joint yaw is the key-point. By solving Eq. (8), the robot rotation angle at hip joint is decided from *0* degree to *90* degree. Meanwhile target position of leg's end-point at *xyz*-axes plane is defined by solving inverse kinematics in Eqs. (14), (15), (16), (20), (21), and interpolation in Eq. (27). At this time the yawing angle θ_{1leg} is fixed at 70°.

During performing the yawing motion, at first the left leg's hip-joint yaw will rotate counterclockwise direction to θ_{1leg}. At the same time, the left leg follows along an ellipse trajectory in regard to *z*-axis direction to move the leg one step. This stepping motion is performed by given the leg's end point position to the robot's control system so that the joint angles of θ_{1leg}~θ_{6leg} could be solved by inverse kinematics calculations. The left leg position is defined by interpolation of the leg end point from its initial position with respect to the *xy*-axes position at a certain calculated distance. At this time the right leg acts as the support axis. Then, the robot corrects its orientation by changing the support axis to the left

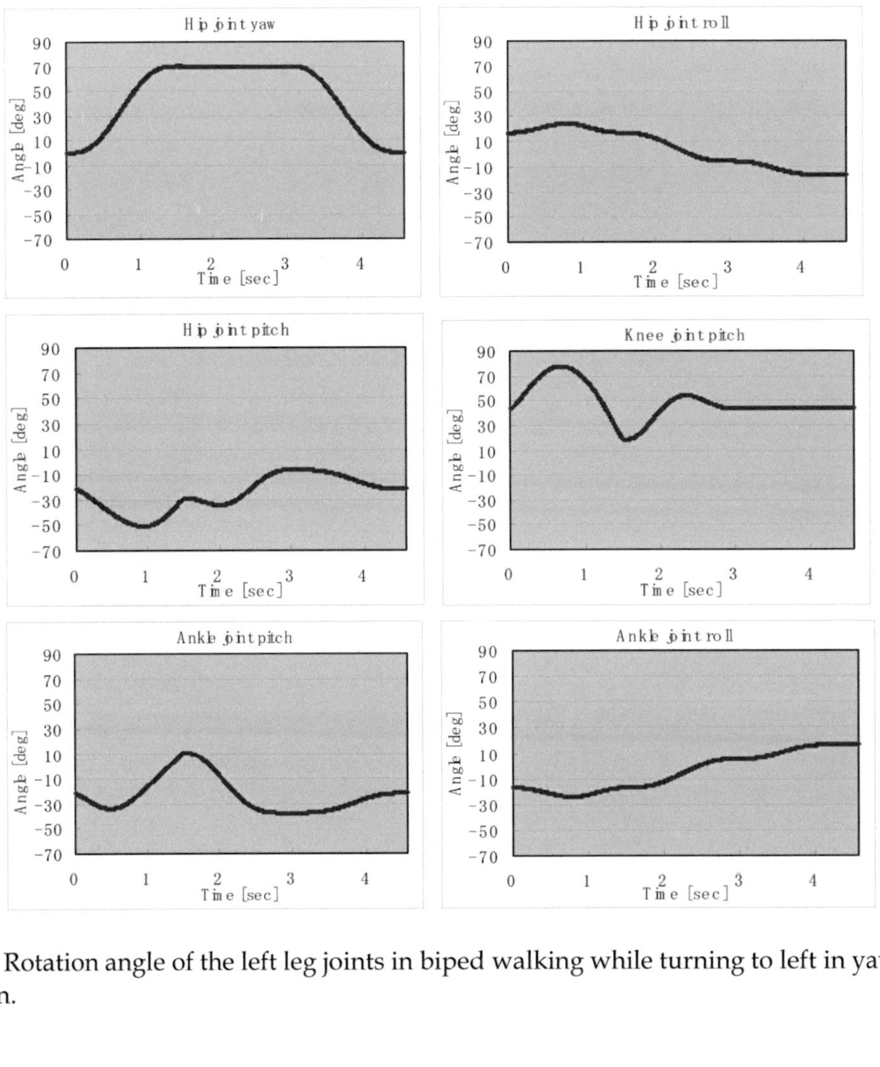

Fig. 7. Rotation angle of the left leg joints in biped walking while turning to left in yawing motion.

leg, while the left leg hip-joint yaw reverses the rotation to clockwise direction to complete the motion.

Each joint's rotation angles were saved and analyzed in a graph structure. For example, the graph for the left leg during yawing motion is plotted in Fig. 7. The graph shows the smooth trajectory of the rotation angles at each leg's joint. These simulation results verified reliability of the proposed kinematics and interpolation formulations to generate smooth and controlled trajectory for humanoid robot *Bonten-Maru II.*

5. Analysis of biped walking speed

5.1 Methodology

The main consideration in a biped humanoid robot is to generate the robot's efficient gait during performing tasks and maintain it in a stable condition until the tasks are completed. The efficiency in biped robots is normally related with how fast and how easy the tasks can be completed. In this research, to increase walking speed without changing the reduction-ratio, we considered three parameters to control the walking speed in biped robot locomotion:

1. ***Step length;*** **s**
2. hip-joint height from the ground; *h*
3. Duty-ratio; *d*

Figure 8 shows initial orientation of *Bonten-Maru II* during performing task which also indicate the step length and hip-joint height of the robot. The step-length is the distance between ankle-joints of a support leg and a swing leg when both of them are settled on the ground during walking motion. The hip-joint height is the distance between intersection point of hip-joint roll and pitch to the ground in walking position. Meanwhile, duty-ratio for biped robot mechanism is described as time ratio of one foot touches the ground when another foot swing to transfer the leg in one cycle of walking motion. In biped gait motion, two steps are equal to one cycle.

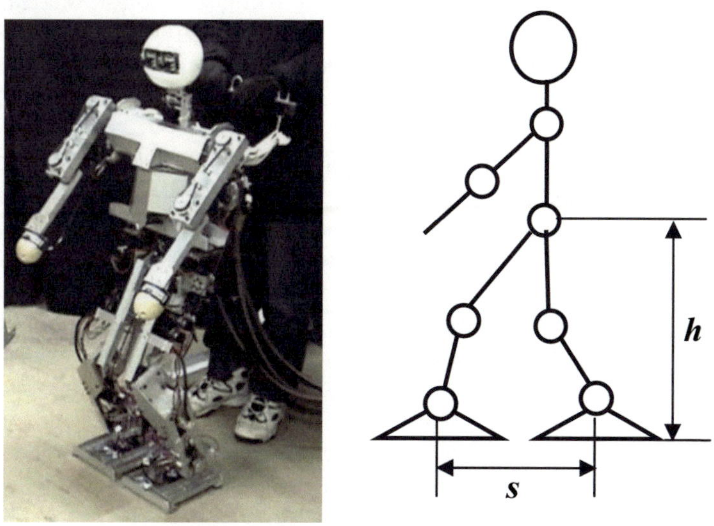

Fig. 8. Orientation of *Bonten-Maru II* to perform motion and parameters of hip-height *h* and step length *s*.

Hip-joint height [mm]	Max. step length in 1 step[mm]	Max. step length in 1 cycle [mm]
h_1=468	350	700
h_2=518	300	600
h_3=568	200	400

Table 3. Relationship of step length against hip-joint height at *Bonten-Maru II*.

Reffering to the *Bonten-Maru II* body and structure of the leg, the link parameters at the legs are used in calculations to define hip-joint height and maximum step length by geometrical analysis. Link parameters of the legs were calculated geometrically to define relation between step-length and hip-joint height. From the geometrical analysis, relation between the step-length and the hip-joint height is defined in Table 3.

At joint-motor system of *Bonten-Maru II*, maximum no-load rotation for the DC servomotor at each joint is 7220 [rpm]. This rotation is reduced by pulley and harmonic drive-reduction system to 1/333, in order to produce high torque output during performing tasks. We considered that the robot required high torque to perform tasks; therefore we do not change the reduction-ratio, which is 333:1. Eventually, these specifications produced maximum joint angular velocity at 130 [deg/s]. However, for safety reason, the joint angular velocity at the motor was reduced to 117 [deg/s]. The step time can be adjusted in the robot control system easily. However, if the step time is too small in order to increase walking speed, the robot motion becomes unstable. Moreover, the maximum step length performed becomes limited. In current condition, the step time for *Bonten-Maru II* to complete one cycle of walking is fixed between 7~10 second at maximum step length 75 [mm]. The duty-ratio d is increased gradually from 0.7 to 0.85. These parameter values are applied in simulation present in the next section.

5.2 Simulation analysis

A simulation analysis of the robot walking velocity using simulation interface that applies GnuPlot was performed based on parameters condition explained at previous section. The time for one circle of walking gait is initially fixed at 10 second. Each joint's rotation angles are saved and analyzed in a graph structure. Based on the joint angle, angular velocity of each joint was calculated.

Figure 9 shows joint angle data for right leg joints when performing 10 steps walk at condition: h=518 [mm], s=100 [mm] and d=0.7. From the angle data, angular velocity for each joint was calculated and presented in Fig. 10. The first and last gait shows acceleration and deceleration of the gait velocity. The three steps in the middle show maximum angular velocity of the legs joint.

Basically, in biped robot the maximum walking gait velocity is calculated from maximum joint angular velocity data by defining minimum step time for one gait. Eventually, by applying the same parameter, even if time for one step is initially different; the final joint angle obtained by the robot is same. Hence, in this analysis we can obtain the minimum step time in one step from the maximum joint angular velocity data that the initial step time was 10 seconds. Basically, the minimum gait time in one step is satisfying following equation:

$$t_{\min} < \frac{v_{\theta \max} \times 10}{V_{\theta \max}}. \tag{31}$$

Here, $V_{\theta\max}$ is the maximum joint angular velocity at the motor, t_{min} is minimum time for one step, and $v_{\theta max}$ is maximum joint angular velocity in each gait. Finally, the maximum walking gait velocity w_{max} is defined by dividing length s with minimum step time t_{min} in each gait, as shown in following equation.

$$w_{\min} = \frac{s}{t_{\min}} \tag{32}$$

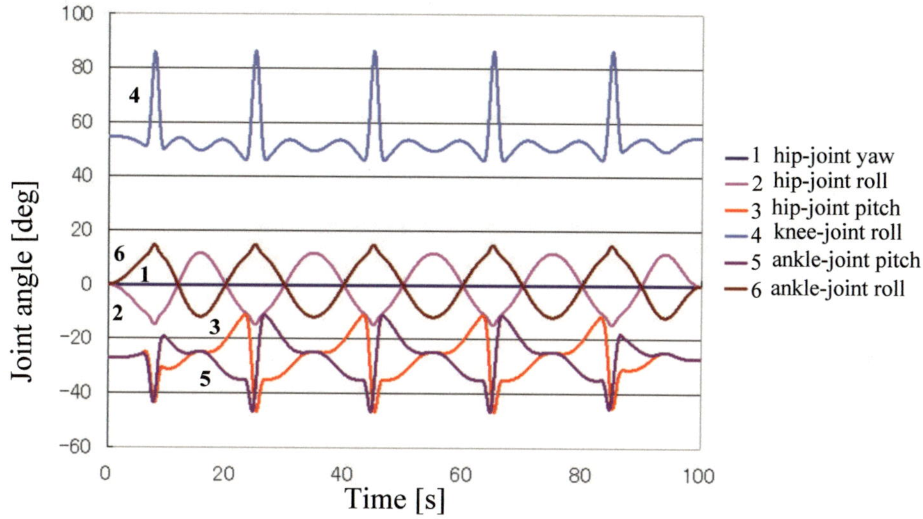

Fig. 9. Graph of joint rotation angle at right leg.

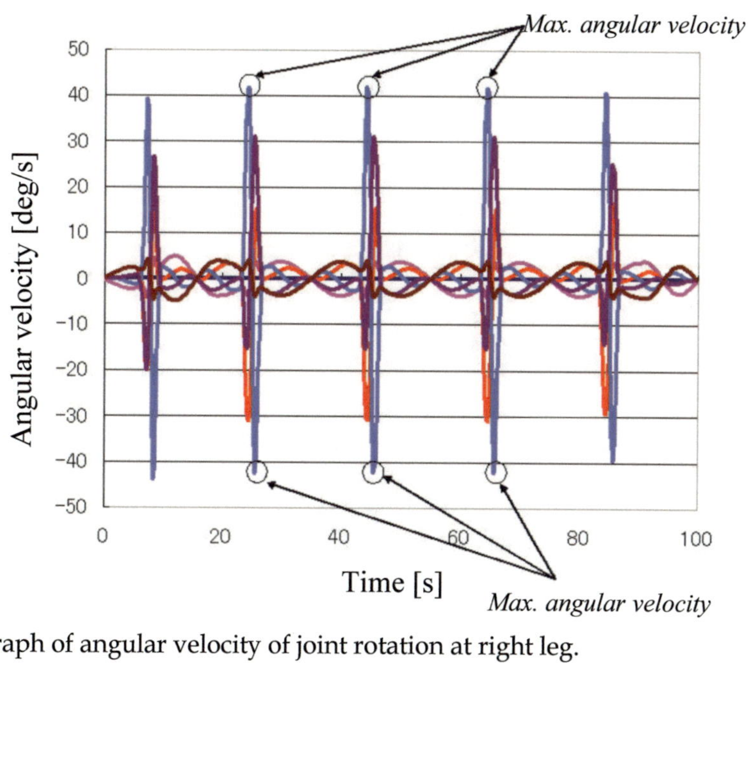

Fig. 10. Graph of angular velocity of joint rotation at right leg.

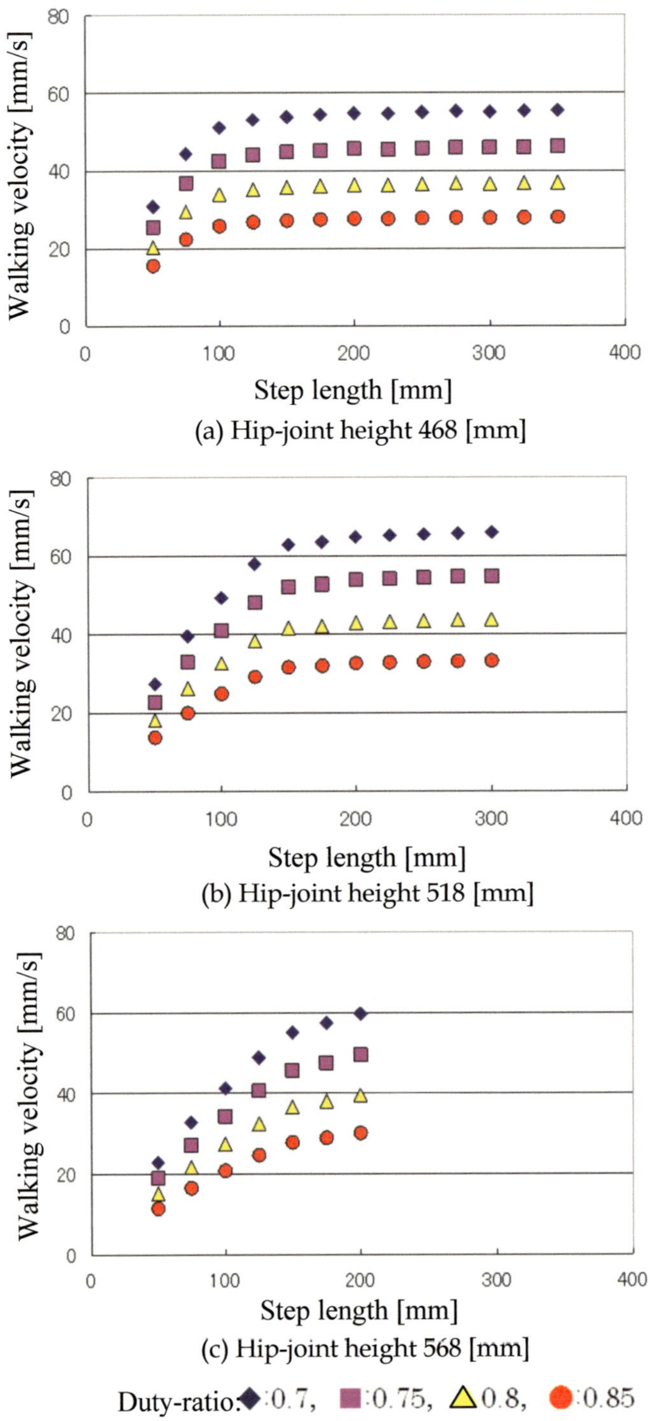

Fig. 11. Analysis results of maximum walking velocity at each gait.

Simulation results of walking gait velocity at each parameters value are compiled in graphs as shown in Fig. 11(a), (b) and (c). According to these graphs, from the relation of walking velocity and step length, the walking velocity was maintain nearly at constant value when it reached certain step length. Moreover, in relation of step length and hip-joint height, the higher hip-joint position is providing wider step length to perform better walking distance. At this point, lower duty-ratio shows the best results in relation of the hip-joint height and the step length for higher walking gait velocity, as shown in Fig. 11(b), where the low duty-ratio shows high walking velocity in relationship between the hip-joint-height and the step-length. It means by shorten the time for the support leg touching the ground will urge swing leg to increase its speed to complete one walking cycle, thus increase the walking velocity. At the same time, by choosing suitable step-length and hip-joint-height parameters, travel distance in each step can be improved. Analysis results revealed that it is possible to control biped walking speed without reducing the reduction-ratio at the joint-motor system.
From the simulation results, we can conclude that lower duty-ratio in suitable hip-joint height comparatively provided higher walking gait velocity. For *Bonten-Maru II*, the maximum walking gait velocity was improved from 30 [mm/s] to 66 [mm/s], which is about two times better than current walking velocity. At this time the hip-joint height is 518 [mm] and the time for one step is 4.5 seconds.

Step 1 Step 2 Step 3 Step 4 Step 5

Fig. 12. Humanoid robot performs biped walking applying the best parameters value from simulation results: h=518 [mm], s=200 [mm] and d=0.7, time per step 4.5 sec.

Step 1 Step 2 Step 3 Step 4 Step 5

Fig. 13. Humanoid robot performs biped walking applying current parameters value: h=568 [mm], s=75 [mm] and d=0.8, time per step 2.5 sec.

5.3 Experiments with *Bonten-Maru II*

Real-time experiments with the biped humanoid robot *Bonten-Maru II* were conducted to evaluate performance of the proposed methodology of optimal biped locomotion to improve walking speed of the humanoid robot. The parameter values that revealed the best result in simulation were applied, in comparison with current walking condition. Figures 12 and 13 respectively show photograph of the actual robot's walking motion in each experiment, which also indicate the parameter values applied. Travel distance was measured during the experiments.

The experimental results show that by applying the best parameters value obtained in the simulation results, the walking speed was improved. At the same time, the travel distance is longer about three times compared with current condition. This result reveals that the travel distance was improved in conjunction with the improvement of walking speed in the biped humanoid robot. The robot performed biped walking in smooth and stable condition.

5.3 Result and discussion

In our approach, we directly analyze geometrically the robot link structures and dimensions, and consider duty-ratio as effective parameters to improve walking speed in biped humanoid robot. Simulation results based on humanoid robot *Bonten-Maru II* parameters reveals that walking speed was improved by applying low duty-ratio at appropriate step length and hip-joint height. The walking speed increased about two times compared to normal condition. Meanwhile, real-time experiments utilizing real biped humanoid robot based on simulation results shows that the robot's travel distance during walking was improved about three times better than current walking condition. This analysis proved that it is possible to improve walking speed in stable biped locomotion without reducing the reduction-ratio.

6. Conclusion

This chapter presented analysis of biped gait locomotion to improve walking speed in humanoid robot without changing reduction-ratio at joint-motor system. Step length, hip-joint height, and duty-ratio were identified as parameters in this analysis. A Relationship between step length and hip-joint height was defined using geometrical calculations. Simulation analysis was conducted followed by real time experiments using humanoid robot *Bonten-Maru II*. Simulation results based on the humanoid robot *Bonten-Maru II* revealed that walking speed was improved by applying low duty-ratio at appropriate step length and hip-joint height. The walking speed increased about two times compared to normal condition. The real-time experiments utilizing *Bonten-Maru II* based on the simulation results shows that the robot's travel distance during walking was improved about three times better than current walking condition. The robot also walked faster in a stable condition compared to current walking condition.

This analysis proved that it is possible to improve walking speed in biped walking robots without reducing the reduction-ratio. The presented optimal gait generation in biped locomotion improved the performance of humanoid robot system towards operation in real world. This was proved by simulation and experimental results. Moreover, analysis results of gait trajectory generation proposed an efficient gait pattern for the biped robot. Since the analysis results revealed that it is possible to control biped locomotion speed without changing reduction-ratio at joint-motor system, the high torque output at robot's manipulator to conduct tasks in various motions is maintained.

7. Acknowledgement

Part of this study was supported by a fiscal 2006 Grant-in-Aid for Scientific Research in Exploratory Research from the Japan Ministry of Education, Culture, Sports, Science and Technology (Grant no. 18656079), and year 2008-2010 Grant-in-Research by the Japan Society for the Promotion of Science (JSPS) no. P08062. The authors are grateful with the support from Universiti Teknologi MARA.

8. References

Althaus, P., Ishiguro, H., Kanda, T., Miyashita T. & Christensen, H. I. (2004). Navigation for Human-Robot Interaction Tasks, *IEEE International Conference on Robotics and Automation (ICRA2004)*, pp. 1894-1900

Bischoff, R. & Graefe, V. (2005). Design principle for dependable robotics assistants, International Journal of Humanoid robotics, Vol. 1 (1), pp. 95-125

Capi, G., Nasu, Y., Mitobe K. & Barolli, L. (2003). Real time gait generation for autonomous humanoid robots: A case study for walking, *Journal Robotics and Autonomous Systems*, Vol. 42, No.2, pp. 169-178

Cheng, G., Nagakubo, A. & Kuniyoshi, Y. (2001). Continuous humanoid interaction: An integrated perspective -gaining adaptivity, redundancy, flexibility- in one, *Journal of Robotics and Autonomous Systems*, Vol. 37, Issues 2-3 pp 161-183

Chevallereau, C. & Aoustin, Y. (2001). Optimal trajectories for walking and running of a biped robot, Robotica, Vol. 19, Issue 5, pp. 557-569

Denavit, J. & Hartenberg, S. (1995). A kinematics notation for lower-pair mechanisms based upon matrices, *Journal of Applied Mechanics*, Vol. 77, pp. 215-221

Goswani, A., Espiau B., & Keramanie, A. (1997). Limit cycles in a passive compass gait biped and passivity-mimicking control laws, *Journal Autonomous Robots*, Vol. 4, pp. 273-286

Hasegawa, Y., Arakawa T. & Fukuda T. (2000). Trajectory generation for biped locomotion robot, *Journal Mechatronics*, Vol. 10, Issue 1-2, pp. 67-89

Hirai, K., Hirose, M., Haikawa, Y. & Takenaka, T. (1998). The development of Honda humanoid robot, *Proceedings of International Conference on Robotics and Automation'98*, pp. 1321-1326

Kajita, S., Nagasaki, T., Kaneko, K., Yokoi, K. & Tanie, K. (2005). A running controller of humanoid biped HRP-2LR, *Proceeding of the 2005 IEEE International Conference on Robotics and Automation (ICRA2005)*, pp. 618-624, Barcelona, Spain

Koker, R. (2005). Reliability-based approach to the inverse kinematics solution of robots using Elman's network, *Journal of Engineering Application of Artificial Intelligence*, Vol. 18, No. 6, pp. 685-693

Konno, A., Nagashima, K., Furukawa, R., Nishiwaki, K., Noda, T., Inaba M. & Inoue, H. (1997). Development of a humanoid robot Saika, *Proceeding of IEEE/RSJ Int. Conf. on Intelligent Robots and Systems (IROS'97)*, Grenoble, France, pp. 805-810

Lim, H., Yamamoto Y. & Takanishi, A. (2000). Control to realize human-like walking of a biped humanoid robot, *Proceeding of IEEE Int. Conf. System, Man and Cybernetics*, Vol. 5, pp. 3271-3276

Nasu, Y., Capi, G., Hanafiah Y., Yamano, M. & Ohka, M. (2007). Development of a CORBA-based humanoid robot and its applications, *Humanoid Robots*, by pro literature Verlag, Chapter 30, pp. 551-590

Takanishi, A., Ishida, M., Yamazaki, Y. & Kato, I. (1985). The realization of dynamic walking robot WL-10RD, *Proceedings of International Conference on Robotics and Automation (ICRA'85)*, pp. 459-466

Vukobratovic, M., Brovac, B., surla D. & Stokic, D. (1990). Biped locomotion, *Springer Verlag*

Yamaguchi, J., Takanishi, A. & Kato, I. (1993). Development of a biped walking robot compensating for three-axis moment by trunk motion, *Proceeding of 2003 EEE/RSJ International Conference on Intelligent Robots and Systems (IROS '93)*

14

A Fast and Smooth Walking Pattern Generator for Biped Robots

Han-Pang Huang and Jiu-Lou Yan
National Taiwan University
Taiwan

1. Introduction

In order to solve inverse kinematics of a multi-DOF (degree of freedom) mechanism, many methods have been proposed with the Jacobian linearization method. When solving inverse kinematics problems of the biped robot with this method, long computation time is required since the Jacobian matrix should be updated in order to solve the configuration for each different end-effector trajectory knot. In this chapter, two smooth trajectories are generated as target positions, one for swing leg's ankle, and the other for center of gravity (COG). These generated knot points in the task space with appropriate distance to each other are used to solve inverse kinematics by the proposed modified Jacobian method—Fixed Leg Jacobian. It can guarantee that only one iteration is required to solve the configuration when it is away from singularity with a small position error (0.0712% of leg length). The proposed algorithm can generate the gait in real-time including singularity avoidance and joint limit avoidance. Simulations have been carried out. The results showed that the proposed method can generate a smooth gait for robot walking on real-time implementation.

Compared with wheeled robots, legged robots have the advantage of being able to traverse uneven or sharp-height-changing environments. Nowadays, many vehicles, buildings and environments are designed for humans. Simple robots cannot enter and adapt to these places. Therefore, we must design complicated humanoid robots to do it. But when the designs become more complicated and with more DOFs, it is getting harder to control and generate the trajectories of them. The proposed algorithm can quickly generate smooth trajectories of the ankle and COG and solve inverse kinematics in order to achieve real-time control of biped robots. In this chapter, the focus is how to coordinate the swing leg, the fixed leg and the COG of the robot, and generate the gait in real-time. In the simulation, the robot has 24 DOFs, 6 in each leg (12 in two legs), 4 in each arm (8 in two arms), 2 in the torso and 2 in the head. The most important DOFs for balancing and walking are the twelve DOFs in the legs. DOFs in the fixed leg dominate the position of the COG, and the position of the ankle of the swing leg is given relative to the position of the ankle of the fixed leg in order to guarantee that the swing leg is in a proper position that it will not hit the fixed leg and touch the ground. The trajectories of the end-effectors planned with desired constraints are inputted to solve inverse kinematics, as shown in Fig. 1.

Many researchers have proposed the solutions to the problem while solving Jacobian linearized inverse kinematics. They include the damped least square method (DLS) (Wampler, 1986) and the robust damped least square method (RDLS) (Nakamura &

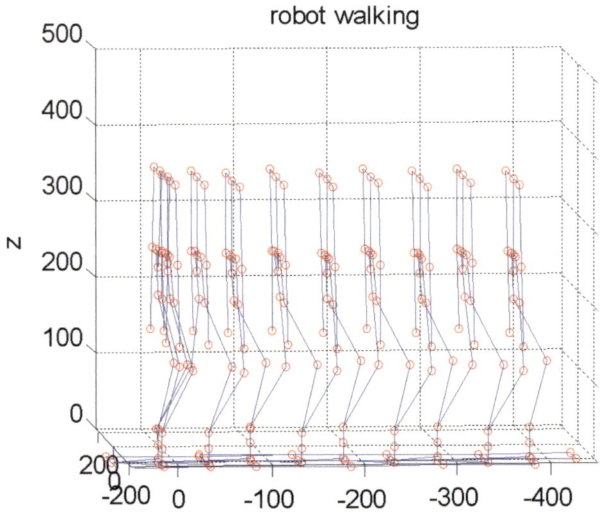

Fig. 1. Solving robot walking gait using inverse kinematics

Hanafusa, 1986), which are used for singularity avoidance. The weighted least-norm method (WLN) (Chan & Dubey, 1995) is used for joint limit avoidance. These two methods (RDLS and WLN) are used in this chapter to improve the performance of the algorithm. Many other methods are proposed, such as the selectively damped least squares methods (SDLS) (Buss & Kim, 2004), the gradient projection method (GPM) (Liegeois, 1997) and the extended Jacobian method EJM (Klein et al.,1995; Tevatia & Schaal, 2000).

On the other hand, the target positions that are one-by-one inputted to solve the inverse kinematics should also be planned appropriately. If the target positions are not planned close enough, many iterations will be wasted since the end-effectors are not in the desired positions and directions. Non-smooth target positions make the robot's joints oscillate. Regarding the generation of the trajectories, many methods were proposed, such as Lagrange interpolations, cubic spline, conventional tension spline (CTS) and modified tension spline (MTS) (Huang & Liu, 2005). Each method has its limitation. In this chapter, MTS is applied in order to generate smooth trajectories in position, velocity, acceleration, and jerk.

2. Robot kinematics

2.1 Inverse kinematics with pseudo inverse

The pseudo inverse method and two improving methods, RDLS, WLN, are used in this chapter to construct the inverse kinematics solver. By using forward kinematics method (DH method), the relationship between the position of the end-effectors and the joint angles can be found as Equation (1).

$$x = f(\theta) \tag{1}$$

The Jacobian linearized relationship between the velocities of the end-effectors and the joint angles are defined as Equation (2).

$$\dot{x} = J\dot{\theta} \tag{2}$$

where J denotes the Jacobian matrix; if there are redundant DOFs in the system, J is a rectangle matrix. Pseudo inverse method can be used to solve $\dot{\theta}$ with given $\dot{x}$:

$$\dot{\theta} = J^{+}\dot{x} \tag{3}$$

2.2 Robust damped least squares methods (RDLS)

If the determinant of JJ^T is zero or close to zero, singularity occurs. In order to avoid the singularity, robust damped least square method (RDLS) is applied. The idea of the damped least square method (DLS) is to minimize $||\dot{x} - J\dot{\theta}||^2 + \alpha||\dot{\theta}||^2$, the sum of the square of the residual error and the joint velocities. Here α is a positive damping factor. Thus, the pseudo inverse with DLS method is shown as Equation (4).

$$J^{+} = J^{T}(JJ^{T} + \alpha I_m)^{-1} \tag{4}$$

where I_m is an identity matrix with the same dimension as JJ^T matrix. The damping factor α helps to avoid singularity, but it also affects the solved $\dot{\theta}$. Thus, α should not be applied at nonsingular configurations. Nakamura et al. proposed a robust DLS method to solve this problem. A factor h is defined as Equation (5).

$$h(\boldsymbol{\theta}) = \sqrt{\det(JJ^{T})} \tag{5}$$

When h approaches to zero, it is getting closer to singularity. Then α is adjusted automatically with Equation (6).

$$\alpha = \begin{cases} \alpha_0(1 - h/h^s), & \textit{if } h < h^s \\ 0, & \textit{otherwise} \end{cases} \tag{6}$$

where h^s denotes the threshold value, α_0 is the value of damping factor at singular points. With the equation above, α is effective only when the configuration is near singular configuration.

2.3 Weighted least-norm method

The weighted least-norm method is designed from the idea of null space. The general solution of $\dot{\theta}$ for solving inverse kinematics can be written as Equation (7).

$$\dot{\boldsymbol{\theta}} = J^{+}\dot{\mathbf{x}} + (I - J^{+}J)\boldsymbol{\varphi} \tag{7}$$

where φ is an arbitrary vector. $J^{+}\dot{x}$ is the particular solution, and $(I-J+J)\varphi$ is the homogeneous solution. Joint limit avoidance is important for humanoid robots in order to act like human beings. A weighted least-norm (WLN) solution based scheme for avoiding joint limits is proposed by Chan & Dubey. In this method, a performance criterion $H(\theta)$ is defined as Equation (8).

$$H(\boldsymbol{\theta}) = \sum_{i=1}^{n} \frac{1}{4} \frac{(\theta_{i,\max} - \theta_{i,\min})^2}{(\theta_{i,\max} - \theta_i)(\theta_i - \theta_{i,\min})} \tag{8}$$

When any joint approaches its limit, the value of $H(\theta)$ grows very fast, and so is its partial differentiation $\partial H(\theta)/\partial\theta_i$. Thus, the weighting matrix is defined as Equations (9) and (10).

$$W = \begin{bmatrix} w_1 & 0 & \cdots & \cdots & 0 \\ 0 & w_2 & & & \vdots \\ \vdots & & \ddots & & \vdots \\ \vdots & & & \ddots & 0 \\ 0 & \cdots & \cdots & 0 & w_n \end{bmatrix} \tag{9}$$

$$w_i = 1 + \left| \frac{\partial H(\boldsymbol{\theta})}{\partial \theta_i} \right| \tag{10}$$

The WLN method can be expressed as Equations (11) and (12).

$$J_W = JW^{-1/2} \tag{11}$$

$$\dot{\boldsymbol{\theta}} = W^{-1/2} J_W{}^{+} \dot{\mathbf{x}} + (W^{-1/2} J_W{}^{+} - J^{+}) \dot{\mathbf{x}} \tag{12}$$

2.4 RWLN method

The RWLN method (Yu, 2006) is the combination of the RDLS method and the WLN method. The equation used to solve inverse kinematics with the RWLN method is shown as Equations (13) and (14).

$$\dot{\boldsymbol{\theta}} = W^{-1/2} J_{W,\alpha}{}^{+} \dot{\mathbf{x}} + (W^{-1/2} J_{W,\alpha}{}^{+} - J_\alpha{}^{+}) \dot{\mathbf{x}} \tag{13}$$

$$J_{W,\alpha}{}^{+} = J_W^T (J_W J_W^T + \alpha_W I_m)^{-1} \tag{14}$$

It can avoid the singularity of J^+, and can also avoid the singularity of J_W^+ with auto-adjusting α and α_W.

3. Gait generation algorithm

The solving process of inverse kinematics is described in the following.

1. The D-H forward kinematics of each limb and head is constructed independently. All limbs have the same base point which is at the middle point of the two hips.
2. Determine the trajectory of the end-effector of the swing leg using MTS method and the trajectory of the COG using preview control method (Kajita et al., 2006). The relationship between the swing leg and fixed leg is discussed in the following discussion "Relative Input".
3. Construct the conventional Jacobian matrix and the COG Jacobian matrix.
4. Construct the proposed F-Jacobian matrix and then combine it into the conventional Jacobian matrix.
5. Solve inverse kinematics using the proposed F-Jacobian method with the inputs, the trajectories of the swing leg and COG.

3.1 Relative input

If the trajectories of the end-effectors of the swing leg and the fixed leg are assigned independently, the swing leg may touch the fixed leg or even be lower than the fixed leg to break the balance of the robot if there is no any other good checking method to check it. So the position of the ankle of the swing leg should be dependent upon the position of the swing leg, as shown in Fig. 2 and Equation (15).

$$\vec{r}_{swing} = \vec{r}_{fixed} + R_n \vec{r}_{bl} + \vec{r}_{traj,i} \tag{15}$$

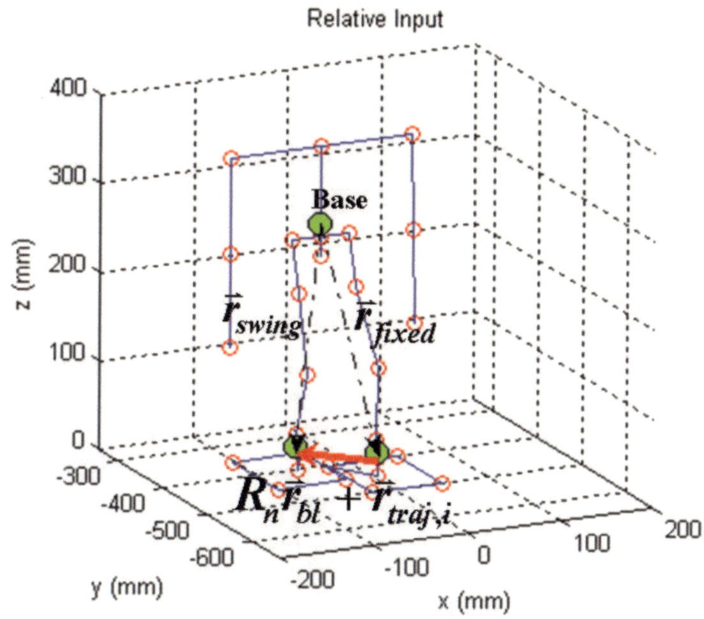

Fig. 2. Relative input trajectory of the swing leg

where $\vec{r}_{swing}$ denotes the position vector from "Base" to the ankle of the swing leg, $\vec{r}_{fixed}$ denotes the position vector from "Base" to the ankle of the fixed leg, $\vec{r}_{bl}$ denotes the position vector from the ankle of the fixed leg to the ankle of the swing leg, R_n denotes the rotation matrix that expresses the rotation between the base coordinate frame and the ground. The magnitude of $\vec{r}_{bl}$ is a constant. And the $\vec{r}_{traj,i}$ denotes the i-th planned trajectory point of the ankle of the swing leg. It can control how the swing leg moves in each stride. The R_n is not an identity matrix when the orientation of the robot is not the same as the world coordinates. In this chapter, in order to simplify the system, the orientation of the robot is the same as the world coordinates and the R_n is an identity matrix. With dependent position input of the ankle of the swing leg, we can give a trajectory easily that the swing leg will not lower than the fixed leg to avoid breaking the balance of the robot and touching the fixed leg.

3.2 Modified Tension Splines (MTS)

It is also important that we should input smooth and well-defined trajectories to our system. MTS method is based on conventional tension splines (CTS), and the tension factor can be arbitrarily assigned in this method. MTS method also generates trajectories with smooth position, velocity, acceleration and jerk. Trajectory planned by MTS is defined as Equation (16).

$$\begin{aligned} q_{j,i}(t) = & \frac{1}{\sigma_{j,i}^2} \frac{\sinh\left[\sigma_{j,i}\left(t_{i+1}-t\right)\right]}{\sinh\left(\sigma_{j,i}h_i\right)} \ddot{q}_{j,i} + \frac{1}{\sigma_{j,i}^2} \frac{\sinh\left[\sigma_{j,i}\left(t-t_i\right)\right]}{\sinh\left(\sigma_{j,i}h_i\right)} \ddot{q}_{j,i+1} \\ & + \left(q_{j,i} - \frac{1}{\sigma_{j,i}^2}\ddot{q}_{j,i}\right)\frac{t_{i+1}-t}{h_i} + \left(q_{j,i+1} - \frac{1}{\sigma_{j,i}^2}\ddot{q}_{j,i+1}\right)\frac{t-t_i}{h_i} \end{aligned} \tag{16}$$

where $h_i = t_{i+1}$-t_i, the time interval of the i-th CTS segment, α_{ij} denotes the tension factor of the ith CTS segment. The acceleration $\ddot{q}_{ij}$ can be found as Equation (17). After solving the $\ddot{q}_{ij}$, the knots q can be found by Equation (16).

$$\begin{aligned} \ddot{q}_{j,i}(t) - \sigma_{j,i}^2 q_{j,i}(t) = & \frac{t_{i+1}-t}{h_i}\left(\ddot{q}_{j,i} - \sigma_{j,i}^2 q_{j,i}\right) \\ & + \frac{t-t_i}{h_i}\left(\ddot{q}_{j,i+1} - \sigma_{j,i}^2 q_{j,i+1}\right), \quad t_i \le t \le t_{i+1} \end{aligned} \tag{17}$$

3.3 Fixed Leg Jacobian (F-Jacobian)

After constructing the DH parameters, the Jacobian matrix can be found by the cross product method, and then the limbs and the head can be controlled independently with the Jacobian matrix. The ankles, the fingertips and the head are chosen as end-effectors. But if we solve inverse kinematics of the limbs and the head of the robot independently, it is very difficult to decide where the positions of the end-effectors should be because DH forward kinematics method constructs the joint positions and orientations in its own coordinates instead of the world coordinates, and all positions of the end-effectors in the world coordinates are influenced by the movements of the fixed leg. Equation (18) describes the conventional Jacobian matrix that is used while solving inverse kinematics independently.

$$\begin{bmatrix} \dot{d}_{LL} \\ \dot{d}_{RL} \\ \dot{d}_{LA} \\ \dot{d}_{RA} \\ \dot{d}_{H} \end{bmatrix} = \begin{bmatrix} J_{LL} & 0 & 0 & 0 & 0 \\ 0 & J_{RL} & 0 & 0 & 0 \\ 0 & 0 & J_{LA} & 0 & 0 \\ 0 & 0 & 0 & J_{RA} & 0 \\ 0 & 0 & 0 & 0 & J_{H} \end{bmatrix} \begin{bmatrix} \dot{\theta}_{LL} \\ \dot{\theta}_{RL} \\ \dot{\theta}_{LA} \\ \dot{\theta}_{RA} \\ \dot{\theta}_{H} \end{bmatrix} \tag{18}$$

where $\dot{d}$ denotes the vector from the current end-effector position to the desired end-effector position in its own coordinates; LL, RL, LA, RA, and H denotes left leg, right leg, left arm, right arm, and head, respectively. If we just want to control the limbs and the head of the robot independently, it is enough to solve inverse kinematics with the equation above in one iteration if $\dot{d}$ is given appropriately (not too large).

In fact, it is not enough to control the end-effectors independently. Relative input should be used to prevent the swing leg from hitting the fixed leg or touching the ground. The velocity of the ankle of the fixed leg in the world coordinates is zero because it is fixed on the ground. The velocities of all other end-effectors in the world coordinates can be found by subtracting the velocity of the fixed leg from them. In world coordinates, the velocities of the end-effectors are shown in Equation (19) - (21).

$$V_{fixed,world} = 0, \quad \omega_{fixed,world} = 0 \tag{19}$$

$$V_{swing,world} = V_{swing,local} - V_{fixed,local}, \quad \omega_{swing,world} = \omega_{swing,local} - \omega_{fixed,local} \tag{20}$$

$$V_{other,world} = V_{other,local} - V_{fixed,local}, \quad \omega_{other,world} = \omega_{other,local} - \omega_{fixed,local} \tag{21}$$

where V_{fixed} and ω_{fixed} denote the velocity and the angular velocity of the fixed leg; the subscript *"world"* denotes the variable is in the world coordinates, and the subscript *"local"* denotes the variable is in its own body-fixed coordinates.
Since the "Base" of the legs and arms are the same, the Jacobian matrix expressing the influence of the fixed leg to all other end-effectors can be calculated as follows. It is calculated using the cross product method, as shown in Equation (22) - (24).

$$J_F = \begin{bmatrix} J_{Translational} \\ J_{Rotational} \end{bmatrix} \tag{22}$$

It is called "F-Jacobian" (Fixed Leg Jacobian) method. Note that $J_{Rotational}$ is given as Equation (23).

$$J_{Rotational} = -\begin{bmatrix} \omega_{e,1} & \omega_{e,2} & \cdots & \omega_{e,j} & \cdots & \omega_{e,n} \end{bmatrix} \tag{23}$$

where n denotes the total number of joints of the fixed leg, $\omega_{e,j}$ are unit normal vectors of the joints of the fixed leg. The minus sign is multiplied since when a joint in the fixed leg rotates clockwise in its coordinates, the body rotates counterclockwise in the world coordinates. On the other hand, the $J_{Translational}$ is calculated as Equation (24).

$$J_{Translational}^{\ T} = -\begin{bmatrix} \omega_{e,1} \times \vec{r}_{1,F\rightarrow end\_eff} \\ \omega_{e,2} \times \vec{r}_{2,F\rightarrow end\_eff} \\ \cdots \\ \omega_{e,j} \times \vec{r}_{j,F\rightarrow end\_eff} \\ \cdots \\ \omega_{e,n} \times \vec{r}_{n,F\rightarrow end\_eff} \end{bmatrix} \tag{24}$$

where $\vec{r}_{j,F\rightarrow end\_eff}$ denotes the vector from the j-th joint of the fixed leg to the end-effector affected by the movement of the motion of the fixed leg.

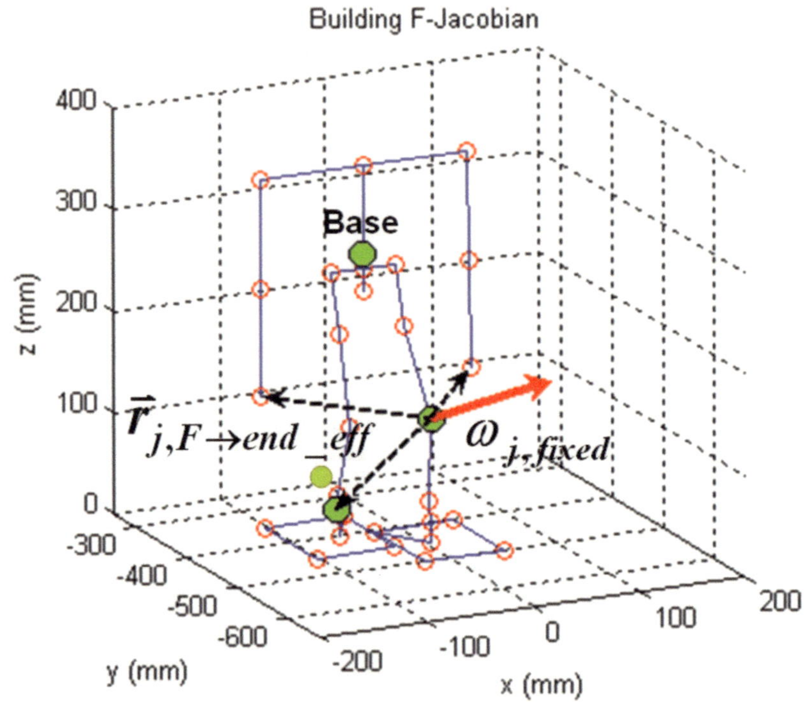

Fig. 3. Build the F-Jacobian

The F-Jacobian matrix is written as Equation (25).

$$\begin{bmatrix} \dot{d}_{fixed} \\ \dot{d}_{swing} \\ \dot{d}_{LA} \\ \dot{d}_{RA} \\ \dot{d}_{H} \end{bmatrix} = \begin{bmatrix} J_{fixed} & 0 & 0 & 0 & 0 \\ J_{f\rightarrow s} & J_{swing} & 0 & 0 & 0 \\ J_{f\rightarrow LA} & 0 & J_{LA} & 0 & 0 \\ J_{f\rightarrow RA} & 0 & 0 & J_{RA} & 0 \\ J_{f\rightarrow H} & 0 & 0 & 0 & J_{H} \end{bmatrix} \begin{bmatrix} \dot{\theta}_{fixed} \\ \dot{\theta}_{swing} \\ \dot{\theta}_{LA} \\ \dot{\theta}_{RA} \\ \dot{\theta}_{H} \end{bmatrix} \tag{25}$$

where $J_{f\rightarrow X}$ denotes the F-Jacobian matrices; the subscript "X" denotes the end-effector affected by the movement of the joints of the fixed leg, such as the swing leg and the fingertips.

Recall the Equation (15).

$$\vec{r}_{swing} = \vec{r}_{fixed} + R_n \vec{r}_{bl} + \vec{r}_{traj,i} \tag{15}$$

The 3-by-1 column vector $\dot{d}_{swing}$ is given by:

$$\dot{d}_{swing} - \vec{r}_{swing,next} - \vec{r}_{swing,now} \tag{26}$$

To achieve the desired position of the fixed leg, the $\vec{r}_{fixed}$ term converges to the target position after iterations of inverse kinematics algorithm. The $\dot{d}_{swing}$ term also changes with the $\vec{r}_{fixed}$ term. Here the F-Jacobian method can be used to describe and compensate the effect of the motion of the fixed leg in the Jacobian matrix in order to reduce the total iterations while solving inverse kinematics. Equation (27) can be obtained by multiplying the second row and the $\dot{\theta}$ column of the matrices in Equation (25) as

$$\dot{d}_{swing} = J_{f\to s}\dot{\theta}_{fixed} + J_{swing}\dot{\theta}_{swing} \tag{27}$$

The $J_{f\to s}\dot{\theta}_{fixed}$ term means the effect of the fixed leg in the world coordinates, and serves as the compensation term. The $J_{swing}\dot{\theta}_{swing}$ term means the affection of the joints of the swing leg itself. Without the compensation of F-Jacobian term, the end-effectors will oscillate and then converge to the desired position slower.

3.4 COG Jacobian (Center of Gravity Jacobian)

The position, velocity and acceleration of COG are highly related with whether the robot falls or not. The position of COG can be computed by averaging the sum of the product of the linkage masses and their position vectors, as shown in Equation (28). Note that the $\vec{r}_{COG}$ is a 3-by-1 vector described in Cartesian coordinates.

$$\vec{r}_{COG} = \frac{1}{n}\sum_{i=1}^{n} m_i \cdot \vec{r}_{m,i} \Big/ \sum_{i=1}^{n} m_i \tag{28}$$

where "Base" denotes the start point of forward kinematics, $\vec{r}_{m,i}$ denotes the vector from base to the COG of linkage i, and m_i denotes the mass of linkage i.

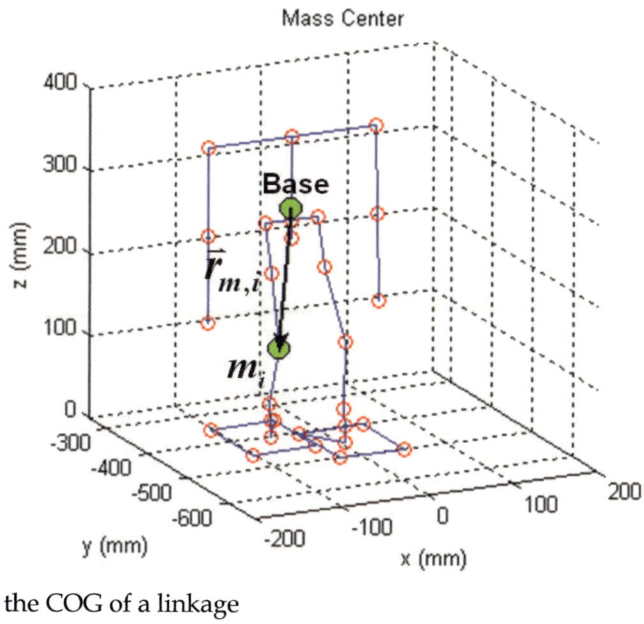

Fig. 4. Position of the COG of a linkage

When a joint rotates, only parts of the whole body are rotated, while the others are not rotated. Separating the rotated parts and the fixed parts (without rotating), we obtain

$$M \cdot \vec{r}_{COG} = \left(\sum_h m_{a,h} \cdot \vec{r}_{a,h} + \sum_k m_{ua,k} \cdot \vec{r}_{ua,k} \right) \Bigg|_{\text{joint}=j} \tag{29}$$

where M denotes the total mass of the robot, subscript "a" denotes the parts that are affected by the rotation, subscript "ua" denotes the parts that are unaffected by the rotation, the $m_{a,h}$ and $m_{ua,k}$ denote the mass that are affected and unaffected by the joint j, and the vectors $\vec{r}_{a,h}$ and $\vec{r}_{ua,k}$ denote the position of the COG of each affected part and each unaffected part. The equation can also be written as

$$M \cdot \vec{r}_{COG} = \left(M_a \cdot \vec{r}_a + M_{ua} \cdot \vec{r}_{ua} \right) \Big|_{\text{joint}=j} = M_{a,j} \cdot \vec{r}_{a,j} + M_{ua,j} \cdot \vec{r}_{ua,j} \tag{30}$$

where $M_{a,j}$ denotes the total mass of the parts affected by joint j, and $M_{ua,j}$ denotes the total mass of the parts unaffected by joint j, $\vec{r}_{a,j}$ and $\vec{r}_{ua,j}$ denote the position vector of the COG of the affected and unaffected parts. Note that the members of affected and unaffected parts change with different joint j, and they also depend upon each different control system. The position change of $\vec{r}_{COG}$ caused by the rotation of joint j can be approximated as

$$\Delta\vec{r}_{COG,j} = \frac{M_{a,j}}{M} \cdot \Delta\vec{r}_{a,j} + \frac{M_{ua,j}}{M} \cdot \Delta\vec{r}_{ua,j} \tag{31}$$

Since $\vec{r}_{ua}$ is unaffected by the rotation, $\Delta\vec{r}_{ua}$ is always equal to zero. $\Delta\vec{r}_{COG,j}$ denotes the displacement of the whole robot's COG caused by the rotation of the j-th joint. Thus, the COG Jacobian can be obtained as

$$\Delta\vec{r}_{COG,j} = \frac{M_{a,j}}{M} \Delta\vec{r}_{a,j} = J_{COG,j} \cdot \dot{\theta}_j \tag{32}$$

where the $J_{COG,j}$ denotes the COG Jacobian of joint j, $\dot{\theta}_j$ denotes the angular speed of joint j. The COG Jacobian matrix of the joints on the limbs except the joints on the fixed leg can be found as Fig. 5 and Equation (33).

$$J_{COG,j} = \frac{M_a}{M} \vec{\omega}_{e,j} \times \vec{r}_{c \to j} \tag{33}$$

where $\vec{\omega}_{a,j}$ denotes the unit vector along the z-direction of the j-th axis, $\vec{r}_{c \to j}$ denotes the vector from the j-th joint to the COG of the affected parts. We can also use this concept to get the COG Jacobian of the fixed leg, as shown in Fig.6 and Equation (34).

$$\Delta\vec{r}_{COG,j} = J_{COG,j} \cdot \dot{\theta}_j = -\frac{M_a}{M} \vec{\omega}_{e,j} \times \vec{r}_{c \to j} \cdot \dot{\theta}_j \tag{34}$$

In the coordinates of the fixed leg, the rotation of a joint rotates the parts that are lower than it. But in the world coordinates, the rotation causes the parts higher than the joint to rotate with the same angular velocity in negative direction.

$$J_{COG,j} = -\frac{M_a}{M} \vec{\omega}_{a,j} \times \vec{r}_{c \to j} \tag{35}$$

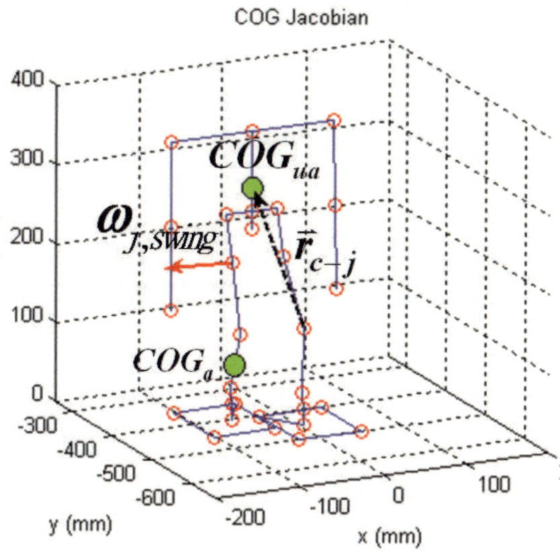

Fig. 5. Construction of COG Jacobian—swing leg

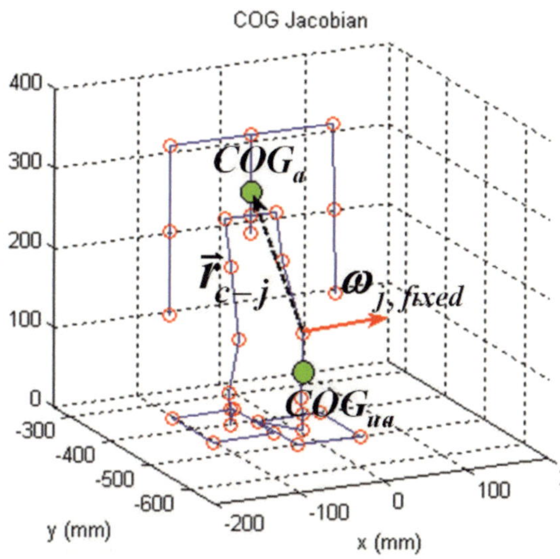

Fig. 6. Construction of COG Jacobian—fixed leg

The Jacobian matrix with COG Jacobian can be written as

$$\begin{bmatrix} \dot{d}_{fixed} \\ \dot{d}_{swing} \\ \dot{d}_{LA} \\ \dot{d}_{RA} \\ \dot{d}_{H} \\ \dot{d}_{COG} \end{bmatrix} = \begin{bmatrix} J_{fixed} & 0 & 0 & 0 & 0 \\ J_{f\to s} & J_{swing} & 0 & 0 & 0 \\ J_{f\to LA} & 0 & J_{LA} & 0 & 0 \\ J_{f\to RA} & 0 & 0 & J_{RA} & 0 \\ J_{f\to H} & 0 & 0 & 0 & J_{H} \\ J_{f\to C} & J_{s\to C} & J_{LA\to C} & J_{RA\to C} & J_{H\to C} \end{bmatrix} \begin{bmatrix} \dot{\theta}_{fixed} \\ \dot{\theta}_{swing} \\ \dot{\theta}_{LA} \\ \dot{\theta}_{RA} \\ \dot{\theta}_{H} \end{bmatrix} \tag{36}$$

where $\dot{d}_{fixed}$ is a 3-by-1 matrix, it controls the three orientations of the fixed leg. The other three DOFs of the fixed leg are used to control the position of the COG, $\dot{d}_{COG}$ is a 3-by-1 matrix that controls the position of COG, $J_{X \to C}$ denotes the COG Jacobian, the "X" denotes the limbs that their movements affect the position of the COM.

4. Simulation

All the input trajectories of the swing leg in the simulations below are generated with "Relative input" method, and the position of the swing leg is given relative to the fixed leg in order to guarantee the swing leg is in a proper position, as shown in Fig. 7. The COG trajectory is generated smoothly and fits the COG/ZMP (zero moment point) inverted pendulum constraint with the preview control method, as shown in Fig. 8.

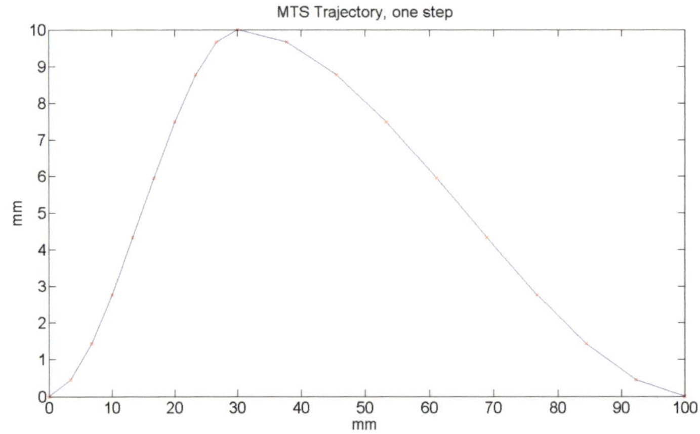

Fig. 7. Planned trajectory of the ankle of the swing leg

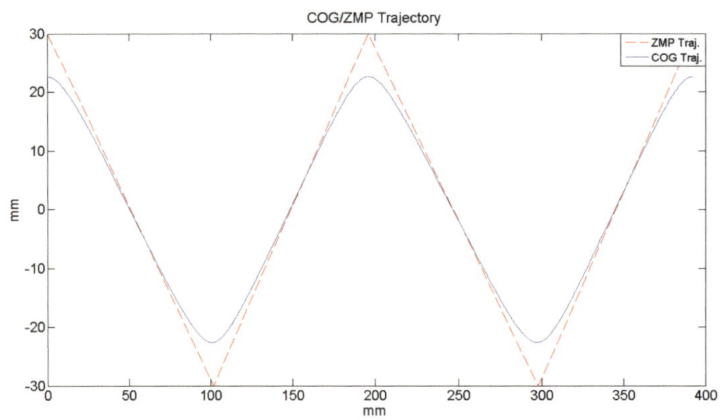

Fig. 8. Planned trajectory of the COG

The inputs above are inputted to the robot walking system in the simulation. The computation times of the same robot model with conventional Jacobian matrix and proposed F-Jacobian matrix will be compared. Each gait configuration should be solved as soon as possible in order to gain more CPU resources in each time period, and then each gait configuration will be sent to the robot within the desired time interval to control the robot walking speed. Human walks about 90 to 110 steps in a minute. It means 0.55 to 0.67 second per step. The robot should walk faster than at least 1.0sec/step to simulate human walking. The simulation results are shown below. All the simulations below are done with a personal computer equipped with Intel Core2Duo E6300 1.83GHz and 2GB RAM.

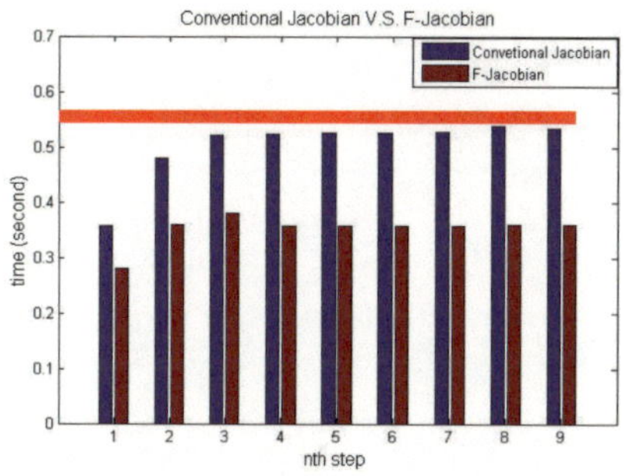

Fig. 9. Simulation results of computation time

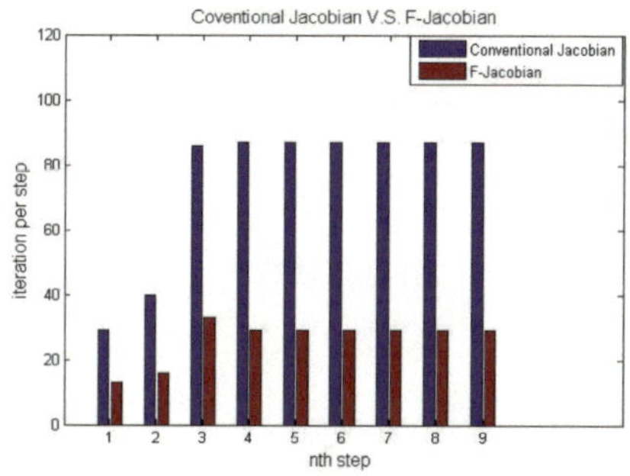

Fig. 10. Simulation results of iterations

The graphics above show the computation time and the iterations per step of the pseudo inverse method with and without F-Jacobian method. Step 1 and step 2 are initial steps, so they are with different computation time and iterations per step. The red bar is the 0.55sec/step reference line of human walking speed. Inverse kinematics with F-Jacobian solves the planned 9 steps in 3.07 second and 232 iterations (average 0.341 second and 25.8 iterations), and inverse kinematics without F-Jacobian solves the planned 9 steps in 4.62 second and 677 iterations (average 0.513 second and 75.2 iterations). Clearly, the proposed method, F-Jacobian, saved 33.55% computation time and 65.73% iterations per step. Except initial steps, each step contains 27 configurations. In the 27 configurations, the first three and last three points need no iteration because they are at the same position. The proposed method, F-Jacobian, can solve each configuration in only one computation when the acceptable error is 0.2mm (0.0712% of the length of legs). "Acceptable error" means the acceptable position error value when solving inverse kinematics. If the position error is smaller than the acceptable error, the next trajectory knot will be inputted to the inverse kinematics solver. If the position error is still larger than the acceptable error, the same trajectory knot will be inputted to the solver again. Since the input to the inverse kinematics are close and smooth enough, the proposed method can get smooth trajectories and solve each configuration in one iteration. But for the same input, inverse kinematics without F-

Jacobian makes the joints oscillate and needs about 3 iterations to solve one configuration. Fig. 11 and Fig. 12 show the solved trajectories of the left ankle (end-effector) with and without F-Jacobian.

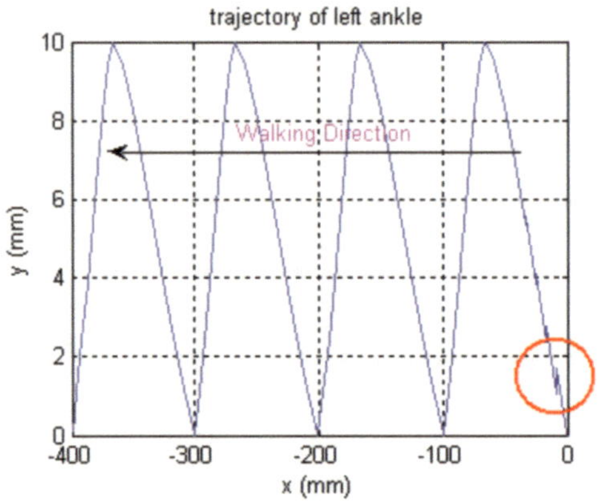

Fig. 11. Solved trajectory with F-Jacobian

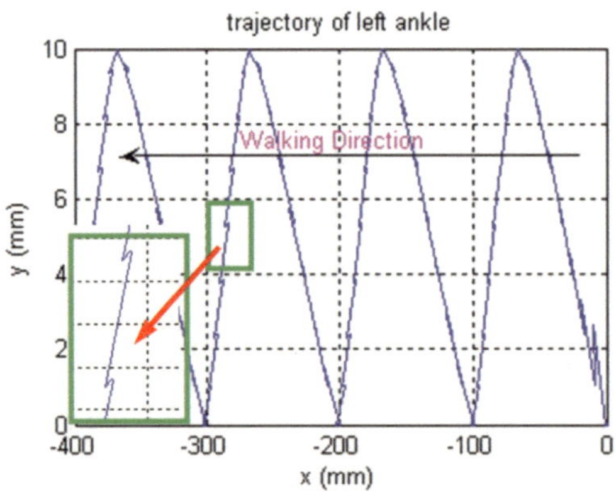

Fig. 12. Solved trajectory without F-Jacobian

Without F-Jacobian, the solved trajectories are not all useable. Only points that are in the acceptable error range are useable. We can reduce the number of useless points with F-Jacobian method. The robot walks from 0 to -400mm in the simulation. Only some configurations in the initial steps are not solved in one computation since the home configuration of the robot is near singularity. After initial steps, the robot has bent its knees, and hence keeps the robot away from the singular configurations. All the configurations after initial steps are solved in one computation. Fig.13 and Fig.14 show the position error after each inverse kinematics computation (conventional Jacobian vs. the proposed F- Jacobian).

In the figures, the max error of conventional Jacobian method is about fifteen times larger than F-Jacobian method. Fig.15 shows the acceptable error versus the total iterations of the inverse kinematics computation (for 232 trajectory knots) with and without F-Jacobian method.

From the figure, we can see the number of total iterations for conventional Jacobian method grows much faster than the F-Jacobian method when we choose the acceptable error from 2mm to 0.0002mm.

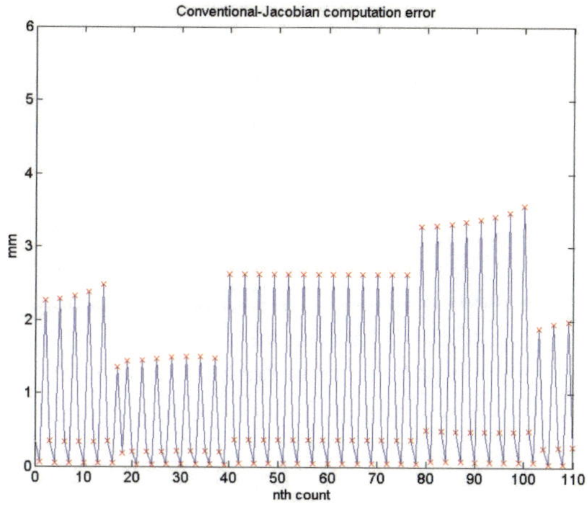

Fig. 13. Position error--conventional Jacobian

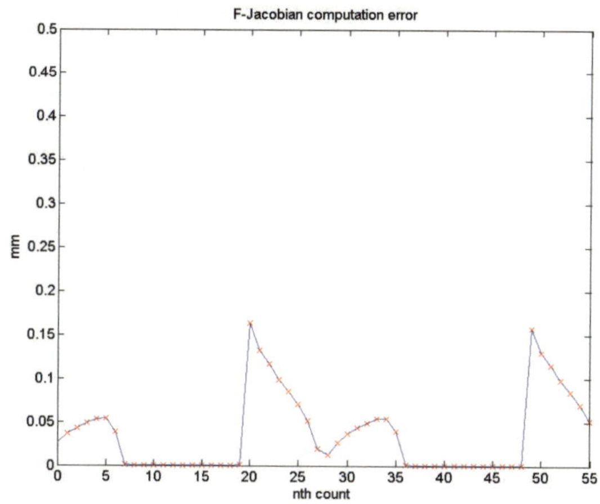

Fig. 14. Position error F-Jacobian

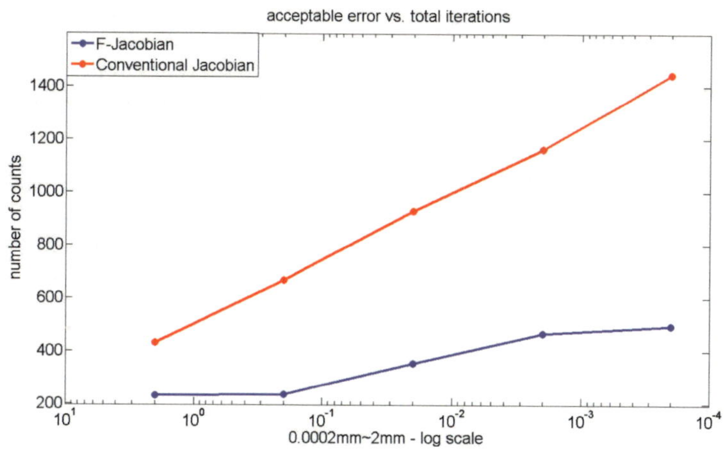

Fig. 15. Acceptable error vs. total iterations

5. Conclusions

Cooperation of the swing leg and the fixed leg is important for a humanoid robot. In this chapter, the position of the swing leg is dependent upon the position of the fixed leg. In order to solve inverse kinematics faster, F-Jacobian method is applied. The F-Jacobian method can compensate the displacements of the other end-effectors that are caused by the rotation of each joint in the fixed leg. F-Jacobian method can save computation time and generate knots more smoothly and effectively than the conventional Jacobian method. F-Jacobian method can be extended to construct COG Jacobian and other Jacobian matrices in order to describe and compensate the effect of the fixed leg. Simulations are done and justify that the F-Jacobian has better performance than conventional Jacobian method.

6. References

Buss, S. R. & Kim, J. S. (2004). Selectively damped least squares for inverse kinematics. *Journal of Graphics Tools*, Vol. 10, pp. 37-49

Chan, T. F. & Dubey, R. V. (1995). A Weighted Least-Norm Solution Based Scheme for Avoiding Joint Limits for Redundant Joint Manipulators. *IEEE Trans. On Robotics and Automation*, Vol. 11, No. 2, pp. 286-292

Huang, H. P. & Liu, C. P. (2005). A Novel Trajectory Optimization and Workspace Boundary Singularity Solution for Industrial Robots. *Proceedings of Automation the Eighth International Conference on Automation Technology Conference*, pp. 1-6

Klein, C. A., Caroline, C. J. & Ahmed, S. (1995). A New Formulation of the Extended Jacobian Method and its Use in Mapping Algorithmic Singularities for Kinematically Redundant Manipulators. *IEEE Trans. on Robotics and Automation*, Vol. 11, No. 1, pp. 50-55

Kajita S., Morisawa M., Harada K., Kaneko K., Kanehiro F., Fujiwara K. & Hirukawa H. (2006). Biped Walking Pattern Generator allowing Auxiliary ZMP Control. *IEEE/RSJ International Conference on Intelligent Robots and Systems*, pp. 2993-2999

Liegeois, A. (1997). Automatic Supervisory Control of the Configuration and Behavior of Multibody Mechanisms. *IEEE Trans. systems, Man, and Cybernetics*, Vol. 7, No. 12

Nakamura, Y. & Hanafusa H. (1986). Inverse Kinematics Solutions with Singularity Robustness for Robot Manipulator Control. *ASME Journal of Dynamic Systems, Measurement and Control*, Vol. 108, pp. 163-171

Tevatia, G. & Schaal S. (2000). Inverse Kinematics for Humanoid Robots. *Proc. IEEE Int. Conf. Robotics and Automation*, pp. 294-299

Wampler, C. W. (1986). Manipulator inverse kinematic solutions based on vector formulations and damped least squares methods. *IEEE Transactions on Systems, Man, and Cybernetics*, Vol. SMC-16, No. 1, pp 93-101

Yu, S. W. (2006). Walking Pattern Analysis and Control of a Humanoid Robot. Master Thesis, Department of Mechanical Engineering, National Taiwan University, pp. 28-30

15

Walking Pattern Generation and Stabilization of Walking for Small Humanoid Robots

Yogo Takada, Tomoki Tajiri, Kiyoshi Ogawa and Tomoyuki Wakisaka
Osaka City University
Japan

1. Introduction

Declining birthrate and a growing proportion of elderly people are closed up as social issues in Japan. Especially, a growing proportion of elderly people is seen as an issue in nations of Europe and North America and so on as well as Japan. New labors engaging in the life support and the nursing for the elderly person will need in the future. To supplement the lack of the manpower, the research that applies the robot technology to the welfare field is important. Especially, because the shape of a humanoid robot looks like human, the person who receives nursing can be relieved. Besides, a humanoid robot can act under person's life environment, and can use the tool that the person uses. In addition, it is possible to avoid colliding with something put on the ground because the biped robot does not move with wheels. The biped robot can be used even in the environment including steps and ruggedness where robots with wheels cannot be used. In a word, the biped humanoid robot is extremely effective as the robot that acts in the environment where we live in daily life.
However, it is necessary to prevent the robot from falling in the case of two-legged locomotion. Moreover, it is also necessary to keep controlling while the robot is only standing with stable posture. Various techniques about the walking of a humanoid robot have been devised up to now in recent years, and the walking pattern generation of a humanoid robot with complex dynamics has become possible.
On the floor where information about the height of the small step was obtained beforehand, steady dynamic walking was achieved by using such as the technique of linear inverse pendulum model where the humanoid robot was expressed with an inverted pendulum of simple single-mass system, and truck model with preview control of ZMP trajectory (Kajita, 2005, 2009). However, in the case that the robot walks in a real environment, it is difficult to obtain the detailed information on the floor beforehand and give the information to the biped robot at any time. There are research examples (Kajita & Tani., 1996) of making the robot measure the shape of the road with sensors while walking. But, it is necessary to install the highly accurate sensor in the robot. The road where the robot walks is not only a smooth road. A small ruggedness and a gradual inclination are contained in many cases. However, it is impossible to record information of detailed shape on the ruggedness onto the robot beforehand.
A biped robot KHR-2HV manufactured by Kondo Kagaku Co., Ltd. in Japan and HOAP-1 manufactured by Fujitsu Automation Ltd. in Japan were used in this study as humanoid robots. The basic walking experiment was conducted by using KHR-2HV. Then, another

biped robot HOAP-1 was used in the attempt of steady gait on the floor including unknown ruggedness. ZMP (Zero Moment Point) that shows dynamic stability is used in a lot of biped robots, and it is important to plan the position trajectory of the ankles and waist beforehand when walking. The locomotory controller for walking, running, swimming and flying animals is based on a Central Pattern Generator (CPG). Models of CPGs have been proposed and many roboticists have adapted these models for the control of robots (Miyakoshi et al., 2000).

In this study, first, by using KHR-2HV, posture control based on ZMP was attempted and the walking posture was stabilized. A simple nervous oscillator circuit was composed based on CPG and the reflex. We tried to stabilize the walking state with posture control by using ZMP, after the walking pattern was generated with CPG.

Next, by using HOAP-1, we tried the dynamic walking of the biped robot when there is ruggedness on the floor. Dynamic walking is more efficient than static walking. But, the dynamic walking is not steady when ruggedness exists on the floor. Therefore, various devices need to control. The walking pattern composed of each joint angle is also necessary for the dynamic walking of biped robot. Linear inverse pendulum mode is used to make the walking pattern. And, the position-based impedance control was conducted to soften the impact caused by the collision of a rugged floor and foot. In this impedance control, the signals from the force sensors installed on the soles were used. In addition, the upper part of the robot was controlled by using the output signals of the gyro sensor and acceleration sensor to correct the inclination when the posture inclined. We verified whether the robot was able to walk stably on the road on which there was unknown ruggedness by applying these methods to the control of robot by using a physical simulator and an actual robot HOAP-1.

2. Stabilization of static walking by use of CPG

Walking of biped robots greatly is distinguished between static walk and dynamic walk. The static walk is described here. At the present time, ZMP (Zero Moment Point) that shows dynamic stability is used in a lot of robots, and the trajectory plans of walking are studied. Moreover, on the other hand, the robots walk with motion generation techniques of rhythm pattern generator CPG (Central Pattern Generator) and the reflex motion by the sense referring creature models.

In this study, first, only the ZMP was used to control the posture. Then, the stabilization with ZMP and the gait pattern generated with the CPG and reflex motion were combined to achieve steadier walking.

2.1 Small humanoid robot

Robot KHR-2HV manufactured by Kondo Kagaku Co. was used for this study. The height is 0.35 m and weight is 1.26 kg. This robot is shown in Fig. 1.

The number of degrees of freedom in the upper part of the body is 7, and that of legs is 10, and the total is 17. The force sensors (made of Inaba Rubber Co, Ltd.) were arranged in each of four corners of the sole of foot, and the back of humanoid robot was equipped with RM-board made of iXs Research Corp. as a controller.

For the simulation of robot motion, open architecture kinetic simulator OpenHRP (Kanehiro et al., 2004), which was an integrated software development environment for humanoid robotics, was used. The simulation model is based on KHR-2HV with the above-mentioned equipments. The sampling period of simulation in this study was set to 140 ms.

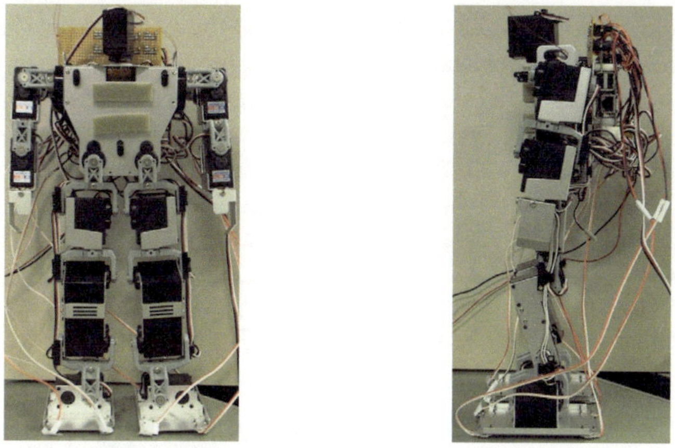

Fig. 1. Photographs of humanoid robot KHR-2HV

2.2 Motion control technique of humanoid robot

The motion control is necessary in order to make the robot KHR-2HV walk. Figure 2 shows the method of the motion control for the robot.

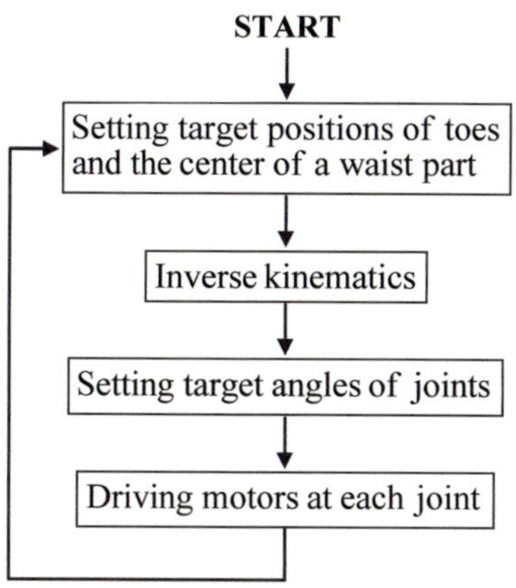

Fig. 2. Motion control technique of humanoid robot

It is important to set target positions of toes and the center of a waist part so as to make the robot move ideally. The target angle of each joint is obtained with the reverse kinematics. Then, each joint is controlled to target joint angle with motor servo control. Consequently, the toes and waist operate ideally, and the stable walking of the robot is achieved. However, in the simulation, the servo control of each motor was not considered. It was assumed that each joint moves in the sampling period to the target angle in this simulation.

2.3 Walking control using ZMP

ZMP is a point on the floor surface where the inertia force and moment produced by gravitation become zero, and it deeply relates to the cause of a tumble while the robot is walking. ZMP is known to be equivalent to a central point of the reaction force from the floor. In this study, ZMP was obtained as the central point calculated from force sensors installed on the sole of foot. The ZMP calculated from the value of the sensor was used for the posture control of the robot shown in Fig. 3. This control system is two degree of freedom PID controller that makes ZMP obtained from the force sensors approach to the target trajectory of robot. The heaviest waist in robot is controlled based on the value of error between the position of ZMP p_{ZMP} while walking and the target of ZMP $\tilde{p}_{ZMP}$. Because ZMP p_{ZMP} approaches the target ZMP, the robot stabilizes its posture. The renewal amount of the waist position Δp_{waist} is given by Equation 1,

$$\Delta p^{n}_{\mathrm{waist}} = K_{\mathrm{P}}\Bigg[\left\{(1-\alpha)\tilde{p}^{n}_{\mathrm{ZMP}} - p^{n}_{\mathrm{ZMP}}\right\} + \frac{1}{T_{\mathrm{I}}}\sum_{i=0}^{n}\left(\tilde{p}^{i}_{\mathrm{ZMP}} - p^{i}_{\mathrm{ZMP}}\right)\Delta T$$

$$+T_{\mathrm{D}}\frac{\left\{(1-\beta)\tilde{p}^{n}_{\mathrm{ZMP}} - p^{n}_{\mathrm{ZMP}}\right\} - \left\{(1-\beta)\tilde{p}^{n-1}_{\mathrm{ZMP}} - p^{n-1}_{\mathrm{ZMP}}\right\}}{\Delta T}\Bigg] \tag{1}$$

where, Δp^{n}_{waist} is renewal amount of waist position at the time $n \cdot \Delta T$, ΔT is sampling period, $\tilde{p}^{n}_{ZMP}$ is target ZMP at $n \cdot \Delta T$, p^{n}_{ZMP} is measured value of ZMP at $n \cdot \Delta T$.
This posture controller was mounted on the robot, and the walking simulation was conducted. The targeted value of right and left foot toes is given beforehand as a time function.

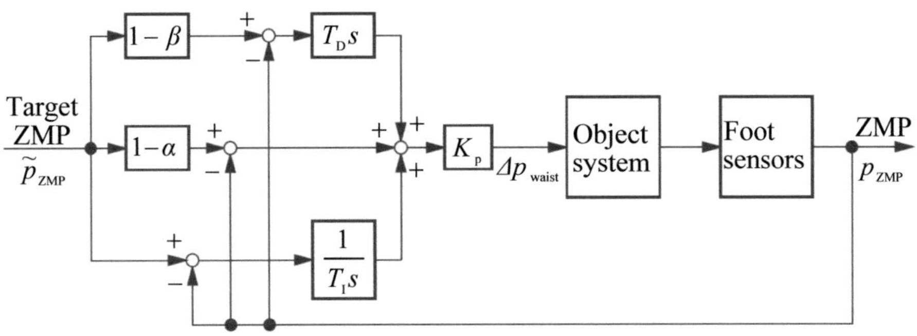

Fig. 3. Posture controller for the robot using ZMP

Figures 4 and 5 show the appearances and trajectories of the walking robot on the simulation. Fig. 4 displays the robot walking for ten seconds from on the left to lower right at intervals of one second. Moreover, the X direction shows travelling direction of robot, and the Y direction shows the lateral direction of robot respectively. The robot did not fall as a result of mounting the posture controller in which the waist center position was controlled to make ZMP approach to the target ZMP trajectory. The robot kept walking stably. The ZMP follows the target in both X direction and Y direction. It is effective in the stabilization of the biped robot walking to operate the heaviest part (waist in this case) in robot.
Though we desperately adjusted the parameters in the posture controller, the parameters that showed a quicker response were not able to be obtained. Therefore, the walking cycle of robot could not shorten any further. Moreover, the waist position of the direction of X has swung to about 15 mm for the period while the idling leg is exchanged with the supporting leg because this control method relies on only the ZMP when the position of the waist is controlled.

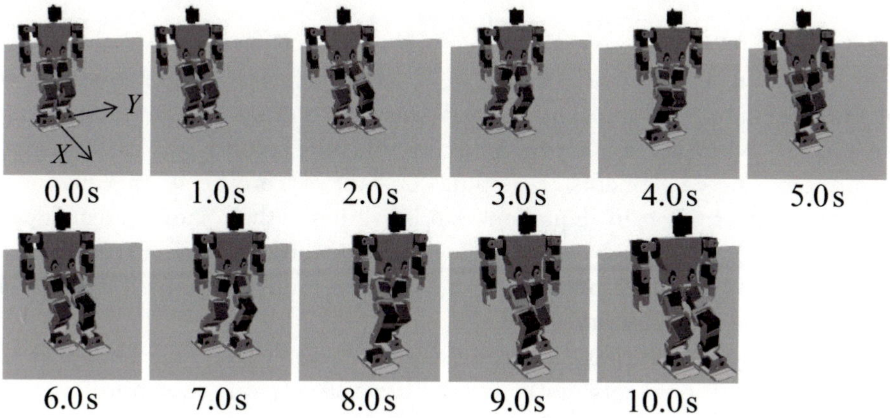

Fig. 4. Appearance of robot walking with posture control by using ZMP

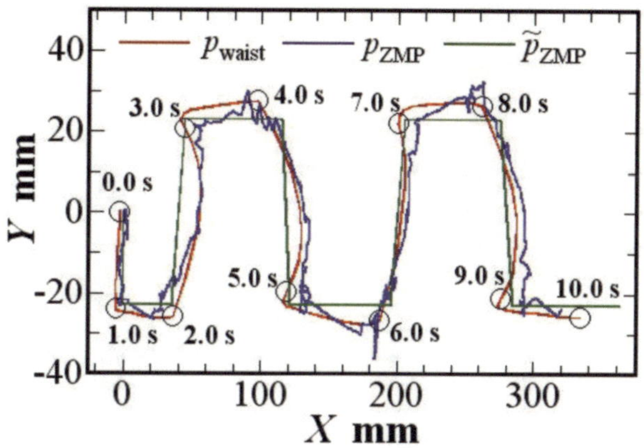

Fig. 5. Trajectories of robot walking with posture control by using ZMP

2.4 Walking pattern generation used CPG and stabilization of walking
2.4.1 Nervous oscillator

In this study, we used a nervous oscillator (Matsuoka, 1985, 1987) composed of adaptable nerve elements (neurons) as the motion generation technique of creature model. The theoretical concept of adaptable nerve elements is given as the following Equations 2 to 5,

$$\tau_{u_i}\dot{u}_i = -u_i - \beta_i v_i + u_i^0 + s_i - \tilde{y}_i \tag{2}$$

$$\tau_{v_i}\dot{v}_i = -v_i + y_i \tag{3}$$

$$y_i = \max(0, u_i) \tag{4}$$

$$\tilde{y}_i = \sum_{j=0,\ldots,N_n-1}^{j\neq i} w_{ij} y_j \tag{5}$$

$$(i = 0, \ldots, N_n - 1) \tag{6}$$

where, N_n is the number of nerve elements, u_i is neuron which shows internal state of nerve element, affixing character i indicates the nerve element of the i th, v_i is variable that shows adaptive effect, y_i is output, s_i is input signal from outside sensor, $\tilde{y}_i$ is variable that shows control from other nerve elements, τ_{ui} is time constant concerning u_i in Equation 2, τ_{vi} is time constant concerning v_i in Equation 3, β_i is variable that shows frequency where adaptation is caused, u_i^0 is steady input, w_{ij} is coupling coefficient from nerve element j th to nerve element i th. When w_{ij} is positive, $\tilde{y}_i$ works as an inhibitory signal. And, when w_{ij} is negative, $\tilde{y}_i$ works as an excitatory signal.

The nervous oscillator has the basic structure where adaptable nerve elements that united to control the other nerve elements mutually. The basic structure is shown in Fig. 6. Because the CPG model with this nervous oscillator changes flexibly the rhythm pattern according to the input signal that come from the outside, the flexible walking can be achieved. Figure 7 shows the behaviour of basic nervous oscillator of Fig. 6. Here, u_i^0 =2.5.

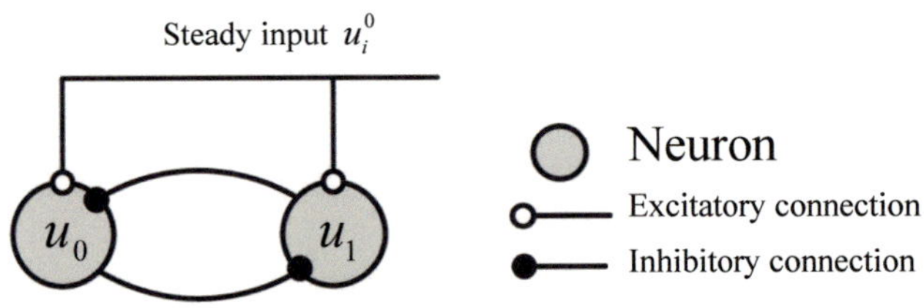

Fig. 6. Structure of nervous oscillator

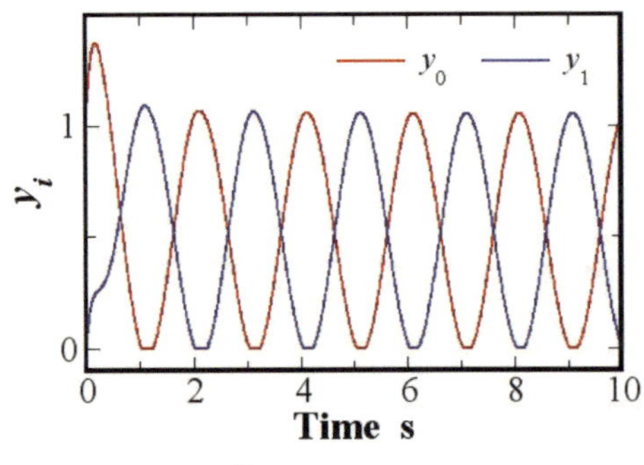

Fig. 7. Output signals from nervous oscillator

Both values y_0 and y_1 rise because of the regular input value at the time of the start. When the nerve element with strong excitement controls the other side by mutually controlling effect, the excitement on the controlled side cools down. However, when the excitement level of the other element is low, the excitement of nerve element becomes stronger. But, the excitement cools down by own adaptive effect. Besides, the other nerve element begins to get excited because the inhibitory power becomes weak when the excitement cools down. Therefore, the phase lag arises between two output signals. Then, a periodic output can be obtained. The gait pattern is generated with the use of such a periodic output to the targeted value of the robot. This is a basic idea of the nervous oscillator with CPG.

2.4.2 Walking pattern generation used CPG

Figures 8 and 9 show the nervous oscillator circuit used for the robot walking. The output signals y_0, y_1 of the nervous oscillator composed of u_0, u_1 in main generator in Fig. 8 were used for the position control of the waist center for the lateral direction. In a word, the walking rhythm of robot was generated with CPG. The outputs y_2, y_3 of the nerve elements u_2, u_3 uniting main generator with excitatory connection were used for the target height of right and left legs' toe. These nerve elements were controlled by the input step signals s_2, s_3 . Working of these nerve elements has improved timing of exchanging the supporting leg and the idling leg. Therefore, the robot steps (does walking motion in a same place) stably.

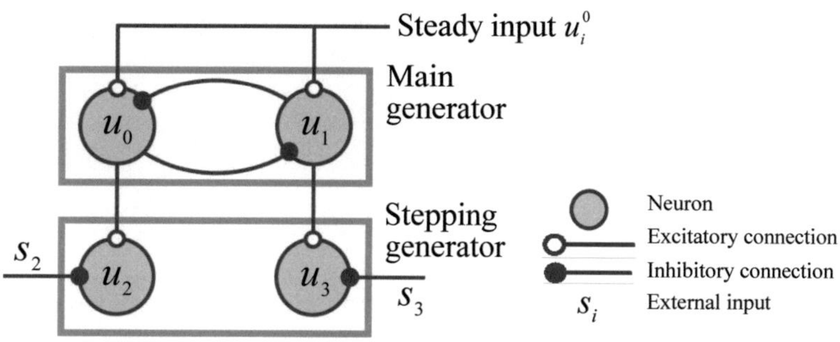

Fig. 8. CPG model for the robot stepping stably

The nervous oscillator composed of u_4, u_5 in Fig. 9, which generates the output signals y_4, y_5 reacting to the input signals s_4, s_5 from force sensors on the sole of the foot, is used for the stepping forward. The robot judges whether the sole of foot touches the ground by using the force sensors. The judged result enters the nervous oscillator as an input signal if the sole of foot touches the ground. The nerve element becomes a state of excitation when the signal enters, and the output signal comes out as reflex action. The robot steps forward by using the signal come from the nervous oscillator according to the contact situation with the ground.

Fig. 9. Reflex model for the robot walking forward

Figures 10 and 11 show the appearances and trajectories of the walking robot with CPG on the simulation. Periodic walking operation was achieved based on the rhythm generated by the CPG model. In addition, the robot was able to walk without falling down by adjusting the timing in which the idling leg was exchanged for the supporting leg by external inputs according to ZMP. Moreover, the motions of ZMP and waist for the travelling direction in this method were calmer than the method in the case that only ZMP signal was used with PID control. Even if the walking cycle shortened, a steady walking could be maintained.

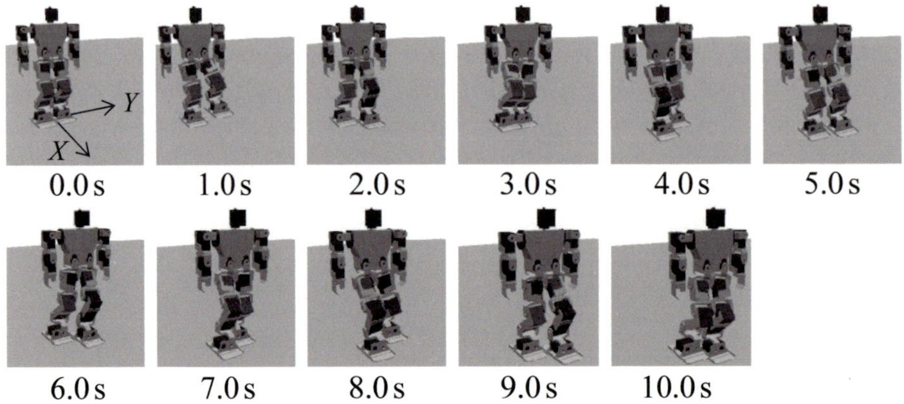

Fig. 10. Appearance of robot walking with CPG and posture control by using ZMP

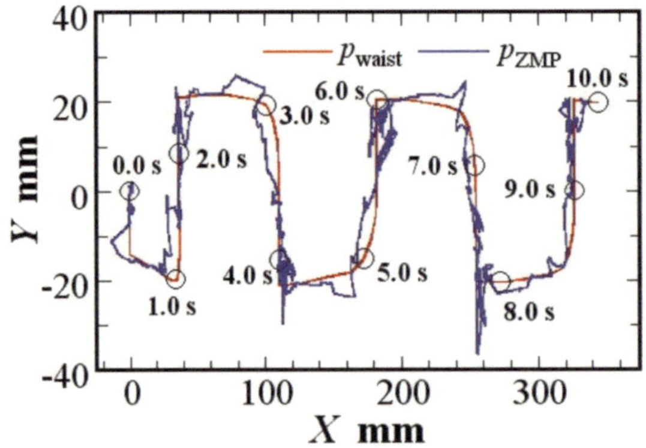

Fig. 11. Trajectories of robot walking with CPG and posture control by using ZMP

2.4.3 Walking experiment by using KHR-2HV

The CPG model used for the walking simulation was applied to a real machine, and the walking experiment was conducted. The gait initiation time of KHR-2HV was after 2.5 s passed from the experiment start. Then, the control period was set to 130 ms. In this time, I/O waiting time was about 80 ms and the computing time was about 50 ms. Figure 12 shows the appearance of KHR-2HV's walking. Moreover, Figs. 13 and 14 show the time behaviors of the waist position and ZMP. The motion of the travelling direction (X direction) is shown in Fig. 13, and that of lateral direction (Y direction) is shown in Fig. 14. Though the trajectory of ZMP was vibrating, a steady walking could be achieved based on the rhythm generated with the CPG model.

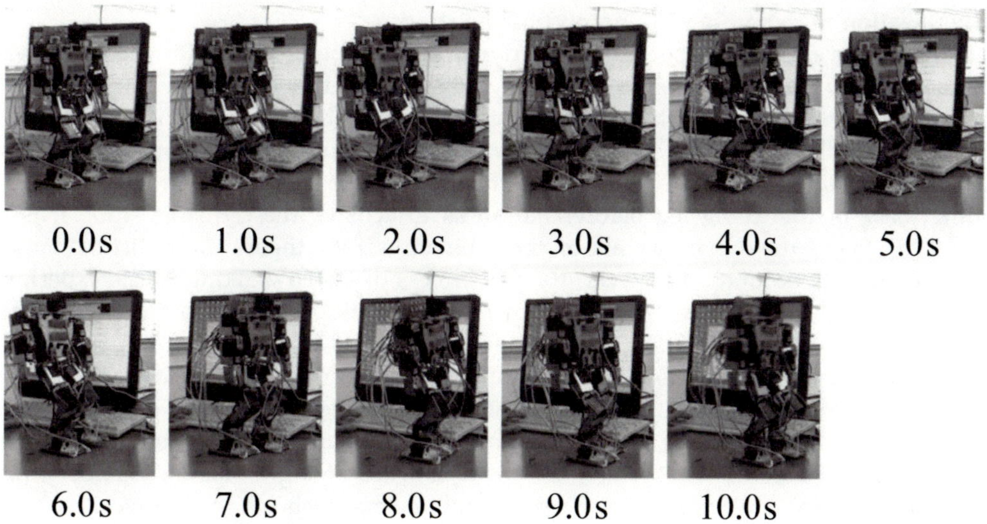

Fig. 12. Appearance of KHR-2HV walking with CPG and posture control by using ZMP

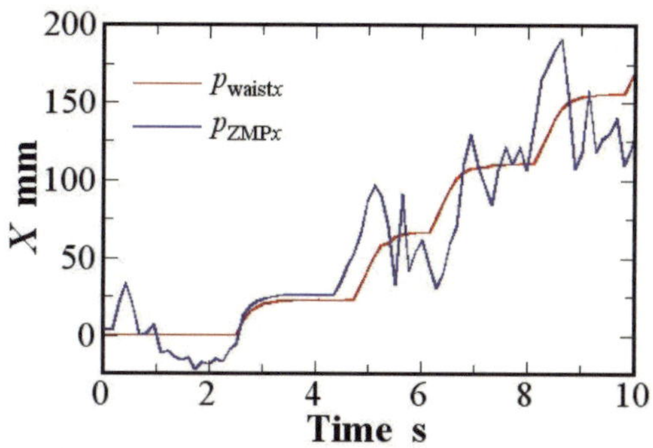

Fig. 13. Trajectories for travelling direction of KHR-2HV walking with CPG and posture control by using ZMP

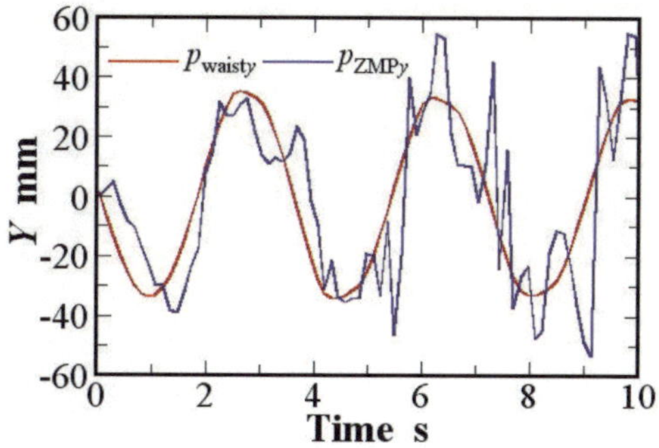

Fig. 14. Trajectories for lateral direction of KHR-2HV walking with CPG and posture control by using ZMP

3. Stabilization of dynamic walking on uneven road

High-speed movement is expected in the dynamic walk compared with the static walk. However, the control system for the dynamic walk is more difficult than that of static walk. We have advanced the research by using more excellent robot HOAP-1 though KHR-2HV had been used till then because the dynamic walk needs a high-integrity controller.

It is necessary to lead the trajectory of the gravity point of a biped robot and predict the landing place of the idling leg beforehand so as to achieve the dynamic walk. It is not difficult to achieve the dynamic walk on the simulation if the theory to predict the landing point is extremely accurate. However, for the success of stable walking in the experiment with a real biped robot, there are conditions that each link of robot is rigid and that floor surface must be completely flat.

In the real environment, minute ruggedness of few millimeters in height exists even if it is a floor surface that looks smooth. The center of gravity of the robot vibrates because of the ruggedness, and the robot falls down. Therefore, it is difficult to achieve the dynamic walk compared with the static walk. Here, the method of achieving a stable dynamic walk is described even if minute ruggedness of several millimeters exists in the floor face.

3.1 Configuration of humanoid robot "HOAP-1"

Biped humanoid robot HOAP-1 was used in this study. Figure 15 shows the photograph of HOAP-1 whose height is 483 mm and weight is 5.9 kg. The number of degrees of freedom in the upper part of the body is 8, and that of legs is 12, and the total is 20. The arrangement of each joint of legs is shown in Fig. 16. Because only 12 motors of the leg were used to walking in this study, the upper part of the body was omitted in this Fig. 16. Two axes in the ankle and three axes in the waist intersect at one point respectively. The parameters of each link are shown in Table 1. The three-axes acceleration sensor and three-axes gyro sensor are installed in the waist. Four force resister sensors are installed in right and left sole of the foot. The resistance of these force resister sensors changes in proportion according to pressure.

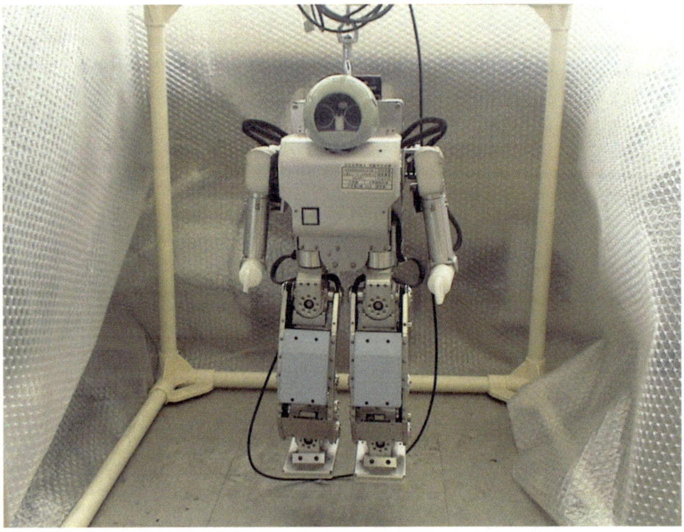

Fig. 15. Photograph of HOAP-1

	Length (m)	Mass (kg)
Leg link 1	L_1 : 0.039	0.04
Leg link 2	L_2 : 0.10	0.46
Leg link 3	L_3 : 0.10	0.46
Leg link 4	L_4 : 0.037	0.14

Table 1. Parameters of each link

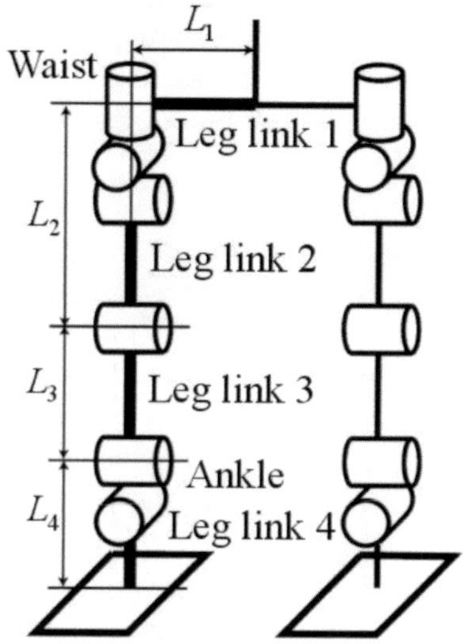

Fig. 16. Arrangement of each joint of legs

3.2 Control system to stabilize the dynamic walk

3.2.1 Position-based impedance control for an idling leg

When the idling leg touches to unknown ruggedness shown in Fig. 17, an actual landing point changes from the one of the planned gait pattern. Then, the impact is generated on the sole of foot when the idling leg touches to the ground. The robot falls down after its posture got worse because this impulsive force becomes very big disturbance in the object system.

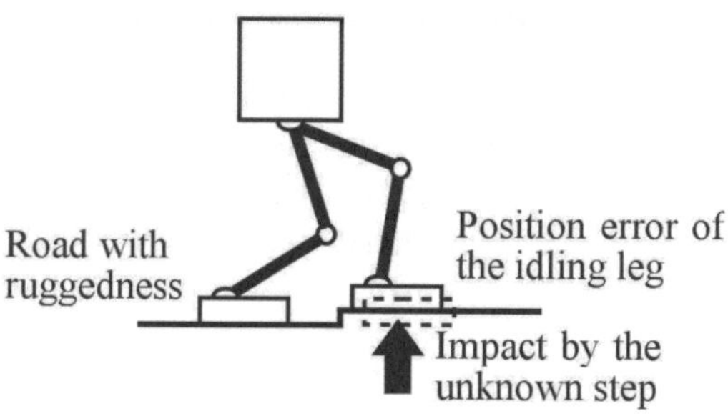

Fig. 17. Walking on a road with unknown step

The impedance control is often used as a technique for controlling the contact strength between the surrounding environment and a part of the robot in the field of the robotic

control. There are some examples in the field of humanoid robot (Kawaji et al., 1996; Sorao et al., 1997; Nisikawa et al., 1999; Hasimoto et al., 2007). In this study, the position-based impedance control was conducted by using the force resister sensors installed on the sole of foot of the robot. In the impedance control, three virtual impedance properties of inertia, rigidity and damping coefficient are given to the part of control object. Then, the amount of the change at the position of ankle is led from the dynamic equation according to the impulsive force measured with the sensors. The impulsive force weakens when the leg moves based on the led position. In a word, it is expected that the impedance control have the same effect as when real springs and dampers are given to the target part.
Four force resister sensors are installed on the each sole of right and left foot of HOAP-1 used in this study, and the arrangement is shown in Fig. 18. In addition, each parameter in Fig. 18 is shown in Table 2.

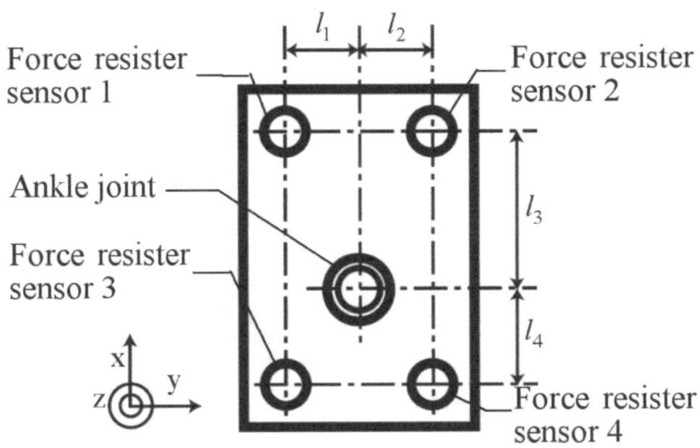

Fig. 18. Arrangement of force resister sensors

	Length (m)
l_1	0.017
l_2	0.017
l_3	0.042
l_4	0.028

Table 2. Parameters for the arrangement of force resister sensors

Only force in the direction of z can be detected with four force resister sensors 1-4, and the forces measured with force resister sensors were defined as f_1, f_2, f_3, f_4 respectively. Torques τ_x, τ_y where the center of gyration is x axis and y axis respectively and force f_z in the direction z can be obtained from Equation 6,

$$\begin{aligned} f_z &= f_1 + f_2 + f_3 + f_4 \\ \tau_x &= (f_1 l_1 + f_2 l_1) - (f_2 l_2 + f_2 l_4) \\ \tau_y &= (f_1 l_3 + f_2 l_3) - (f_3 l_2 + f_4 l_4) \end{aligned} \quad (6)$$

and the virtual spring and damper mechanism shown in Fig. 19 can be utilized for the control system to stabilize the dynamic walk by using the force and torque on the toe of idling leg obtained by Equation 6. The dynamic equation of this toe of idling leg model is given as Equation 7,

$$\boldsymbol{F} = \boldsymbol{M}_{\text{imp}}\ddot{\boldsymbol{p}}_{\text{leg}} + \boldsymbol{C}_{\text{imp}}\dot{\boldsymbol{p}}_{\text{leg}} + \boldsymbol{K}_{\text{imp}}\boldsymbol{p}_{\text{leg}}$$
$$\boldsymbol{F} = \begin{bmatrix}\tau_{\text{x}} & \tau_{\text{y}} & f_{\text{z}}\end{bmatrix}^{\text{T}},\ \boldsymbol{p}_{\text{leg}} = \begin{bmatrix}\theta_{\text{x}} & \theta_{\text{y}} & p_{\text{z}}\end{bmatrix}^{\text{T}} \tag{7}$$

where, $\mathbf{M}_{\text{imp}}$ (diagonal matrix: $\text{diag}(J_{\text{x}},J_{\text{y}},M_{\text{z}})$) is virtual inertia and mass, $\mathbf{C}_{\text{imp}}$ ($\text{diag}(C_{\text{x}},C_{\text{y}},C_{\text{z}})$) is virtual viscous coefficient, $\mathbf{K}_{\text{imp}}$ ($\text{diag}(K_{\text{x}},K_{\text{y}},K_{\text{z}})$) is virtual coefficient of rigidity, θ_{x} is rotation angle around x axis, θ_{y} is rotation angle around y axis, p_{z} is moving distance in the z direction. In addition, the angle θ_{x} , θ_{y} and moving distance p_{z} are defined to zero on standing upright as shown in Fig. 16, and the arrow direction in Fig. 19 is defined as positive direction.

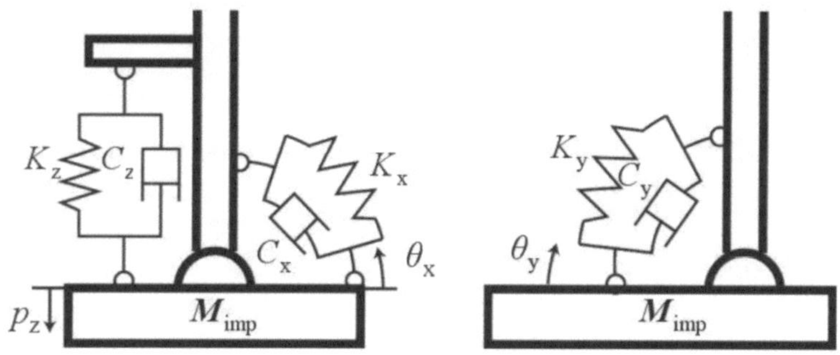

(a) Front view (b) Side view

Fig. 19. Model of impedance control

When the ideal joint angle $\theta_{\text{x}}^{\text{n}}$ around x axis of toe was led by using Equation 7, Equation 8 was obtained,

$$A = K_{\text{x}} + C_{\text{x}} / \Delta t + J_{\text{x}} / \Delta t^2$$
$$\theta_{\text{x}}{}^{n} = \frac{\tau_{\text{x}} + K_{\text{x}}\theta_{x}{}^{0}}{A} + \frac{C_{\text{x}}\theta_{\text{x}}{}^{n-1}}{A\Delta t} + \frac{J_{\text{x}}(2\theta_{\text{x}}{}^{n-1} - \theta_{\text{x}}{}^{n-2})}{A\Delta t^2} \tag{8}$$

where, θ_{x}^{0} is initial angle (momentary value that the idling leg touch to the ground with unknown ruggedness), $\theta_{\text{x}}^{\text{n}}$ is current angle, $\theta_{\text{x}}^{\text{n-1}}$ is angle before time for 1 period of control period, $\theta_{\text{x}}^{\text{n-2}}$ is angle before time for 2 periods of control period. The rotation angle of toe around y axis and the moving distance in the z direction were controlled as well as x axis.

The impact caused by touching to unknown ruggedness was weakened by impedance control in which the force f_z and torque τ_x , τ_y in Equation 6 were controlled to become small. We verified the effect of impedance control. The parameters in Table 3 were used in this verification. The height of an unknown step on the ground was set to 20 mm. The time behaviors of vertical distance from the waist of the robot to the toe of foot are shown in Fig. 20 as a result. Here, only f_z was applied in the impedance control in this verification. The idling leg's position of the z direction has changed according to the force f_z as shown in Fig. 20. Therefore, it is confirmed that landing point in the z direction of idling leg was corrected in response to unknown step on the ground.

	Value
M_z	1.4 kg
C_z	70.0 Ns/m
K_z	100.0 N/m

Table 3. Parameters used in impedance control

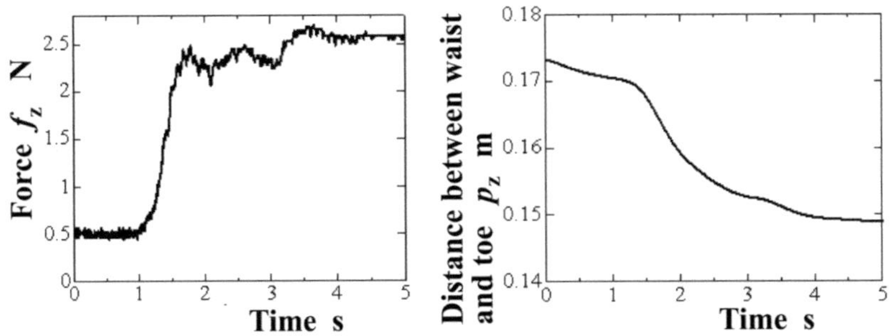

Fig. 20. Time behaviors of force measured with force sensors and distance between waist and toe of foot touching the unknown step on the ground in impedance control

3.2.2 Posture control for upper part of robot

Vertical posture of the upper part in the body is good for steady walking of the biped robot while walking. However, the upper part of the body occasionally inclines as shown in Fig. 21 by the small inclination of floor surface on uneven road. In this case, there is a possibility that ZMP goes out from the stable zone of the foot-ankle assembly. Therefore, when the posture inclined, the upper part of the body must be controlled to keep vertical posture by using the inclination of upper part measured with the gyro sensor and the acceleration sensor. In this posture stabilization control, the motor of the ankle rotated according to the value of the measured inclination.

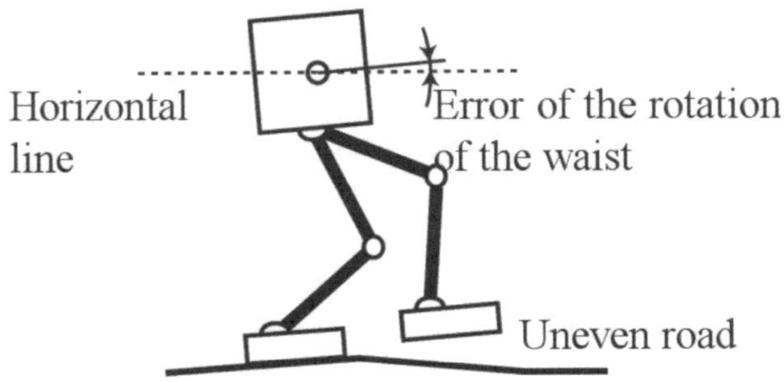

Fig. 21. Walking on a road with slopes

A concrete content is described here. The current inclination of upper part of the body θ_{gyro} was obtained by integrating the output of the gyro sensor installed in the waist of the robot. Then, the upper part of the body was kept vertical by the servo control of the ankle joint according to obtained inclination of the upper part θ_{gyro} and target posture, which was target inclination of upper part θ_{target}^{body} .

At this time, target inclination the upper part $\theta_{\text{target}}^{\text{body}}$ was obtained by using the output values a_x, a_y and a_z of the 3-dimensional acceleration sensor installed in the waist of the robot as shown in Equation 9,

$$\theta_{\text{target}}^{\text{body}} = \left[\tan^{-1}(a_z / a_x) \quad \tan^{-1}(a_z / a_y)\right]^{\mathrm{T}}. \tag{9}$$

However, the influence of robot's acceleration is included in the output of this acceleration sensor. Therefore, while the robot has accelerated to the travelling direction, the upper part of the body does not become perpendicular but leans toward the travelling direction a little. This inclination preferably prevents the robot from falling behind, and has stabilized the robot walking more.

The PI controller was used for this posture control as shown in Equation 10,

$$\theta_{\text{ref}}^{\text{leg}} = K_{\text{P}}\left\{\left(\theta_{\text{gyro}} - \theta_{\text{target}}^{\text{body}}\right) + \frac{1}{T_{\text{I}}}\sum\left(\theta_{\text{gyro}} - \theta_{\text{target}}^{\text{body}}\right)\Delta t\right\} \tag{10}$$

where, K_{P} and T_{I} are parameters of PI controller, and the current inclination of upper part θ_{gyro} is obtained from Equation 11,

$$\theta_{\text{gyro}} = \sum \omega_{\text{gyro}} \Delta t \tag{11}$$

where, ω_{gyro} is angular velocity of the waist obtained from the output of the gyro sensor. The initial posture angle of the robot needed in Equation 11 was obtained after the acceleration sensor outputs when the robot was fixed to flat surface floor were substituted for Equation 9. The posture of robot is controlled so that θ_{gyro} can approach $\theta_{\text{target}}^{\text{body}}$ by the PI controller shown in Equation 10.

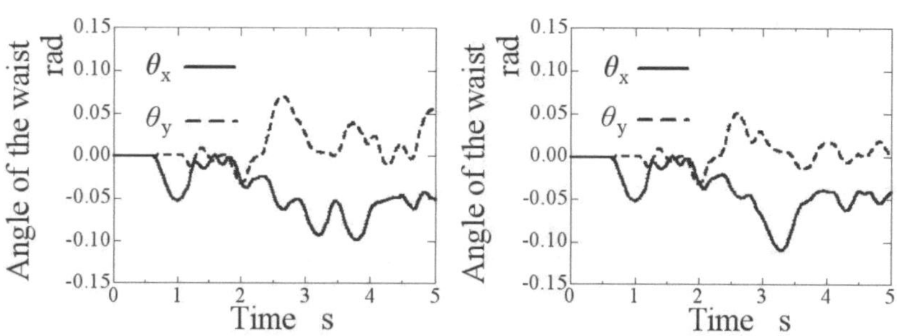

(a) Without PI controller (b) With PI controller

Fig. 22. Inclination angle of the waist

Figure 22 shows the time behaviors about the inclination of the upper part (waist) of the body. The robot walked on the sloping road that is inclining on 2.0°(0.035 rad) in the direction around y axis. The inclination of the upper part was obtained by integrating the output of the gyro sensor. In Fig. 22(a), the time behaviors of robot walking without feedback by the PI control are shown. In Fig. 22(b), the time behaviors in case where inclination angle of upper part was controlled by the PI controller are shown. The robot began walking at 0.5 seconds, and it began to go up on the slope at 2.0 s in both cases.

When there was no posture control, inclination angle in the travelling direction θ_y was 0.02 rad on the average, and the dispersion was 0.024 rad. When there was the posture control, the average was 0.009 rad and the dispersion was 0.017 rad. It was confirmed that the posture of robot was improved as a result of the PI controller.

3.3 Walking on road with ruggedness

3.3.1 Generation of gait pattern with inverse pendulum model

Linear inverse pendulum model (Kajita, 2005) was used for the walking pattern generation in this study. For the linear inverse pendulum model, a robot is assumed to be a single-mass system as shown in Fig. 23, and the dynamic equation of the mass system inverse pendulum model is set up. The trajectory of gravity point in a robot for a steady walking is calculated by solving the dynamic equation about the mass system of inverse pendulum. Adjustment of kick force f prevents the gravity point from falling, and gravity point can be fixed to arbitrary height (Z_c) in Fig. 23. For the linear inverse pendulum model, the trajectory of gravity point while walking can be obtained by setting the walking cycle and the landing position of idling leg beforehand. In this study, relative values as shown in Fig. 24 from a present landing position to the next landing position must be decided. The next landing points are calculated by using Equations 12 and 13 whenever the foot lands on ground,

$$p_x^{(i+1)} = p_x^{(i)} + S_x \tag{12}$$

$$p_y^{(i+1)} = p_y^{(i)} + (-1)^i S_y \tag{13}$$

where, $p_x^{(i)}$ and $p_y^{(i)}$ show x coordinate and y coordinate at the landing point of i th step.

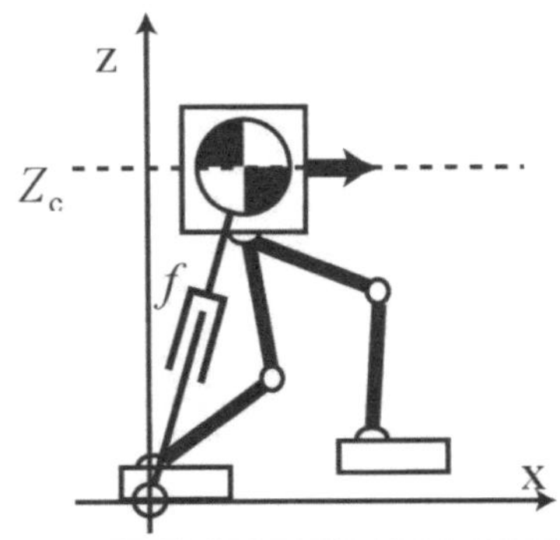

Fig. 23. Liner inverse pendulum model (LIPM) for humanoid robot walking

It is possible to divide into two time period (one-legged support period and two-legged support period) in the walking cycle of human. However, the linear inverse pendulum model is a trajectory plan method only for the one-legged supporting period. Therefore, it is necessary to consider another trajectory plan of gravity point for two-legged supporting period. In this study, in order to connect two-legged supporting period to other one-legged supporting period smoothly and continuously, the positions, velocities and accelerations of gravity point were interpolated by the fifth function.

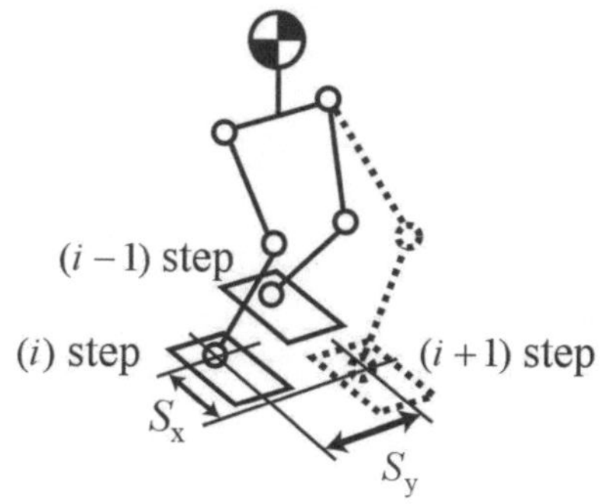

Fig. 24. Parameters used in LIPM

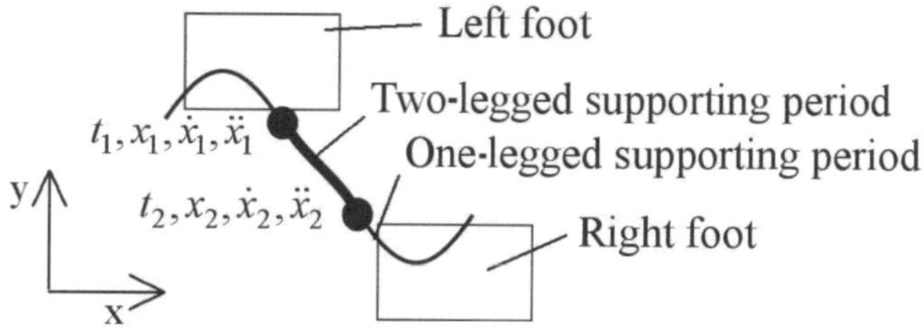

Fig. 25. Connection of two-legged supporting period and one-legged supporting period concerning trajectory of gravity point

The positions, velocities, accelerations and times of gravity point at the start and end of two-legged supporting period were decided beforehand. The trajectory of gravity point $x(t)$ for the two-legged supporting period was obtained by interpolating them with the polynomial equation shown in Equation 14,

$$x(t) = a_0 t^5 + a_1 t^4 + a_2 t^3 + a_3 t^2 + a_4 t + a_5 \tag{14}$$

where, the coefficients from a_1 to a_5 are obtained by Equation 15,

$$\begin{bmatrix} a_0 \\ a_1 \\ a_2 \\ a_3 \\ a_4 \\ a_5 \end{bmatrix} = \begin{bmatrix} t_0{}^5 & t_0{}^4 & t_0{}^3 & t_0{}^2 & t_0 & 1 \\ t_1^5 & t_1^4 & t_1^3 & t_1^2 & t_1 & 1 \\ 5t_0^4 & 4t_0^3 & 3t_0^2 & 2t_0 & 1 & 0 \\ 5t_1^4 & 4t_1^3 & 3t_1^2 & 2t_1 & 1 & 0 \\ 20t_0^3 & 12t_0^2 & 6t_0 & 2 & 0 & 0 \\ 20t_1^3 & 12t_1^2 & 6t_1 & 2 & 0 & 0 \end{bmatrix}^{-1} \begin{bmatrix} x_1 \\ x_2 \\ \dot{x}_1 \\ \dot{x}_2 \\ \ddot{x}_1 \\ \ddot{x}_2 \end{bmatrix}. \tag{15}$$

Moreover, the gravity point trajectory of y-axis $y(t)$ was similarly interpolated and obtained. The trajectories of right and left idling legs were created based on Equation 16,

$$x(t) = x_1 + \frac{S_x}{2}\left(1 - \cos\frac{\pi t}{T_{one}}\right)$$
$$z(t) = \frac{h}{2}\left(1 - \cos\frac{2\pi t}{T_{one}}\right) \tag{16}$$

where, it is necessary to decide the maximum value of height h, the stride S_x and the time for one-legged supporting period T_{one} to obtain $x(t)$ and $z(t)$. Each trajectory was calculated every time, when a present leg left the ground, and it moved to the following planned landing position. By using the Equation 16, the movement of foot, where the acceleration of the idling leg is gradual at beginning and at landing, can be realized as shown in Fig. 26.

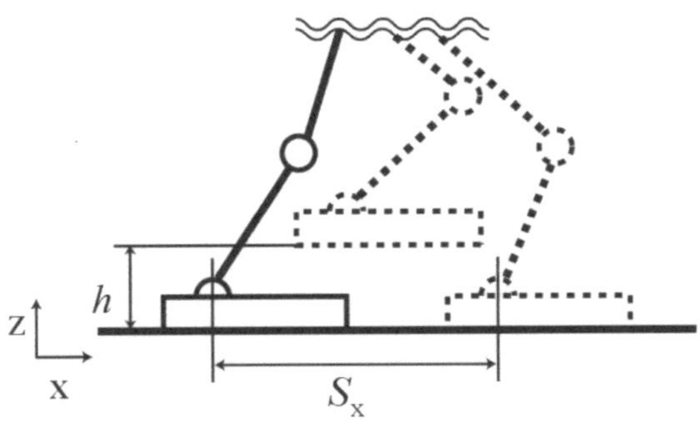

Fig. 26. Movement of idling foot

3.3.2 Walking on ruggedness road in simulation

In the dynamic walk, stable walking can be achieved by applying the gait pattern based on the theory to predict the landing point by using liner inverse pendulum model. However, there are conditions that each link of robot is rigid and that floor surface must be completely flat. In addition, it is necessary to use high-precision actuator systems that operate by target signals. In the real environment, minute ruggedness of few millimeters in height exists even if it is a floor surface that looks smooth. When robots walk on the ruggedness, the gravity point of the robot vibrates because of the ruggedness, and the robot falls down. Therefore, the gait pattern has been improved by using the stabilization control system that describes ahead. Figure 27 shows the outline of the walking control system. Trajectory $\boldsymbol{p}^{COG}$ of the gravity point of robot and trajectory $\boldsymbol{p}^{leg}_{ref}$ of the idling leg were generated with the walking pattern generator. The trajectory of the idling leg was calculated by the impedance control system and becomes $\boldsymbol{p}^{leg}_{target}$. Then, θ_{ref} is calculated with the inverse kinematics. After θ^{leg}_{ref} calculated by the PI controller is added to θ_{ref}, angle θ_{target} becomes target angle vector of all of motors of joints.

The trial of robot walking on the road including unknown ruggedness was conducted with using these walking-pattern generation and stabilization control system on the simulation. OpenHRP (Kanehiro et al., 2004), which was an integrated software development environment for humanoid robotics, was used for the simulator. The time interval of simulator was set to 1 msec, and the control cycle of robot was set to 5 msec. Then other parameters in this simulator are shown in Table 4. Figure 28 shows the size concerning an

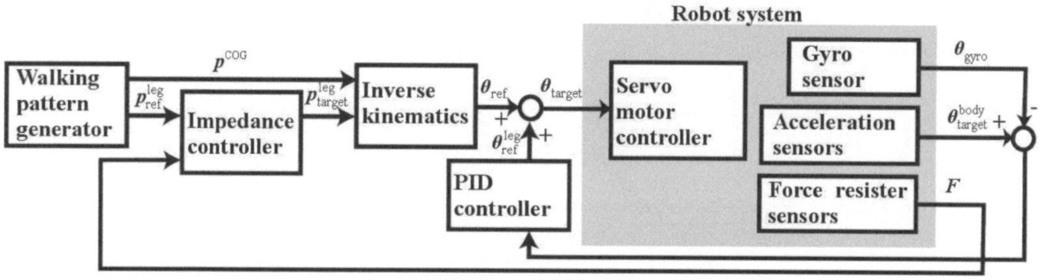

Fig. 27. Outline of walking control system for HOAP-1

LIPM			Impedance controller	
	Value			Value
T_{one}	0.3 s		J_x, J_y	1.0 kgm^2
T_{two}	0.2 s		M_z	1.0 kg
S_x	0.06 m		K_x, K_y	100.0 Nm/rad
S_y	0.04 m		K_z	1.0 N/m
h	0.02 m		C_x, C_y	70.0 Nms/rad
Z_c	0.13 m		C_z	10.0 Ns/m
			K_P	2.5 s^{-1}
			T_I	0.8 s

Table 4. Parameters used in liner inverse pendulum model (LIPM) and impedance controller

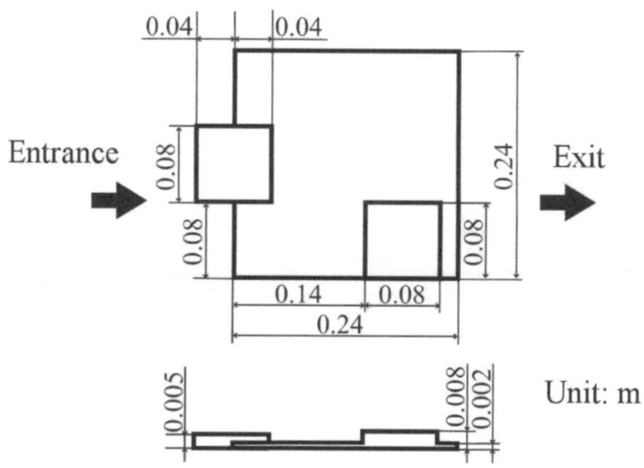

Fig. 28. Unknown ruggedness road used in simulation

unknown ruggedness of the road used in the simulation. The result of walking simulation about the trajectory of the gravity point is shown in Fig. 29 and the trajectory of toe of idling leg is shown in Fig. 30. In this figure, the position of toe of idling leg is appropriately controlled according to the height of unknown step while walking. Figure 31 shows the walking appearance on the simulation with the snap shot of each 0.3 sec. The left leg of the robot stepped on the ruggedness of 5 mm in height as shown in Fig. 31(e). The robot kept walking on the ruggedness afterwards.

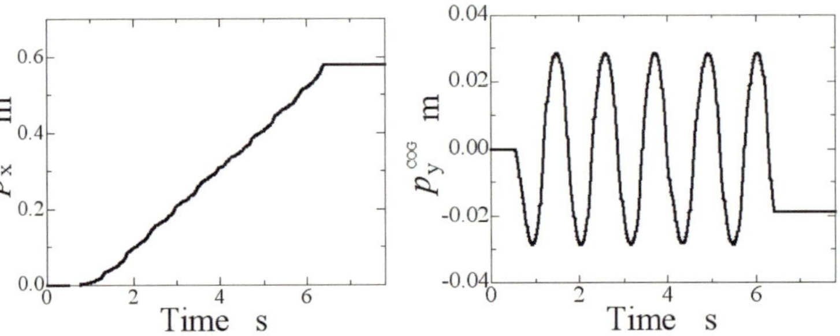

Fig. 29. Time behaviors of COG (center of gravity)

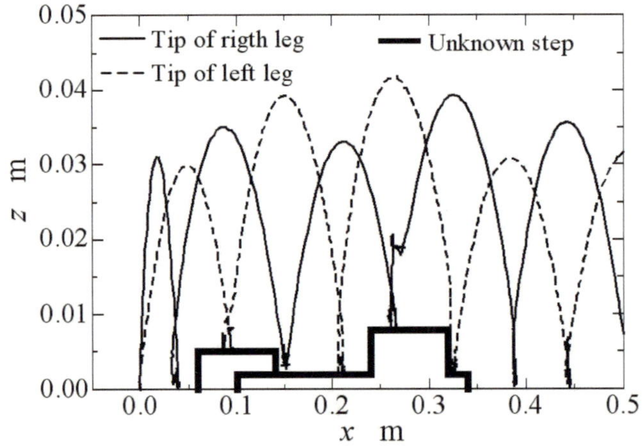

Fig. 30. Trajectory of legs

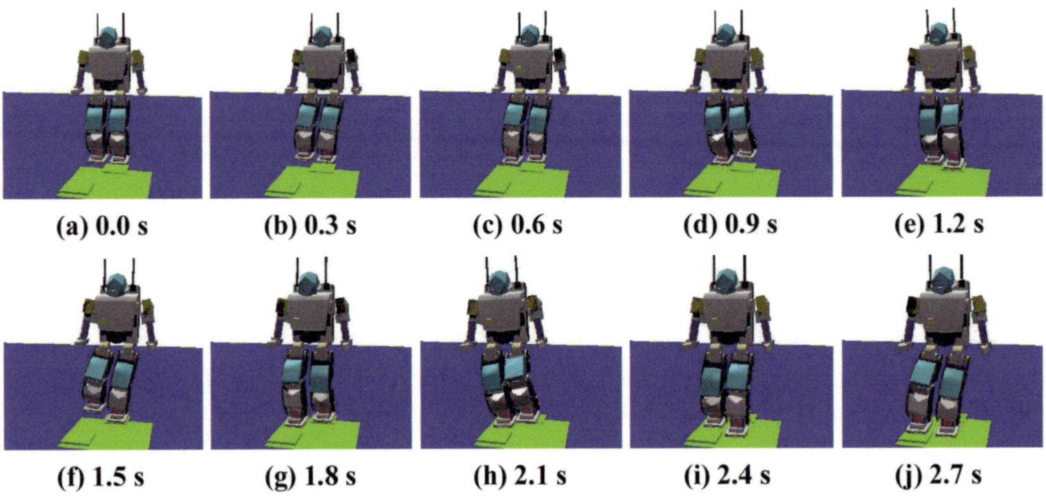

Fig. 31. Snapshots of the walking motion of HOAP-1 on the simulaiton

However, it can be confirmed that the behavior of the right leg did unexpected motion when x reached 0.25 m in Fig. 30. The unexpected motion was generated because the stabilization control system was not able to weaken the impact from an unknown step of 6 mm in height. This walking control is very effective in a real environment because most ruggedness of floor on a smooth road is almost less than 6 mm. Moreover, ruggedness that is higher than 6 mm can be recognized beforehand as an obstacle.

3.4 Experimental verification

This section shows the experimental results of which the previously described control techniques were applied for an actual humanoid robot.

The control cycle was set to 5 msec same as on the simulation, and other control parameters are shown in Table 5. The value of the control parameter is larger than that of the parameters used on the simulation. This is because it was necessary to reduce the value of the parameters in the simulation environment where impact is greater than that of real environment. Because the elastic deformation by the outside force is not considered in the simulation, the calculated impact becomes greater. The value of parameters in the inverse pendulum model was same as the parameters used in the simulation.

	Value
J_x, J_y	1.0 kgm^2
M_z	1.0 kg
K_x, K_y	5.0 Nm/rad
K_z	200.0 N/m
C_x, C_y	10.0 Nms/rad
C_z	140.0 Ns/m
K_P	3.0 s^{-1}
T_I	1.1 s

Table 5. Parameters used in controller

Moreover, the acrylic boards of 4 mm and 6 mm in thickness were arranged as unknown ruggedness like Fig. 32. The experimental result concerning the trajectory of the gravity point is shown in Fig. 33. The time behavior of the length between the toe of idling leg and waist is shown in Fig. 34. The distance from the toe of idling leg to the waist changes in the range from 0.14 to 0.18 m. The vertical movement of idling foot of the robot is remarkably seen in this Figure. Moreover, points from a to d in this Figure indicate the moment of landing of idling foot. In a word, it can be confirmed that the value of P_z changes a little by landing of the idling foot.

The walking appearance of the real robot HOAP-1 is shown in Fig. 35 as snap shot of photographs. In Fig. 35(a), the right leg is on the acrylic board of 6mm, and the left leg is on another acrylic board of 4mm. Then, in Fig. 35(d), the right leg stepped acrylic board of 4mm on which the left leg was stepping. Afterwards, the robot kept walking stably.

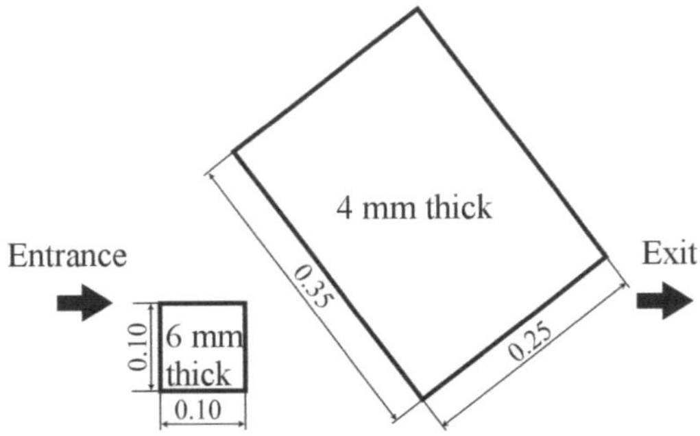

Fig. 32. Arranged acrylic boards of 4 mm and 6 mm as unknown ruggedness

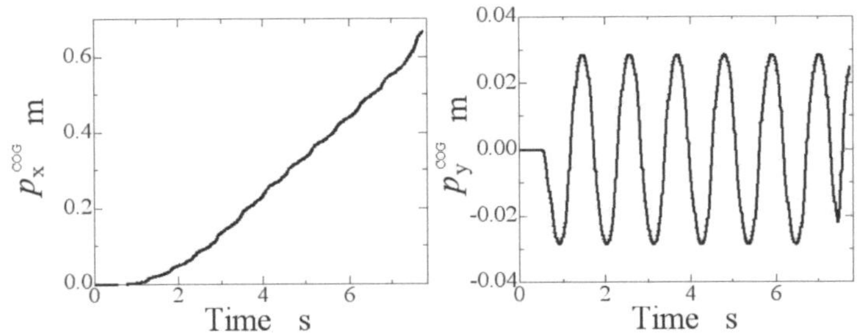

Fig. 33. Trajectory of gravity point of HOAP-1 obtained by experiment

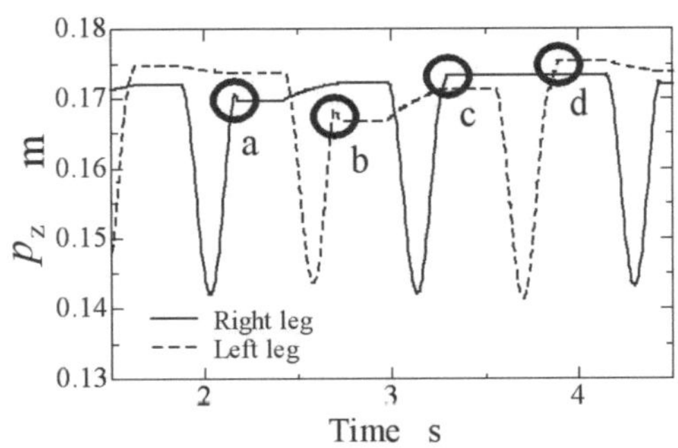

Fig. 34. Trajectory of legs obtained by experiment

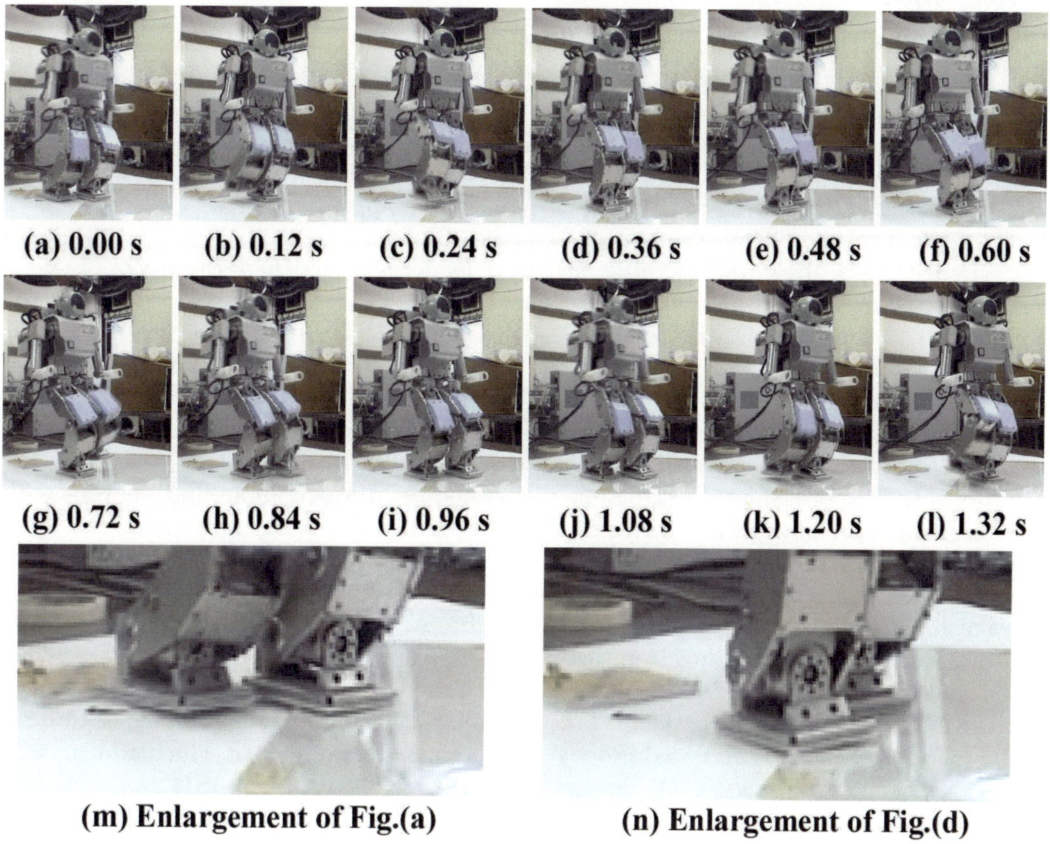

Fig. 35. Arranged acrylic boards of 4 mm and 6 mm as unknown ruggedness

4. Concluding remarks

By using KHR-2HV, posture control based on ZMP was attempted and the walking posture was stabilized. As a result, the robot voluntarily adjusted the exchange timing of supporting leg and idling leg according to the state of robot by using the nerve oscillator circuit. Then, the robot walked controlling posture by using ZMP without the tumble in the simulation. In addition, the CPG model used for the walking simulation was applied to a real biped robot, and the walking pattern was generated. The biped robot KHR-2HV kept walking and maintaining steady balance for a few seconds. But, in this case, KHR-2HV walked slowly with static walking.

Then, by using HOAP-1, we tried the dynamic walking of the biped robot when there is ruggedness on the floor. The biped robot maintained the stability even if the foot collided to the small step from which information was not given, by correcting the position and turning angle of tip of idling leg with the impedance control. In addition, walking steady has been possible as a result of the experiment with a real biped robot HOAP-1 by these control techniques.

5. References

Hasimoto, K., Sugahara, Y., Kawase, M., Hayashi, A., Tanaka, C., Ohta, A., Sawato, T., Endo, N., Lim, H. & Takanishi, A. (2007). *Journal of the Robotics Society of Japan*, Vol.25, No.6, pp. 53-60

Kajita, S., (2005). *Humanoid Robot*, Ohmsha,ISBN4-274-20058-2, Japan

Kajita, S. (2009). From Inverted Pendulums to Biped Locomotion -ZMP and Control Theory-, *Journal of the Robotics Society of Japan*, Vol.27, No.4, pp. 392-395

Kajita, S. & Tani, K. (1996). Control of Dynamic Biped Locomotion Based on Realtime Sensing of the Groung Profile, *Journal of the Robotics Society of Japan*, Vol.14, No.7, pp.1062-1069

Kanehiro, F., Hirukawa, H. & Kajita, S. (2004). OpenHRP: Open Architecture Humanoid Robotics Platform, *The International Journal of Robotics Research*, Vol.23, No.2, pp. 155-165

Kawaji, S., Ogasawara,K. & Iimori, J. (1996). Biped Locomotion Control with Compliance, *IEEJ Transaction on Industrial Applications*, Vol.116-D, No.1, pp. 11-18

Matsuoka, K. (1985). Sustained Oscillations Generated by Mutually Inhibiting Neurons with Adaptation, *Biological Cybernetics*, Vol.52, pp. 367-376

Matsuoka, K. (1987). Mechanisms of Frequency and Pattern Control in the Neural Rhythm Generations, *Biological Cybernetics*, Vol.56, pp. 345-353

Miyakoshi, S., Taga, G., Kuniyoshi, Y., & Nagakubo, A. (2000). Three Dimensional Bipedal Stepping Motion using Neural Oscillators, *Journal of the Robotics Society of Japan*, Vol.18, No.1, pp. 87-93

Nisikawa, N., Fujimoto, Y., Murakami, T. & Ohnishi, K. (1999). Variable Compliance Control for 3 Dimensional Biped Robot considering Environmental Fluctuations, *IEEJ Transaction on Industrial Applications*, Vol.119-D, No.12, pp. 1507-1514

Sorao, S., Murakami, T. & Ohnishi, K. (1997). Walking Control of a Biped Robot by Impedance Control, *IEEJ Transaction on Industrial Applications*, Vol.117-D, No.10, pp. 1227-1233